AF389094

Novel Approaches to Minimising Mycotoxin Contamination

Novel Approaches to Minimising Mycotoxin Contamination

Special Issue Editors

Mar Rodríguez Jovita
Félix Núñez

MDPI • Basel • Beijing • Wuhan • Barcelona • Belgrade • Manchester • Tokyo • Cluj • Tianjin

Special Issue Editors
Mar Rodríguez Jovita
University of Extremadura
Spain

Félix Núñez
Universidad de Extremadura
Spain

Editorial Office
MDPI
St. Alban-Anlage 66
4052 Basel, Switzerland

This is a reprint of articles from the Special Issue published online in the open access journal *Toxins* (ISSN 2072-6651) (available at: https://www.mdpi.com/journal/toxins/special_issues/approach_minising_mycotoxin).

For citation purposes, cite each article independently as indicated on the article page online and as indicated below:

LastName, A.A.; LastName, B.B.; LastName, C.C. Article Title. *Journal Name* **Year**, *Article Number*, Page Range.

ISBN 978-3-03928-937-0 (Pbk)
ISBN 978-3-03928-938-7 (PDF)

© 2020 by the authors. Articles in this book are Open Access and distributed under the Creative Commons Attribution (CC BY) license, which allows users to download, copy and build upon published articles, as long as the author and publisher are properly credited, which ensures maximum dissemination and a wider impact of our publications.

The book as a whole is distributed by MDPI under the terms and conditions of the Creative Commons license CC BY-NC-ND.

Contents

About the Special Issue Editors

Mar Rodríguez Jovita graduated in Veterinary Science from the University of Extremadura (Spain, 1988). She obtained her Ph.D. in Food Science and Technology from the University of Extremadura (1995), and continued as a Postdoctoral Fellow in the Department of Molecular Biology of Institute of Food Research of Reading (U.K., 1997–1998). In addition, she has been a visiting researcher at the University of Davis-California (USA, 2004). She is currently a senior researcher in the Meat and Meat Products Research Institute. All her research has mainly focused on the technological and toxicological evaluation of microorganisms in meat products, the molecular techniques for their detection, and identification and production of toxins and the procedures for their prevention and/or elimination.

Félix Núñez graduated in Veterinary Science from the University of Extremadura (Spain, 1988), and obtained his Ph.D. in Food Science and Technology from the University of Extremadura (1995). He continued his formation as a Postdoctoral Fellow in the Department of Food Science and Human Nutrition at the Michigan State University (USA, 1996–1997). He is currently a senior researcher in the Meat and Meat Products Research Institute and the coordinator of the Ph.D. Programme in Food Science at the University of Extremadura. His research interests are mainly focused on the biocontrol of toxigenic microorganisms in dry-cured foods from animal origin, including protective cultures and antifungal proteins.

 *toxins*

Editorial

Novel Approaches to Minimizing Mycotoxin Contamination

Mar Rodríguez * and Félix Núñez *

Food Hygiene and Safety, Meat and Meat Products Research Institute, Faculty of Veterinary Science, University of Extremadura, Avda. de las Ciencias, s/n, 10003 Cáceres, Spain
* Correspondence: marrodri@unex.es (M.R.); fnunez@unex.es (F.N.)

Received: 17 January 2020; Accepted: 26 March 2020; Published: 29 March 2020

Contamination of foods and agricultural commodities by various types of toxigenic fungi is a concerning issue for human and animal health. Molds naturally present in foods can produce mycotoxins and contaminate foodstuffs under favorable conditions of temperature, relative humidity, pH, and nutrient availability. Mycotoxins are, in general, stable molecules, which are difficult to remove from foods once they have been produced. Therefore, the prevention of mycotoxin contamination is one of the main goals of the agriculture and food industries.

Chemical control or decontamination techniques may be quite efficient. However, the more sustainable and restricted use of fungicides, the lack of efficiency in some foods, and the consumer demand for chemical residue-free foods require new approaches to control this hazard. Therefore, food safety demands permanent research efforts for exploring new strategies to reduce mycotoxin contamination.

This Special Issue contains original contributions and reviews that advance the knowledge about the most current promising approaches to minimize mycotoxin contamination, including biological control agents (BCAs), phytochemical antifungal compounds, enzyme detoxification, and the use of novel technologies. Most of the studies focus on *Fusarium* toxins, but the toxicity of aflatoxins and ochratoxin A is also addressed. In addition, a few studies are focused on the control of plant pathogenic fungi.

Several studies examined the potential of biological control strategies as alternative methods in both plant pathogenic and mycotoxigenic fungi. Within the group of articles about biological control strategies, Ntushelo et al. [1] reported that *Bacillus* species adopt various mechanisms, including the production of bioactive compounds, to inhibit the growth of the mycotoxin-producing plant pathogenic fungus *Fusarium graminearium* and provided a perspective of the techniques used to study antagonist metabolites. Moraes-Bazioli et al. [2] review the potential of BCAs, including bacteria, yeasts, and natural plant products for the control of *Penicillium digitatum*, *Penicillium italicum*, and *Geotrichum citriaurantii*, which are responsible for postharvest citrus diseases. Zearalenone (ZEA) is an estrogenic mycotoxin which can cause loss in animal production. Chen et al. [3] selected a *Bacillus* strain with a strong esterase activity that exhibited a high ZEA detoxification capability in maize using a fermentation process to validate their potential application in the feed industry. In the same sense, Liu et al. [4] tested the combination of probiotic strains from *Bacillus subtilis* and *Candida utilis* with cell-free extracts from *Aspergillus oryzae* to degrade ZEA. Interestingly, the authors showed the BCA effect of alleviating the negative impact of ZEA on normal growth performance in pig keeping. Aflatoxin M1 (AFM1) is secreted in the milk of lactating mammals through the ingestion of feedstuff contaminated by aflatoxin B1, being a health concern for dairy industries and consumers of dairy products. Assaf et al. [5] review AFM1 decontamination methods including different bio-adsorbents such as bacteria, yeasts, or mixtures of both. The efficiency of these decontamination methods in addition to their plausible experimental variants, advantages, limitations, and prospective applications are broadly discussed. Ochratoxin A (OTA) is the mycotoxin most commonly found in meat products, which contribute significantly to human exposure to this toxin. Cebrian et al. [6] propose the use of a combined protective culture containing selected strains of *P. chrysogenum* in combination with *Debaryomyces hansenii* as a promising strategy to reduce OTA production by *P. nordicum* in dry-cured hams. The efficacy of BCAs was tested in dry-cured hams under industrial ripening, resulting in a significant reduction of OTA contamination.

Three papers are focused on exploring phytochemical compounds as potential antifungal agents, two of them dealing with the use of garlic-derived compounds to control toxigenic molds in cereals and the other one with apple pomace. Quiles et al. [7] evaluate the antifungal activity of allyl isothiocyanate against aflatoxigenic *Aspergillus flavus* and ochratoxigenic *Penicillium verrucosum* on cereals in small-scale silos, obtaining a significant reduction of the *A. flavus* and *P. verrucosum* growth as well as an important reduction of the OTA. Mylona et al. [8] reported that the efficacy of treatment with propyl propane thiosulfonate and propyl propane thiosulfinate to reduce *Fusarium* growth and mycotoxin production was dependent on the specific "*Fusarium* species-toxin" pathosystem and the a_w of cereals. The work of Oleszek et al. [9] showed that apple pomace could be a good source of natural bio-fungicides inhibiting the growth of crop pathogens, mycotoxigenic molds, being the strongest antifungal activity exerted by a fraction containing phloridzin.

The use of mycotoxin-degrading methods is a promising approach to control this hazard and to counteract their toxic effects in livestock. With respect to enzymatic detoxification, Alberts et al. [10] developed an innovative method using the commercial fumonisin esterase FumD to reduce fumonisin B in whole maize, resulting in the formation of the hydrolyzed breakdown product, HFB1, associated with the aqueous phase to be discarded. Fruhauf et al. [11] concluded that the cleavage of ZEA by the zearalenone-lactonase Zhd101p reduces its estrogenicity in piglets, providing an important basis for the further evaluation of ZEA-degrading enzymes.

Finally, four papers explore the effectiveness of new technologies in preventing or reducing mycotoxin contamination of foods and feeds. Wang et al. [12] synthesized a light-responsive dendritic-like α-Fe$_2$O$_3$ that showed good activity for the photocatalytic degradation of deoxynivalenol (DON) in aqueous solution under visible light irradiation with a significant decrease of the toxicity of this mycotoxin. Casas-Junco et al. [13] propose a treatment of OTA-contaminated roasted coffee with cold plasma, achieving a decrease of OTA concentration and a reduction in toxicity of the treated coffee. Liu et al. [14] proved that the treatment of wheat kernels with superheated steam is an effective method to reduce the content of DON by thermal degradation, and the reduction rate increased significantly with the steam temperature. In addition, the treatment improves the qualities of crisp biscuits made from processed wheat. García-Díaz et al. [15] proposed a novel niosome-based encapsulated essential oil product applied to polypropylene bags that significantly reduced the development of *A. flavus* and aflatoxin contamination in maize up to 75 days. Thus, the correct application of this product may be a sustainable way to avoid the occurrence of aflatoxins in stored maize.

Therefore, the results contained in this Special Issue represent several interesting advances in different methodologies for the control of mycotoxins in food and feeds, providing useful information for the development of further research in this field.

Funding: This research received no external funding.

Acknowledgments: The editors are grateful to all the authors who contributed their work to this Special Issue. Special thanks go to the rigorous evaluations of all of the submitted manuscripts by the expert peer reviewers who contributed to this Special Issue. Lastly, the valuable contributions, organization, and editorial support of the MDPI management team and staff are greatly appreciated.

Conflicts of Interest: The authors declare no conflict of interest.

References

1. Ntushelo, K.; Ledwaba, L.K.; Rauwane, M.E.; Adebo, O.A.; Njobeh, P.B. The mode of action of *Bacillus* species against *Fusarium graminearum*, tools for investigation, and future prospects. *Toxins* **2019**, *11*, 606. [CrossRef] [PubMed]

2. Moraes-Bazioli, J.; Belinato, J.R.; Costa, J.H.; Akiyama, D.Y.; de Moraes Pontes, J.G.; Kupper, K.C.; Augusto, F.; de Carvalho, J.E.; Pacheco-Fill, T. Biological control of citrus postharvest phytopathogens. *Toxins* **2019**, *11*, 460. [CrossRef] [PubMed]

3. Chen, S.; Wang, H.; Shih, W.; Ciou, Y.; Chang, Y.; Ananda, L.; Wang, S.; Hsu, J. Application of Zearalenone (ZEN)-Detoxifying *Bacillus* in animal feed decontamination through fermentation. *Toxins* **2019**, *1*, 330. [CrossRef] [PubMed]

4. Liu, C.; Chang, J.; Wang, P.; Yin, Q.; Huang, W.; Dang, X.; Lu, F.; Gao, T. Zearalenone biodegradation by the combination of probiotics with cell-free extracts of *Aspergillus oryzae* and its mycotoxin-alleviating effect on pig production performance. *Toxins* **2019**, *11*, 552. [CrossRef] [PubMed]

5. Assaf, J.C.; Nahle, S.; Louka, N.; Chokr, A.; Atoui, A.; El Khoury, A. Assorted methods for decontamination of aflatoxin. *Toxins* **2020**, *11*, 304. [CrossRef] [PubMed]

6. Cebrián, E.; Rodríguez, M.; Peromingo, B.; Bermúdez, E.; Núñez, F. Efficacy of the combined protective cultures of *Penicillium chrysogenum* and *Debaryomyces hansenii* for the control of Ochratoxin A hazard in dry-cured ham. *Toxins* **2019**, *11*, 710. [CrossRef] [PubMed]

7. Quiles, J.M.; de Melo Nazareth, T.; Luz, C.; Luciano, F.B.; Mañes, J.; Meca, G. Development of an antifungal and antimycotoxigenic device containing allyl isothiocyanate for silo fumigation. *Toxins* **2019**, *11*, 137. [CrossRef] [PubMed]

8. Mylona, K.; Garcia-Cela, E.; Sulyok, M.; Medina, A.; Magan, N. Influence of two garlic-derived compounds, propyl propane thiosulfonate (PTS) and propyl propane thiosulfinate (PTSO), on growth and mycotoxin production by *Fusarium* species in vitro and in stored cereals. *Toxins* **2019**, *11*, 495. [CrossRef] [PubMed]

9. Oleszek, M.; Pecio, Ł.; Kozachok, S.; Lachowska-Filipiuk, Ż.; Oszust, K.; Frąc, M. Phytochemicals of apple pomace as prospect bio-fungicide agents against mycotoxigenic fungal species—in vitro experiments. *Toxins* **2019**, *11*, 361. [CrossRef] [PubMed]

10. Alberts, J.; Schatzmayr, G.; Moll, W.D.; Davids, I.; Rheeder, J.; Burger, H.M.; Shephard, G.; Gelderblom, W. Detoxification of the fumonisin mycotoxins in maize: An enzymatic approach. *Toxins* **2019**, *11*, 523. [CrossRef] [PubMed]

11. Fruhauf, S.; Novak, B.; Nagl, V.; Hackl, M.; Hartinger, D.; Rainer, V.; Labudová, S.; Adam, G.; Aleschko, M.; Moll, W.-D.; et al. Biotransformation of the mycotoxin zearalenone to its metabolites hydrolyzed zearalenone (HZEN) and decarboxylated hydrolyzed zearalenone (DHZEN) diminishes its estrogenicity in vitro and in vivo. *Toxins* **2019**, *11*, 481. [CrossRef] [PubMed]

12. Wang, H.; Mao, J.; Zhang, Z.; Zhang, Q.; Zhang, L.; Zhang, W.; Li, P. Photocatalytic degradation of deoxynivalenol over dendritic-like α-Fe$_2$O$_3$ under visible light irradiation. *Toxins* **2019**, *11*, 105. [CrossRef] [PubMed]

13. Casas-Junco, P.P.; Solís-Pacheco, J.R.; Ragazzo-Sánchez, J.A.; Aguilar-Uscanga, B.R.; Bautista-Rosales, P.U.; Calderón-Santoyo, M. Cold plasma treatment as an alternative for ochratoxin a detoxification and inhibition of mycotoxigenic fungi in roasted coffee. *Toxins* **2019**, *11*, 337. [CrossRef] [PubMed]

14. Liu, Y.; Li, M.; Bian, K.; Guan, E.; Liu, Y.; Lu, Y. Reduction of deoxynivalenol in wheat with superheated steam and its effects on wheat quality. *Toxins* **2019**, *11*, 414. [CrossRef] [PubMed]

15. García-Díaz, M.; Patiño, B.; Vázquez, C.; Gil-Serna, J. A novel niosome-encapsulated essential oil formulation to prevent *Aspergillus flavus* growth and aflatoxin contamination of maize grains during storage. *Toxins* **2019**, *1*, 646. [CrossRef] [PubMed]

© 2020 by the authors. Licensee MDPI, Basel, Switzerland. This article is an open access article distributed under the terms and conditions of the Creative Commons Attribution (CC BY) license (http://creativecommons.org/licenses/by/4.0/).

Perspective

The Mode of Action of *Bacillus* Species against *Fusarium graminearum*, Tools for Investigation, and Future Prospects

Khayalethu Ntushelo [1,*], Lesiba Klaas Ledwaba [1,2], Molemi Evelyn Rauwane [1], Oluwafemi Ayodeji Adebo [3] and Patrick Berka Njobeh [3]

1   Department of Agriculture and Animal Health, Science Campus, University of South Africa, Corner Christiaan De Wet and Pioneer Avenue, Private Bag X6, Florida 1709, Guateng, South Africa; ledwabal@arc.agric.za (L.K.L.); rauwaneme@gmail.com (M.E.R.)
2   Agricultural Research Council-Vegetable and Ornamental Plants, Private Bag X293, Pretoria 0001, Tshwane, South Africa
3   Department of Biotechnology and Food Technology, University of Johannesburg, Corner Siemert and Louisa Street, Doornfontein 2028, Gauteng, South Africa; oadebo@uj.ac.za (O.A.A.); pnjobeh@uj.ac.za (P.B.N.)
*   Correspondence: ntushk@unisa.ac.za; Tel.: +27-011-471-3648

Received: 5 August 2019; Accepted: 23 August 2019; Published: 18 October 2019

Abstract: *Fusarium graminearum* is a pervasive plant pathogenic fungal species. Biological control agents employ various strategies to weaken their targets, as shown by *Bacillus* species, which adopt various mechanisms, including the production of bioactive compounds, to inhibit the growth of *F. graminearum*. Various efforts to uncover the antagonistic mechanisms of *Bacillus* against *F. graminearum* have been undertaken and have yielded a plethora of data available in the current literature. This perspective article attempts to provide a unified record of these interesting findings. The authors provide background knowledge on the use of *Bacillus* as a biocontrol agent as well as details on techniques and tools for studying the antagonistic mechanism of *Bacillus* against *F. graminearum*. Emphasizing its potential as a future biological control agent with extensive use, the authors encourage future studies on *Bacillus* as a useful antagonist of *F. graminearum* and other plant pathogens. It is also recommended to take advantage of the newly invented analytical platforms for studying biochemical processes to understand the mechanism of action of *Bacillus* against plant pathogens in general.

Keywords: *Bacillus*; *Fusarium graminearum*; antagonism; mode of action

Key Contribution: This article reports the overall mode of action of the bacteria *Bacillus* against the mycotoxin-producing plant pathogenic fungus *Fusarium graminearium* and provides a perspective of the techniques used to study antagonist metabolites. The goal is to illustrate research done so far and recommend study directions for the future.

---

1. Introduction

Biotic stresses such as plant pests and pathogens are the major factors threatening global crop production. Proliferation in plants of these pathogens can cause devastating epidemics, which can cause severe food shortages especially in countries with limited resources. The current state of crop losses due to pathogenic diseases is alarming, with an estimated 8–40% of crop yield losses caused by plant pathogens worldwide [1,2]. One of such prominent pathogens of health and economic importance is *Fusarium graminearum*. This fungus is a known causative agent of Fusarium head blight (FHB), which is an economically important cereal crop disease that accounts for worldwide losses estimated between 20 and 100% [3–6]. According to Dean et al. [7], *F. graminearum* is ranked the fourth most important plant fungal pathogen, on the basis of its scientific and economic importance. This filamentous ascomycete infects floral tissues of cereal plants and contaminates food grains [7]. Infection is associated with premature bleaching symptoms, which mainly reduce grain quality and, less often, yield [3,7–9].

In addition to grain quality reduction, *F. graminearum* also produces various types of mycotoxins, which if ingested in huge amounts, cause various toxicoses in animals and humans [9]. The major mycotoxin produced by *F. graminearum* is deoxynivalenol (DON) together with other mycotoxins including the trichothecene nivalenol (NIV) and its derivatives, 3- and 15-acetyldeoxynivalenol (3-ADON and 15-ADON). These mycotoxins are reported to contaminate grain food products, thereby posing a threat to humans and animals by causing neurological disorders and immunosuppression [10–12] amongst other dysfunctions. However, these health complications vary from one animal species to the other and according to several factors such as trichothecene type, level, and route of exposure. This assembled body of evidence justifies the need for the biocontrol of *F. graminearum* in several foodstuffs [12]. In the past three decades, control strategies against this devastating plant pathogen have been based solely on fungicide application, which has resulted in long-term undesirable environmental pollution [13]. Herbicides and insecticides have also been used over the years to suppress the activity of this pathogenic microorganism causing FHB, amongst other diseases, in crops. Coupled with fungicides are various control practices such as sanitation, good agricultural practices, as well as the use of resistant cultivars. With the increase in awareness of the danger of chemical control applications, fungicides are beginning to take a back foot, with the use of biocontrol products being exploited.

With the increased desire for environmental friendliness and sustainability, the biocontrol of pathogens is equally receiving attention. Biocontrol is defined as the use of natural products and living organisms to suppress pathogen populations. The use of biocontrol agents either as an alternative to other forms of plant disease control or as a supplement has attracted worldwide attention to be included in an integrated pathogen management strategy in various food systems. However, to prevent an irrational selection of plant pathogen antagonists to be adopted as commercial products, the modes of the antagonists' activities and effects need to be fully understood. Bacterial antagonists are commonly used, and many of them belong to the genus *Bacillus* [14].

In this perspective manuscript, we summarize the current knowledge about the mode of action of *Bacillus* species against the pervasive plant pathogen *F. graminearum*. Background information about *Bacillus* is provided, the antagonism of *Bacillus* and its mode of action, tools and techniques to uncover the mechanisms of the antagonism are described, and future prospects are presented.

2. Overview of *Bacillus* Species as a Protective Agent against Pathogens

Bacillus is one of the largest genera of bacteria that produce aerobically dormant endospores under diverse growth conditions [15]. Species belonging to this genus can play a role as human pathogens, whilst others promote plant health and development [16]. Due to their different genetic characteristics, *Bacillus* species are ideal candidates as biocontrol agents. *Bacillus* species play a role as bacterial antagonists to pathogens due to their ability to reproduce actively and their resistance to unfavorable environmental conditions [14]. The species' antagonistic activities are associated with the production of metabolites with antibiotic properties [17]. Particularly, volatile metabolites produced by these microorganisms also play an important role in the activation of plant defense mechanisms by triggering induced systemic resistance (ISR) in plants [18]. In addition, plant host defense responses can also be activated during the production of metabolites by *Bacillus* species [19]. As documented in the literature, *Bacillus* spp. also directly antagonize fungal pathogens by competing and depriving them of essential nutrients, by producing fungitoxic compounds, and by inducing systemic acquired resistance in plants [20–23].

A wide range of pathogenic microorganisms have been controlled using *Bacillus*-based biocontrol agents [17,24–26]. Several disease control products produced from various strains have also been registered and are commercially available. A broad spectrum of resistance mechanisms against plant diseases have been reported to be induced by *Bacillus* strains in many studies [17,27]. Furthermore, the activity of other *Bacillus* strains was also investigated in different crops and found to be effective against various fungal plant pathogens and diseases, including *Fusarium* wilt in tomato [28] as well as FHB in wheat and barley [19,27].

2.1. Biological Activity of Bacillus in General and Against F. graminearum

Bacillus species can produce different antimicrobial substances that confer protection and act as biological agents [29]. Such substances include subtilin [30], bacilysin [31], mycobacillin [32], bacillomycin [33,34], mycosubtilin [35,36], iturins, fengycins, and surfactins [37]. These substances have been reported to exert antibacterial and/or antifungal activities against pathogenic microorganisms [17,19,26–38]. As noted in the literature, among these antimicrobial substances produced by *Bacillus*, the most studied with regard to *F. graminearum* are surfactin, fengycin, and iturin. For this reason, the literature reported herein focuses on these three *Bacillus*-produced antimicrobial agents.

The antagonism of these antimicrobial substances has been reported against *F. graminearum* [26,27], *Fusarium oxysporum* [39,40], *Fusarium solani*, and *Rhizoctonia solani* [40], amongst many other plant pathogenic fungi. In a study by Földes and colleagues [29], antagonistic compounds produced by *Bacillus subtilis* IFS-01 exhibited antimicrobial effects against phytopathogenic, food-borne, and spoilage microorganisms. In an agar diffusion assay, some of the filamentous fungi and yeasts tested showed no visible growth within the inhibition zone (about 10 mm from the colony) due to the antagonistic effect of *B. subtilis* IFS-01. These findings confirmed the biological control ability of this *Bacillus* strain against these fungi and yeasts. The iturin family of the lipopeptides produced by *Bacillus amyloliquefaciens* PPCB044 strain showed antagonism against pathogenic fungi from seven citrus plants during postharvest [41]. All the fungal pathogens were deterred by the *B. amyloliquefaciens* PPCB004 strain, as the strain produced compounds related to iturin A, fengycin, and surfactin [41]. Similar results were also described by Gong et al. [26], who reported the antagonism of iturin A and plipastatin A from *B. amyloliquefaciens* S76-3 in wheat inoculated with *F. graminearum*. The data obtained from both the growth chamber and the field plot assays revealed a strong antagonistic activity of strain S76-3 against the growth and development of *F. graminearum*. Iturin A killed the conidia at the minimal inhibitory concentration of 50 μg/mL, while plipastatin A exhibited a strong fungal activity at 100 μg/mL.

Zalila-Kolsi et al. [19] studied the FZB42 strain of *B. amyloliquefaciens* and found that the commercial bacterial strain produces the lipopeptide bacillomycin D, which contributes to its antimicrobial activity. Bacillomycin D showed a strong antagonism against *F. graminearum* at 30 μg/mL, which is its 50% effective concentration. The plasma membrane morphology and cell wall of *F. graminearum* were affected by bacillomycin D, while inducing the accumulation of reactive oxygen species (ROS) [19]. Furthermore, this lipopeptide caused cell death of the tested *F. graminearum*. Lipopeptide-type compounds from the iturin, fengycin, and surfactin families, synthesized by various strains of *Bacillus*, effectively suppressed the growth of pathogenic microorganisms [26–28,39–41]. These lipopeptides have different residues at specific positions but consist of variants with the same peptide length. Molecules of the iturin lipopeptide family are linked to a β-amino fatty acid of variable length (C_{14}–C_{17}), those of the surfactin family to a β-hydroxyl fatty acid (C_{12}–C_{16}), while fengycin decapeptides are linked to a β-hydroxyl fatty acid chain (C_{14}–C_{18}) [42]. These nonribosomal peptide synthetase-mediated compounds are surface-active and have emulsifying and foaming properties and haemolytic activity [43–45].

Different strains of *Bacillus* produce different groups of lipopeptides [46], and their role in suppressing/controlling plant pathogens may vary. Similar lipopeptides produced by various *Bacillus* strains can also suppress and control other pathogens of economic importance. In a study by Guo et al. [47], the antagonistic effect of the *B. subtilis* strain NCD-2, a fengycin-deficient mutant, was strong against *R. solani* in vitro and suppressed cotton damping-off disease in vivo. In addition, *B. amyloliquefacien* CM-2 and T-5 showed their antagonistic activities against the bacterium *Ralstonia solanacearum* in tomato [28]. The disease incidences were reduced by over 70% by both strains in comparison to the control. On the other hand, crude lipopeptide extracts of *B. amyloliquefaciens* SS-12.6 successfully suppressed leaf spot disease severity on sugar beet plants [48]. These studies showed significant antagonism of the various *Bacillus* strains against various pathogens. Many studies have reported the success of *Bacillus* as a biological control agent against *F. graminearum* in various crops and diseases. However, the potential of these biocontrol agents has not been fully exploited to control other pathogens. Therefore, different strains of *Bacillus* species should be studied further as potential biocontrol agents against other pathogenic microorganisms.

2.2. Surfactins, Fengycins, and Iturins in Bacillus Species

The production of surfactins, fengycins, and iturins by various strains of *B. subtilis* has been reported by numerous researchers [49–58], and a crude lipopeptide mixture of the supernatant of *B. subtilis* was once found to contain these polypeptides [59]. The main congener structures of these cyclic lipopeptide families are shown in Figure 1. Among the most studied in the surfactin family are surfactin linchenysin, pumilacidin WH1, and fungin; for the iturin family, the various iturin isomers—bacillomycins, mycosubtilin—are the best known, while for fengycin, the main compounds are feycin, plipastatin, and agrastatin 1 [60]. An overview of the activity of these three lipopeptides against fungi, with emphasis on *F. graminearum*, is provided in the following sections of this review.

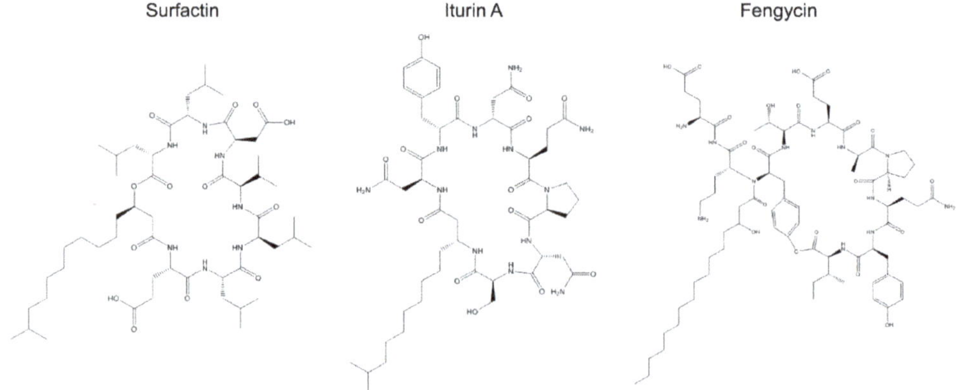

Figure 1. Congener structures of the cyclic lipopeptides; surfactin, iturin A, and fengycin. source: [61].

2.2.1. Surfactins

Surfactins are natural lipopeptides that have been reported to possess antifungal activity [42,61]. They include β-hydoxy hepta cyclic depsipeptides with possibile alanine, valine, leucine, or isoleucine amino acid variations at positions 2, 4, and 7 in the cyclic depsipeptide moiety and C_{13} to C_{16} variation in the β-hydroxy fatty acid chains [62–64]. Surfactin is amphiphilic, with a polar amino acid head and a hydrocarbon chain. This molecular structure makes surfactin a strong biosurfactant, which is at the basis of its antifungal properties. It is assumed that its antibiotic properties are due to its ability to produce selective cationic channels in the membrane phospholipid bilayer [65]. Several studies have been conducted to determine the effect of surfactin on fungi. Qi et al. [66] found a new surfactin, WH1fungin (Figure 2), which induces apoptosis in fungal cells. The same surfactin has also been reported in other studies as an oral immunoadjuvant that could be used for the development of vaccines [67,68]. Surfactin was also found to be effective against the plant pathogenic fungus *Colletotrichum gloeosporiodes* [57]. Another surfactin, Leu[7]-surfactin, produced by *Bacillus mojavensis*, was found to be effective against *Fusarium verticillioides* [69]. A similar inhibitory activity of surfactin was discovered against *F. graminearum* [17], *F. oxysporum* [70], and *Fusarium moniliforme* (presently

F. verticillioides) [71]. This effect on *F. graminearum* can be culture condition-dependent [17,19], with iron concentration being the most important determinant [19].

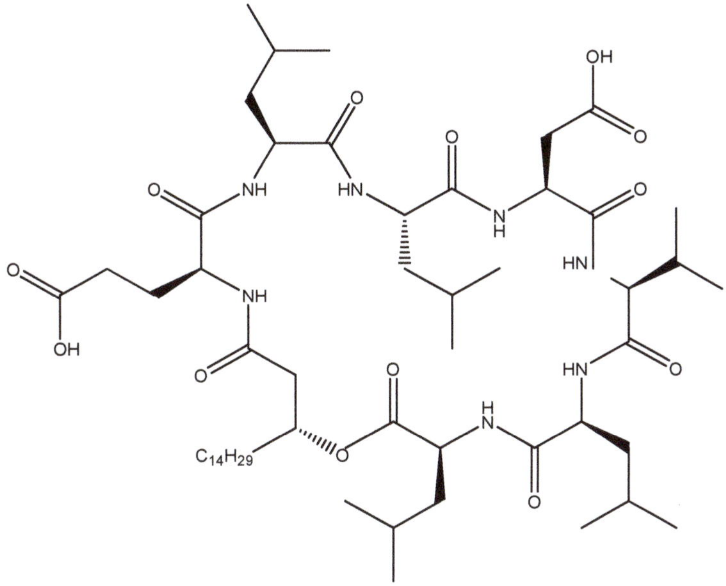

Figure 2. Structure of WH1fungin; source: [72].

2.2.2. Fengycin

The antimicrobial activity of *Bacillus*-produced lipopeptides is based on their chemistry. This is also the case with fengycin, which is a cyclic lipodecapeptide that contains a β-hydroxy fatty acid with a side chain consisting of 16–19 carbon atoms [73]. Fengycin is particularly active against filamentous fungi and inhibits the functions of the enzymes phospholipase A2 and aromatase [73]. It has various isoforms, which differ in length and branching of the β-hydroxy fatty acid moiety, as well as in the amino-acid composition of the peptide ring [50]. For instance, position 6 D-alanine (as in fengycin A) can be replaced by D-valine (as in fengycin B) [73,74]. Fengycin A presents 1 D-Ala, 1 L-Ile, 1 L-Pro, 1 D-allo-Thr, 3 L-Glx, 1 D-Tyr, 1 L-Tyr, 1 D-Orn, whereas in fengyicn B, D-Ala is replaced by D-Val.

Fengycin affects the integrity of biological membranes in a molar-ratio-dependent manner. The effects of fengycin on biological membranes depend on the concentration, but ultimately high concentrations completely disrupt membranes [75]. Fengycins are elicitors of plant defense [76] and have been found to be effective against many fungi including *Magnaporthe grisea* [77], *Plasmodiophora brassicae* [78], *Botryosphaeria dothidea* [79], *C. gloeosporiodes* [57], and a number of other fungi [80]. A cluster of fengycin homologues were found to be effective against *F. verticillioides* [80], *F. solani* [81], *F. solani* f. sp. *radicicola* [80], *F. oxysporum* [25,39], *F. oxysporum* f. sp. *spinaciae* [27], fumonisin production by *F. verticillioides* [82] and proliferation of *F. graminearum* [17,27,80,83–85]. On *F. graminearum*, fengycin causes structural deformations of the hyphae and suppresses in planta proliferation and mycotoxin production [27,84], permeabilization of hyphae [85], and in planta arrest of ear rot development of maize [83]. The study of Liu et al. [86] also revealed that fengycin could block the growth of *F. graminearum*, disrupt cell membrane structure increasing permeability, and create primary lesions in the membrane of fungal cells, thus compromising cell integrity. While the efficacy of fengycin cannot be disputed, its effect on *F. graminearum* can be concentration-dependent [80,86].

2.2.3. Iturin

Iturins exhibit strong fungitoxic properties by forming ion-conducting pores upon contact with fungal membranes. These amphiphilic compounds possess a heptapeptide backbone connected to a C_{13}-to-C_{17} β-amino fatty acid chain [56,87]. Iturins vary in structure, their differences consisting in the type of amino acid residues and in the length and branching of the fatty acid chain. Some examples include iturins A, C, D, and E, bacillomycins D, F, and L, bacillopeptin, and mycosubtilin, all of which are arranged in an LDDLLDL configurational sequence [88]. Length and fatty acid chain branching heterogeneity is clearly demonstrated by iturin A, which has up to 8 isomers with between the 10 to 14 carbons and branching with *n-*, *iso-*, or *anteiso* configurations of the fatty acid chain [89]. Members of the iturin family bacillomycin and bacillopeptin have different amino acids at the third, fourth, and fifth positions. Mycosubtilin, a *B. subtilis*-produced iturin family member, targets, through its sterol group, ergosterol present in the membranes of sensitive fungi [90]. Bacillomycin L is presumed to act by inducing membrane permeabilization and disruption, as well as by targeting intercellular structrues [91]. Iturins have been found to be effective against a number of plant pathogenic fungi, which include *Botrytis, Penicillium, Monilinia* [92], *R. solani* [93], *Colletotricum* [94], *F. oxysporum* [95–97], and *F. graminearum* [19,26].

On *F. graminearum*, iturin causes morphological distortions in conidia and hyphae and severe damage to the plasma membrane, which lead to leakage of the cell contents [26]. Figure 3 illustrates the effects of iturin on *F. graminearum* conidia. Co-cultured with *Bacillus, F. graminearum* is not able to decrease the germination ability of wheat seed [19].

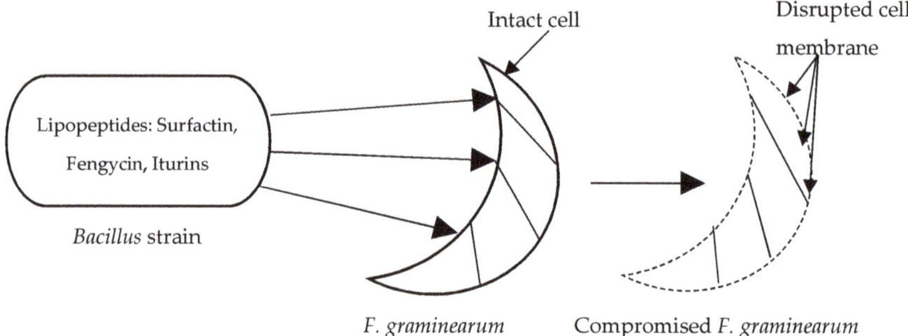

Figure 3. Graphical illustration of *Fusarium graminearum* cell disruption by *Bacillus*.

3. Techniques Applied to Establish Potential Modes of Action of *Bacillus* against *Fusarium graminearum*

The effect of an organism or a substance against the growth of a target organism is traditionally studied by means of bioassays. In a bioassay, the organism is grown in the presence of the antagonist, and its growth monitored over time in comparison to that of an experimental control. Characteristically, a zone of growth inhibition is formed around the inhibited microbe. Various bioassays have been conducted to assess the effect of *B. subtilis* on the growth of *F. graminearum*. Notable is the study of Zhao et al. [27], which clearly demonstrated an antagonism of *Bacillus* against *F. graminearum*, whose mechanism still remains not fully elucidated. If a polypeptide is suspected to be a growth deterrent against target microorganisms, genes (their presence or relative expression) which code for the polypeptide can be detected in the growth culture by means of the polymerase chain reaction (PCR) technique. This was the case in the studies of Arrebola et al. [41], Velho et al. [98], and He et al. [99].

The questions needing answers would then be: What are the antagonistic compounds and how do these antagonistic compounds inhibit growth? Studies based on bioassays analyze the growth medium in which the antagonistic microbe and its target are grown. As part of the biochemical analysis, this growth medium is compared with a control growth medium, and inhibitory compounds are detected. Detection is done using techniques such as liquid chromatography–mass spectrometry (LC–MS). Examples of these studies are those which were conducted to detect and/or analyze surfactin, fengycin, and iturin produced by *Bacillus* against various plant pathogenic fungi [77,100–107]. Two initial scenarios may require this type of testing. The first is when the presence of a specific compound responsible for the antagonistic effect is supposed. This is a targeted analysis, which seeks to confirm the presence of the 'suspected' compound. Alternatively, if the presence of a specific molecule is not presumed, an untargeted analysis to assess culture conditions in comparison to the control is performed. A target analysis follows this untargeted analysis. Studying the antagonistic effect of *Bacillus* against *F. graminearum* for the protection of durum wheat, Zalisa-Kolsi et al. [19] performed an in vitro bioassay, which was followed by an in planta growth inhibitory test. Similarly, in studying the effect of three *Bacillus* strains against *Fusarium*, Dunlap et al. [17] followed a radial diffusion assay with analysis of candidate lipopeptides using high-performance liquid chromatography (HPLC) and a matrix-assisted laser desorption/ionization time-of-flight (MALDI-TOF) system. Zhao et al. [27] performed a similar experiment and discovered an antagonistic effect of *B. subtilis* strain SG6 on *F. graminearum*, as many other similar studies [26,108].

4. Tools for the Detection of Surfactin, Fengycin, and Iturin Genes in *Bacillus* Strains and Culture with Biological Activity against *Fusarium graminearum*

Genomic analysis of *Bacillus* has shown that these bacteria possess genes which code for metabolites associated with biological control [38,109–113]. Genetic information made available by genomic sequencing has led to a better understanding of *Bacillus* biocontrol features. Chen et al. [112] characterized the genome of the *Bacillus velezensis* LM2303 strain, known for its strong biocontrol potential against *F. graminearum*. This strain presented the largest number of biocontrol genes and gene clusters when compared with strains studied earlier. Thirteen biosynthetic gene clusters associated with biocontrol activity were identified using an integrated approach of genome mining and chemical analysis, including the three antifungal metabolites fengycin B, iturin A, and surfactin A [112]. Another strain, *B. velezensis* LM2303, which has antimicrobial activity against *F. graminearum* in addition to other plant pathogenic fungi, also presented a plethora of genes encoding antimicrobial compounds. These findings demonstrated the value of genomic analysis in both biocontrol strain characterization and understanding of the basis of biocontrol activity. A plethora of co-culturing studies have utilized PCR to detect genes involved in biological control in culture, to later identify the basis of their biological control activity. However, biological control genes are sometimes detected in pure *Bacillus* strains undergoing characterization [46]. The study by Adeniji et al. [46] analyzed seven isolates of *Bacillus* with bio-suppressive effects against *F. graminearum* and found them to have valuable gene clusters encoding biocontrol agents. The fingerprint of the combination of genes detected by PCR indicates that strain differentiation and selection are important to identify the strain demonstrating the highest antimicrobial activity as a candidate biocontrol agent. Studies to identify surfactin, fengycin, and iturin in culture are routinely carried out and have uncovered a myriad of antimicrobial substances able to act against plant pathogenic fungi, including *F. graminearum*. These studies make use of combined chromatography and mass spectrometry to identify the compounds which have antagonistic activity. Using reverse-phase high-performance liquid chromatography and electrospray ionization mass spectrometry (RP-HPLC/ESI–MS) analyses, Gong et al. [26] identified iturin and surfactin in a culture of *B. amyloliquefaciens* isolated from wheat infected with *F. graminearum*. Further characterization of iturin showed that it causes leakage and/or inactivation of *F. graminearum* cellular contents. Using thin-layer chromatography–bioautography, Lee et al. [96] identified iturin A in a butanol extract of a culture of *B. amyloliquefaciens* strain DA12, which was found to be active against *F. graminearum*. The same study also attributed this activity to volatile heptanones, some of which were detected using gas chromatography–mass spectrometry (GC–MS). A similar study was performed using ultra-high-performance liquid chromatography coupled with mass spectrometry (UHPLC–MS) to confirm the presence of fengycin B, iturin A, and surfactin A in *B. velezensis* [112]. Also, the study of

Adeniji et al. [46] used electrospray ionization–quadrupole mass spectrometry (ESI–Q-ToF-MS) to detect surfactin, fengycin, and iturin in the *F. graminearum*-supressing *B. velezensis* strain NWUMFkBS10.5. The power of these analytical techniques lies on their sensitivity and accuracy of detection, and their application is critical for, amongst other things, the detection of toxins in food to ensure compliance with food safety standards based on critical threshold values. Moreover, their application to detect bioactive components of *Bacillus* against *F. graminearum* is particularly relevant.

5. Future Prospects and Conclusions

The evidence that *Bacillus* species can act as biocontrol agents against *F. graminearum* encourages the exploitation of *Bacillus* in crop protection and their potential use for organic farming to supplement the despised control measures that pose various environmental hazards and health risks. Ideally, their use may completely replace the current strategies for the control of *F. graminearum* in wheat and other crops. This is supported by various studies conducted to assess the suitability of *Bacillus* to control wheat diseases, in particular FHB. The biofungicide, *B. subtilis* strain QST 713 suspension concentrate (Serenade®ASO) was tested against yellow rust in wheat and showed promising applicability for the control of this fungal infection. However, control tests proved that this biofungicide can be more effective as part of an integrated control strategy than as a standalone remedy [114]. Further work is, therefore, necessary to design an integrated control strategy which utilizes Serenade®ASO together with other organic disease control methods. *B. amyloliquefaciens* CC09 was also reported to have great potential as a biocontrol agent for wheat powdery mildew [115]. The same CC09 strain was found to be effective against take-all disease caused by *Gaeumannomyces graminis* and against a myriad of symptoms caused by *Bipolaris sorokiniana*. This strain effectively colonized the wheat tissue and was found to express genes encoding iturin A synthetase, thereby gaining the name "potential vaccine" [116]. Through its ability to also form spores, *Bacillus* can be an effective biological control agent against *F. graminearum* in wheat. With spore formation, *Bacillus* can overwinter and protect wheat against FHB over several growing seasons. Although the use of biocontrol agents must be extensively tested, ensuring they have a reasonable shelf life, compatibility with other treatments and affordability must be ascertained. Such is not the case with *Bacillus*, which seems to have passed many of these hurdles to become an effective commercial biocontrol product against *F. graminearum*. This is evident in available patents registered, such as those for *Bacillus* species against FHB in cereals [117,118]. The widespread adoption of these patented products to control FHB can benefit organic farming with a healthier and more sustainable wheat product.

Massive screening of various *Bacillus* strains against a wide array of crop pathogens is still nonetheless necessary to identify new antagonistic species. Furthermore, the application of new tools and techniques for assessing the efficacy of biocontrol agents against crop pathogens can accelerate the discovery of new biocontrol strains of *Bacillus*. Equally important is the study of the mechanism of action of *Bacillus* against *F. graminearum*, which should be analyzed more accurately using the new tools of genome-wide studies and the sensitive and accurate platforms of metabolomics. High-resolution techniques of chromatography and mass spectrometry can make the detection of new antagonistic molecules possible even at traceable levels. Specifically, if explored extensively, *Bacillus* may replace in the control *F. graminearum* most of the current widely applied control agents, such as fungicides, and cultural practices which impact negatively on health and the environment.

Author Contributions: Conceptualization, K.N.; writing, K.N., L.K.L., M.E.R., O.A.A., P.B.N.; editing, K.N., L.K.L., M.E.R., O.A.A., P.B.N.; supervision, K.N., P.B.N.

Funding: This study was funded by the National Research Foundation of South Africa under grant number Reference: TTK170413227119.

Conflicts of Interest: The authors declare no conflict of interest.

References

1. Savary, S.; Ficke, A.; Aubertot, J.N.; Hollier, C. Crop losses due to diseases and their implications for global food production losses and food security. *Food Secur.* **2012**, *4*, 519–537. [CrossRef]
2. Savary, S.; Willocquet, L.; Pethybridge, S.J.; Esker, P.; McRoberts, N.; Nelson, A. The global burden of pathogens and pests on major food crops. *Nat. Ecol. Food Evol.* **2019**, *3*, 430–439. [CrossRef] [PubMed]
3. McMullen, M.P.; Jones, R.; Gallenberg, D. Scab of wheat and barley: A re-emerging disease of devastating impact. *Plant Dis.* **1997**, *81*, 1340–1348. [CrossRef] [PubMed]
4. Manning, B.; Southwell, R.; Hayman, P.; Moore, K. *'Fusarium head blight in Northern NSW*; NSW Agriculture: Orange, Australia, 2000.
5. Nganje, W.E.; Bangsund, D.A.; Leistritx, F.L.; Wilson, W.W.; Tlapo, N.M. Regional economic impacts of *Fusarium* head blight in wheat and barley. *Rev. Agric. Econ.* **2004**, *26*, 332–347. [CrossRef]
6. Dweba, C.C.; Figlan, S.; Shimelis, H.A.; Motaung, T.E.; Sydenham, S.; Mwadzingeni, L.; Tsilo, T.J. *Fusarium* head blight of wheat: Pathogenesis and control strategies. *Crop Prot.* **2017**, *91*, 114–122. [CrossRef]
7. Dean, R.; van Kan, J.A.L.; Pretorius, Z.A.; Hammond-Kosack, K.E.; Di Pietro, A.; Spanu, P.D.; Rudd, J.J.; Dickman, M.; Kahmann, R.; Ellis, J.; et al. The top-10 fungal pathogens in molecular plant pathology. *Plant Pathol.* **2012**, *13*, 414–430.
8. Pestka, J.J.; Smolinski, A.T. Deoxynivalenol: Toxicology and potential effects on humans. *J. Toxicol. Environ. Heal. B Crit. Rev.* **2005**, *8*, 39–69. [CrossRef]
9. Mahmoud, A.F. Genetic variation and biological control of *Fusarium graminearum* isolated from wheat in Assiut-Egypt. *Plant Pathol.* **2015**, *32*, 145–156. [CrossRef]
10. Desjardins, A.E.; Hohn, T.M.; McCormick, S.P. Trichothecene biosynthesis in *Fusarium* species: Chemistry, genetics, and significance. *Microbiol. Mol. Biol. Rev.* **1993**, *157*, 595–604.
11. Desjardins, A.E.; Hohn, T.M. Mycotoxins in plant pathogenesis. *Mol. Plant Microbe Interact.* **1997**, *10*, 147–152. [CrossRef]
12. Goswami, R.S.; Kistler, H.C. Heading for disaster: *Fusarium graminearum* on cereal crops. *Mol. Plant Pathol.* **2004**, *5*, 515–525. [CrossRef] [PubMed]
13. Zhang, Y.J.; Yu, J.J.; Zhang, Y.N.; Zhang, X.; Cheng, C.J.; Wang, J.X.; Hollomon, D.W.; Fan, P.S.; Zhou, M.G. Effect of carbendazim resistance on trichothecene production and aggressiveness of *Fusarium graminearum*. *Mol. Plant Microbe Interact.* **2009**, *22*, 1143–1150. [CrossRef] [PubMed]
14. Shafi, J.; Tian, H.; Ji, M. *Bacillus* species as versatile weapons for plant pathogens: A review. *Biotechnol. Biotech. Equip.* **2017**, *31*, 446–459. [CrossRef]
15. Zeigler, D.R.; Perkins, J.B. The genus *Bacillus*. In *Practical Handbook of Microbiology*; Goldman, E., Green, L.H., Eds.; CRC Press: Boca Raton, FL, USA, 2018; pp. 309–326.
16. Chitlaru, T.; Altboum, Z.; Reuveny, S.; Shafferman, A. Progress and novel strategies in vaccine development and treatment of Anthrax. *Immunol. Rev.* **2011**, *239*, 221–236. [CrossRef]
17. Dunlap, C.A.; Schisler, D.A.; Price, N.P.; Vaughn, S.F. Cyclic lipopeptide profile of three *Bacillus subtilis* strains; antagonists of *Fusarium* head blight. *J. Microbiol.* **2011**, *49*, 603–609. [CrossRef]
18. Compant, S.; Duffy, B.; Nowak, J.; Clement, C.; Barka, E.A. Use of plant growth promoting bacteria for biocontrol of plant diseases: Principles, mechanisms of action, and future prospects. *Appl. Environ Microbiol.* **2005**, *71*, 4951–4959. [CrossRef]
19. Zalila-Kolsi, I.; Mahmoud, A.B.; Ali, H.; Sellami, S.; Nasfi, Z.; Tounsi, S.; Jamoussi, K. Antagonist effects of *Bacillus* spp. strains against *Fusarium graminearum* for protection of durum wheat (*Triticum turgidum* L. subsp. *durum*). *Microbiol. Res.* **2016**, *192*, 148–158. [CrossRef]
20. Whipps, J.M. Microbial interactions and biocontrol in the rhizosphere. *J. Exp. Bot.* **2001**, *511*, 487–511. [CrossRef]
21. Cawoy, H.; Debois, D.; Franzil, L.; De Pauw, E.; Thonart, P.; Ongena, M. Lipopeptides as main ingredients for inhibition of fungal phytopathogens by *Bacillus subtilis/amyloliquefaciens*. *Microb. Biotechnol.* **2015**, *8*, 281–295. [CrossRef]
22. Khan, N.; Maymon, M.; Hirsch, A.M. Combating *Fusarium* infection using *Bacillus*-based antimicrobials. *Microorganisms* **2017**, *5*, 75. [CrossRef]
23. Radhakrishnan, R.; Hashem, A.; Abd Allah, E.F. *Bacillus*: A biological tool for crop improvement through bio-molecular changes in adverse environments. *Front. Physiol.* **2017**, *8*, 667. [CrossRef] [PubMed]

24. Ayed, H.B.; Hmidet, N.; Béchet, M.; Chollet, M.; Chataigné, G.; Leclère, V.; Jacques, P.; Nasri, M. Identification and biochemical characteristics of lipopeptides from *Bacillus mojavensis* A21. *Process Biochem.* **2014**, *49*, 1699–1707. [CrossRef]

25. Cao, Y.; Xu, Z.; Ling, N.; Yuan, Y.; Yang, X.; Chen, L.; Shen, B.; Shen, Q. Isolation and identification of lipopeptides produced by *B. subtilis* SQR 9 for suppressing *Fusarium* wilt of cucumber. *Sci. Hortic.* **2012**, *135*, 32–39. [CrossRef]

26. Gong, A.D.; Li, H.P.; Yuan, Q.S.; Song, X.S.; Yao, W.; He, W.J.; Zhang, J.B.; Liao, Y.C. Antagonistic mechanism of iturin A and plipastatin A from *Bacillus amyloliquefaciens* S76-3 from wheat spikes against *Fusarium graminearum*. *PLoS ONE* **2015**, *10*, e0116871. [CrossRef] [PubMed]

27. Zhao, Y.; Selvaraj, J.N.; Xing, F.; Zhou, L.; Wang, Y.; Song, H.; Tan, X.; Sun, L.; Sangare, L.; Folly, Y.M.E.; et al. Antagonistic action of *Bacillus subtilis* strain SG6 on *Fusarium graminearum*. *PLoS ONE* **2014**, *9*, e92486. [CrossRef]

28. Tan, S.; Dong, Y.; Liao, H.; Huang, J.; Song, S.; Xu, Y.; Shen, Q. Antagonistic bacterium *Bacillus amyloliquefaciens* induces resistance and controls the bacterial wilt of tomato. *Pest Manag. Sci.* **2013**, *69*, 1245–1252.

29. Földes, T.; Bánhegyi, I.; Herpai, Z.; Varga, L.; Szigeti, J. Isolation of *Bacillus* strains from the rhizosphere of cereals and in vitro screening for antagonism against phytopathogenic, food-borne pathogenic and spoilage micro-organisms. *J. Appl. Microbiol.* **2000**, *89*, 840–846. [CrossRef]

30. Gross, E.; Kiltz, H.H.; Nebelin, E.; Subtilin, V.I. Die Struktur des Subtilins. Hoppe-Seyler Z. *Physiol. Chem.* **1973**, *354*, 810–812.

31. Walker, J.E.; Abraham, E.P. The structure of bacilysin and other products of *Bacillus subtilis*. *Biochem. J.* **1970**, *118*, 563–570. [CrossRef]

32. Sengupta, S.; Banerjee, A.B.; Bose, S.K. γ-Glutamyl and D-or L-peptide linkages in mycobacillin, a cyclic peptide antibiotic. *Biochem. J.* **1971**, *121*, 839–846. [CrossRef]

33. Besson, F.; Peypoux, F.; Michel, G.; Delcambe, L. The structure of bacillomycin L, an antibiotic from *Bacillus subtilis*. *Eur. J. Biochem.* **1977**, *77*, 61–67. [CrossRef] [PubMed]

34. Peypoux, F.; Marion, D.; Maget-Dana, R. Structure of bacillomycin F, a new peptidolipid antibiotic of the iturin group. *Eur. J. Biochem.* **1985**, *153*, 335–340. [CrossRef] [PubMed]

35. Peypoux, F.; Michel, G.; Delcambe, L. The structure of mycosubtilin, an antibiotic isolated from *Bacillus subtilis*. *Eur. J. Biochem.* **1976**, *63*, 391–398. [CrossRef] [PubMed]

36. Peypoux, F.; Pommier, M.T.; Marion, D.; Ptak, M.; Das, B.C.; Michel, G. Revised structure of mycosubtilin, a peptidolipid antibiotic from *Bacillus subtilis*. *J. Antibiot.* **1986**, *39*, 636–641. [CrossRef] [PubMed]

37. Zeriouh, H.; Romero, D.; Garcia-Gutierrez, L.; Cazorla, F.M.; de Vicente, A.; Perez-Garcia, A. The iturin-like lipopeptides are essential components in the biological control arsenal of *Bacillus subtilis* against bacterial diseases of cucurbits. *Mol. Plant Microbe Interact.* **2011**, *24*, 1540–1552. [CrossRef] [PubMed]

38. Dunlap, C.A.; Bowman, M.J.; Schisler, D.A. Genomic analysis and secondary metabolite production in *Bacillus amyloliquefaciens* AS 43.3: A biocontrol antagonist of *Fusarium* head blight. *Biol. Control* **2013**, *64*, 166–175. [CrossRef]

39. Yuan, J.; Raza, W.; Huang, Q.; Shen, Q. The ultrasound-assisted extraction and identification of antifungal substances from *B. amyloliquefaciens* strain NJN-6 suppressing *Fusarium oxysporum*. *J. Basic Microbiol.* **2012**, *52*, 721–730. [CrossRef]

40. Kumar, P.; Dubey, R.C.; Maheshwari, D.K. *Bacillus* strains isolated from rhizosphere showed plant growth promoting and antagonistic activity against phytopathogens. *Microbiol. Res.* **2012**, *167*, 493–499. [CrossRef]

41. Arrebola, E.; Jacobs, R.; Korsten, L. Iturin A is the principal inhibitor in the biocontrol activity of *Bacillus amyloliquefaciens* PPCB004 against postharvest fungal pathogens. *J. Appl. Microbiol.* **2010**, *108*, 386–395. [CrossRef] [PubMed]

42. Stein, T. *Bacillus subtilis* antibiotics: Structure, syntheses and specific functions. *Mol. Microbiol.* **2005**, *56*, 845–857. [CrossRef]

43. Roongsawang, N.; Washio, K.; Morikawa, M. Diversity of nonribosomal peptide synthetases involved in the biosynthesis of lipopeptide biosurfactants. *Int. J. Mol. Sci.* **2011**, *12*, 141–172. [CrossRef] [PubMed]

44. Deleu, M.; Razafindralambo, H.; Popineau, Y.; Jacques, P.; Thonard, P.; Paquot, M. Interfacial and emulsifying properties of lipopeptides from *Bacillus subtilis*. *Coll. Surf. A Physicochem. Eng. Asp.* **1999**, *152*, 3–10. [CrossRef]

45. Peypoux, F.; Bonmatin, J.; Wallach, J. Recent trends in the biochemistry of surfactin. *Appl. Microbiol. Biotechnol.* **1999**, *51*, 553–563. [CrossRef] [PubMed]

46. Adeniji, A.A.; Aremu, O.S.; Babalola, O.O. Selecting lipopeptide-producing, *Fusarium*-suppressing *Bacillus* spp.: Metabolomic and genomic probing of *Bacillus velezensis* NWUMFkBS10. 5. *MicrobiologyOpen* **2019**, *8*, e00742. [CrossRef] [PubMed]

47. Guo, Q.; Dong, W.; Li, S.; Lu, X.; Wang, P.; Zhang, X.; Wang, Y.; Ma, P. Fengycin produced by *Bacillus subtilis* NCD-2 plays a major role in biocontrol of cotton seedling damping-off disease. *Microbiol. Res.* **2014**, *169*, 533–540. [CrossRef]

48. Nikolić, I.; Berić, T.; Dimkić, I.; Popović, T.; Lozo, J.; Fira, D.; Stanković, S. Biological control of *Pseudomonas syringae* pv. *aptata* on sugar beet with *Bacillus pumilus* SS-10.7 and *Bacillus amyloliquefaciens* (SS-12.6 and SS-38.4) strains. *J. Appl. Microbiol.* **2019**, *126*, 165–176.

49. Winkelmann, H.; Allgaier, H.; Lu, R.; Jung, G. Iturin AL—A new long chain iturin A possessing an unusual high content of C16 β amino acids. *J. Antibiot.* **1983**, *11*, 1451–1457. [CrossRef]

50. Vanittanakom, N.; Loeffler, W.; Koch, U.; Jung, G. Fengycin—A novel antifungal lipopeptide antibiotic produced by *Bacillus subtilis* F-29-2. *J. Antibiot.* **1986**, *7*, 888–901. [CrossRef]

51. Vollenbroich, D.; Ozel, M.; Vater, J.; Kamp, R.M.; Pauli, G. Mechanism of inactivation of enveloped viruses by the biosurfactant surfactin from *Bacillus subtilis*. *Biologicals* **1997**, *25*, 289–297. [CrossRef]

52. Vollenbroich, D.; Pauli, G.; Ozel, M.; Vater, J. Antimycoplasma properties and application in cell culture of surfactin, a lipopeptide antibiotic from *Bacillus subtilis*. *Appl. Environ. Microbiol.* **1997**, *63*, 44–49.

53. Kracht, M.; Rokos, H.; Ozel, M.; Kowall, M.; Pauli, G.; Vater, J. Antiviral and hemolytic activities of surfactin isoforms and their methyl ester derivatives. *J. Antibiot.* **1999**, *52*, 613–619. [CrossRef] [PubMed]

54. Kim, S.Y.; Kim, J.M.; Kim, S.H.; Bae, H.J.; Yi, H.; Yoon, S.H.; Koo, B.S.; Kwon, M.; Cho, J.Y.; Lee, C.E.; et al. Surfactin from *Bacillus subtilis* displays anti-proliferative effect via apoptosis induction, cell cycle arrest and survival signalling suppression. *FEBS Lett.* **2007**, *581*, 865–871. [CrossRef] [PubMed]

55. Nagorska, K.; Bikowski, M.; Obuchowki, M. Multicellular behaviour and production of a wide variety of toxic substance support usage of *Bacillus subtilis* as powerful biocontrol agent. *Acta. Biochim. Pol.* **2007**, *54*, 495–508. [PubMed]

56. Ongena, M.; Jacques, P. *Bacillus* lipopeptides: Versatile weapons for plant disease biocontrol. *Trends Microbiol.* **2008**, *16*, 115–124. [CrossRef]

57. Kim, P.I.; Ryu, J.; Kim, Y.H.; Chi, Y.T. Production of biosurfactant lipopeptides iturin A, fengycin and surfactin from *Bacillus subtilis* CMB32 for control of *Colletotrichum gloeosporides*. *J. Microbiol. Biotechnol.* **2010**, *20*, 138–145.

58. Geetha, L.; Manonmani, A.M.; Paily, K.P. Identification and characterization of a mosquito pupicidal metabolite of *Bacillus subtilis* subsp. *subtilis* strain. *Appl. Microbiol. Biotechnol.* **2010**, *86*, 1737–1744. [CrossRef]

59. Pathak, K.V.; Keharia, H. Identification of surfactins and iturins produced by potent fungal antagonist, *Bacillus subtilis* K1 isolated from aerial roots of banyan (*Ficus benghalensis*) tree using mass spectrometry. *3 Biotech.* **2014**, *4*, 283–295. [CrossRef]

60. Beltran-Gracia, E.; Macedo-Raygoza, G.; Villafaña-Rojas, J.; Martinez-Rodriguez, A.; Chavez-Castrillon, Y.Y.; Espinosa-Escalante, F.M.; Di Mascio, P.; Ogura, T.; Beltran-Garcia, M.J. Production of lipopeptides by fermentation processes: Endophytic bacteria, fermentation strategies and easy methods for bacterial selection. In *Fermentation Processes*, 1st ed.; Menestrina, G., Serra, M.D., Jozala, A.F., Eds.; Intech Open Science: London, UK, 2017; pp. 260–271.

61. Geissler, M.; Oellig, C.; Moss, K.; Schwack, W.; Henkel, M.; Hausmann, R. High-performance thin-layer chromatography (HPTLC) for the simultaneous quantification of the cyclic lipopeptides surfactin, iturin A and fengycin in culture samples of Bacillus species. *J. Chromatogr. B* **2017**, *1044*, 214–224. [CrossRef]

62. Peypoux, F.; Bonmatin, J.M.; Labbe, H.; Grangemard, I.; Das, B.C.; Ptak, M.; Wallach, J.; Michel, G. [Ala4] surfactin, a novel isoform from *Bacillus subtilis* studied by mass and NMR spectroscopies. *Eur. J. Biochem.* **1994**, *224*, 89–96. [CrossRef]

63. Kowall, M.; Vater, J.; Kluge, T.; Stein, P.; Ziessow, D. Separation and characterization of surfactin isoforms produced by *Bacillus subtilis* OKB 105. *J. Coll. Interface Sci.* **1998**, *204*, 1–8. [CrossRef]

64. Hue, N.; Serani, L.; Laprevote, O. Structural investigation of cyclic peptidolipids from *Bacillus subtilis* by high energy tandem mass spectrometry. *Rapid Commun. Mass Spectrom.* **2001**, *15*, 203–209. [CrossRef]

65. Sheppard, J.D.; Jumarie, C.; Cooper, D.G.; Laprade, R. Ionic channels induced by surfactin in planar lipid bilayer membranes. *Biochim. Biophys. Acta* **1991**, *1064*, 13–23. [CrossRef]

66. Qi, G.; Zhu, F.; Du, P.; Yang, X.; Qiu, D.; Yu, Z.; Chen, J.; Zhao, X. Lipopeptide induces apoptosis in fungal cells by a mitochondria-dependent pathway. *Peptides* **2010**, *31*, 1978–1986. [CrossRef] [PubMed]

67. Gao, Z.; Wang, S.; Qi, G.; Pan, H.; Zhang, L.; Zhou, X.; Liu, J.; Zhao, X.; Wu, J.A. Surfactin cyclopeptide of WH1 fungin used as a novel adjuvant for intramuscular and subcutaneous immunization in mice. *Peptides* **2012**, *38*, 163–171. [CrossRef] [PubMed]

68. Gao, Z.; Zhao, X.; Lee, S.; Li, J.; Liao, H.; Zhou, X.; Wu, J.; Qi, G. WH1fungin a surfactin cyclic lipopeptide is a novel oral immunoadjuvant. *Vaccine* **2013**, *31*, 2796–2803. [CrossRef] [PubMed]

69. Snook, M.E.; Mitchell, T.; Hinton, D.M.; Bacon, C.W. Isolation and characterization of Leu7-surfactin from the endophytic bacterium *Bacillus mojavensis* RRC 101, a biocontrol agent for *Fusarium verticillioides*. *J. Agric. Food Chem.* **2009**, *57*, 4287–4292. [CrossRef]

70. Vitullo, D.; Di Pietro, A.; Romano, A.; Lanzotti, V.; Lima, G. Role of new bacterial surfactins in the antifungal interaction between *Bacillus amyloliquefaciens* and *Fusarium oxysporum*. *Plant Pathol.* **2012**, *61*, 689–699. [CrossRef]

71. Jiang, J.; Gao, L.; Bie, X.; Lu, Z.; Liu, H.; Zhang, C.; Lu, F.; Zhao, H. Identification of novel surfactin derivatives from NRPS modification of *Bacillus subtilis* and its antifungal activity against *Fusarium moniliforme*. *BMC Microbiol.* **2016**, *16*, 31. [CrossRef]

72. Nielsen, D.S.; Shepherd, N.E.; Xu, W.; Lucke, A.J.; Stoermer, M.J.; Fairlie, D.P. Orally absorbed cyclic peptides. *Chem. Rev.* **2017**, *117*, 8094–8128. [CrossRef]

73. Steller, S.; Vater, J. Purification of the fengycin synthetase multienzyme system from *Bacillus subtilis* b213. *J. Chromatogr. B Biomed. Sci. Appl.* **2000**, *737*, 267–275. [CrossRef]

74. Vater, J.; Kablitz, B.; Wilde, C.; Franke, P.; Mehta, N.; Cameotra, S.S. Matrix-assisted laser desorption ionization-time of flight mass spectrometry of lipopeptide biosurfactants in whole cells and culture filtrates of *Bacillus subtilis* C-1 isolated from petroleum sludge. *Appl. Environ. Microbiol.* **2002**, *68*, 6210–6219. [CrossRef] [PubMed]

75. Deleu, M.; Paquot, M.; Nylander, T. Fengycin interaction with lipid monolayers at the air-aqueous interface-implications for the effect of fengycin on biological membranes. *J. Coll. Interface Sci.* **2005**, *283*, 358–365. [CrossRef] [PubMed]

76. Ongena, M.; Jourdan, E.; Adam, A.; Paquot, M.; Brans, A.; Joris, B.; Arpigny, J.L.; Thonart, P. Surfactin and fengycin lipopeptides of *Bacillus subtilis* as elicitors of induced systemic resistance in plants. *Environ. Microbiol.* **2007**, *9*, 1084–1090. [CrossRef] [PubMed]

77. Zhang, L.; Sun, C. Fengycins, cyclic lipopeptides from marine *Bacillus subtilis* strains, kill the plant-pathogenic fungus *Magnaporthe grisea* by inducing reactive oxygen species production and chromatin condensation. *Appl. Environ. Microbiol.* **2018**, *84*, e00445-18. [CrossRef] [PubMed]

78. Li, X.Y.; Yang, J.J.; Mao, Z.C.; Ho, H.H.; Wu, Y.X.; He, Y.Q. Enhancement of biocontrol activities and cyclic lipopeptides production by chemical mutagenesis of *Bacillus subtilis* XF-1, a biocontrol agent of *Plasmodiophora brassicae* and *Fusarium solani*. *Indian J. Microbiol.* **2014**, *54*, 476–479. [CrossRef]

79. Fan, H.; Ru, J.; Zhang, Y.; Wang, Q.; Li, Y. Fengycin produced by *Bacillus subtilis* 9407 plays a major role in the biocontrol of apple ring rot disease. *Microbiol. Res.* **2017**, *199*, 89–97. [CrossRef]

80. Li, L.; Ma, M.; Huang, R.; Qu, Q.; Li, G.; Zhou, J.; Zhang, K.; Lu, K.; Niu, X.; Luo, J. Induction of chlamydospore formation in *Fusarium* by cyclic lipopeptide antibiotics from *Bacillus subtilis* C2. *J. Chem. Ecol.* **2012**, *38*, 966–974. [CrossRef]

81. Li, B.; Li, Q.; Xu, Z.; Zhang, N.; Shen, Q.; Zhang, R. Responses of beneficial *Bacillus amyloliquefaciens* SQR9 to different soilborne fungal pathogens through the alteration of antifungal compounds production. *Front. Microbiol.* **2014**, *5*, 636. [CrossRef]

82. Hu, L.B.; Zhang, T.; Yang, Z.M.; Zhou, W.; Shi, Z.Q. Inhibition of fengycins on the production of fumonisin B1 from *Fusarium verticillioides*. *Lett. Appl. Microbiol.* **2009**, *48*, 84–89. [CrossRef]

83. Chan, Y.K.; Savard, M.E.; Reid, L.M.; Cyr, T.; McCormick, W.A.; Seguin, C. Identification of lipopeptide antibiotics ogyjuygthkjf a *Bacillus subtilis* isolate and their control of *Fusarium graminearum* diseases in maize and wheat. *BioControl* **2009**, *54*, 567. [CrossRef]

84. Hanif, A.; Zhang, F.; Li, P.; Li, C.; Xu, Y.; Zubair, M.; Zhang, M.; Jia, D.; Zhao, X.; Liang, J.; et al. Fengycin produced by *Bacillus amyloliquefaciens* FZB42 inhibits *Fusarium graminearum* growth and mycotoxins biosynthesis. *Toxins* **2019**, *11*, 295. [CrossRef] [PubMed]

85. Wang, J.; Liu, J.; Chen, H.; Yao, J. Characterization of *Fusarium graminearum* inhibitory lipopeptide from *Bacillus subtilis* IB. *Appl. Microbiol. Biotechnol.* **2007**, *76*, 889–894. [CrossRef] [PubMed]

86. Liu, Y.; Lu, J.; Sun, J.; Bie, X.; Lu, Z. Membrane disruption and DNA binding of *Fusarium graminearum* cell induced by C16-Fengycin A produced by *Bacillus amyloliquefaciens*. *Food Cont.* **2019**, *102*, 206–213. [CrossRef]

87. Aranda, F.J.; Teruel, J.A.; Ortiz, A. Further aspects on the haemolytic activity of the antibiotic lipopeptide iturin A. *Biochem. Biophys. Acta* **2005**, *1713*, 51–56. [CrossRef]

88. Bland, J.M. The first synthesis of a member of the iturin family, the antifungal cyclic lipopeptide, iturin-A2. *J. Org. Chem.* **1996**, *61*, 5663–5664. [CrossRef]

89. Isogai, A.; Takayama, S.; Murakoshi, S.; Suzuki, A. Structure of β-amino acids in antibiotics iturin A. *Tetrahedron Lett.* **1982**, *23*, 3065–3068. [CrossRef]

90. Nasir, M.N.; Besson, F. Interactions of the antifungal mycosubtilin with ergosterol-containing interfacial monolayers. *Biochim. Biophys. Acta* **2012**, *1818*, 1302–1308. [CrossRef]

91. Zhang, B.; Dong, C.; Shang, Q.; Han, Y.; Li, P. New insights into membrane-active action in plasma membrane of fungal hyphae by the lipopeptide antibiotic bacillomycin L. *Biochim. Biophys. Acta* **2013**, *1828*, 2230–2237. [CrossRef]

92. Calvo, H.; Mendiara, I.; Arias, E.; Blanco, D.; Venturini, M.E. The role of iturin A from *B. amyloliquefaciens* BUZ-14 in the inhibition of the most common postharvest fruit rots. *Food Microbiol.* **2019**, *82*, 62–69. [CrossRef]

93. Zohora, U.S.; Ano, T.; Rahman, M.S. Biocontrol of *Rhizoctonia solani* K1 by iturin A producer *Bacillus subtilis* RB14 seed treatment in tomato plants. *Adv. Microbiol.* **2016**, *6*, 424–431. [CrossRef]

94. Arroyave-Toro, J.J.; Mosquera, S.; Villegas-Escobar, V. Biocontrol activity of *Bacillus subtilis* EA-CB0015 cells and lipopeptides against postharvest fungal pathogens. *Biol. Control* **2017**, *114*, 195–200. [CrossRef]

95. Fujita, S.; Yokota, K. Disease suppression by the cyclic lipopeptides iturin A and surfactin from *Bacillus* spp. against *Fusarium* wilt of lettuce. *J. Gen. Plant Pathol.* **2019**, *85*, 44–48. [CrossRef]

96. Lee, T.; Park, D.; Kim, K.; Lim, S.M.; Yu, N.H.; Kim, S.; Kim, H.Y.; Jung, K.S.; Jang, J.Y.; Park, J.C.; et al. Characterization of *Bacillus amyloliquefaciens* DA12 showing potent antifungal activity against mycotoxigenic *Fusarium* species. *Plant Pathol. J.* **2017**, *33*, 499–507. [CrossRef] [PubMed]

97. Cao, Y.; Pi, H.; Chandrangsu, P.; Li, Y.; Wang, Y.; Zhou, H.; Xiong, H.; Helmann, J.D.; Cai, Y. Antagonism of two plant-growth promoting *Bacillus velezensis* isolates against *Ralstonia solanacearum* and *Fusarium oxysporum*. *Sci. Rep.* **2018**, *8*, 4360. [CrossRef] [PubMed]

98. Velho, R.V.; Medina, L.F.C.; Segalin, J.; Brandelli, A. Production of lipopeptides among *Bacillus* strains showing growth inhibition of phytopathogenic fungi. *Folia Microbiol.* **2011**, *56*, 297–303. [CrossRef] [PubMed]

99. He, Y.; Zhu, M.; Huang, J.; Hsiang, T.; Zheng, L. Biocontrol potential of a *Bacillus subtilis* strain BJ-1 against the rice blast fungus *Magnaporthe oryzae*. *Can. J. Plant Pathol.* **2019**, *41*, 47–59. [CrossRef]

100. Koumoutsi, A.; Chen, X.H.; Henne, A.; Liesegang, H.; Hitzero th, G.; Franke, P.; Vater, J.; Borriss, R. Structural and functional characterization of gene clusters directing nonribosomal synthesis of bioactive cyclic lipopeptides in *Bacillus amyloliquefaciens* strain FZB42. *J. Bacteriol.* **2004**, *186*, 1084–1096. [CrossRef]

101. Nanjundan, J.; Ramasamy, R.; Uthandi, S.; Ponnusamy, M. Antimicrobial activity and spectroscopic characterization of surfactin class of lipopeptides from *Bacillus amyloliquefaciens* SR1. *Microb. Pathog.* **2019**, *128*, 374–380. [CrossRef]

102. Ding, L.; Guo, W.; Chen, X. Exogenous addition of alkanoic acids enhanced production of antifungal lipopeptides in *Bacillus amyloliquefaciens* Pc3. *Appl. Microbiol. Biotechnol.* **2019**, *103*, 5367–5377. [CrossRef]

103. Saechow, S.; Thammasittirong, A.; Kittakoop, P.; Prachya, S.; Thammasittirong, S.N.R. Antagonistic activity against dirty panicle rice fungal pathogens and plant growth-promoting activity of *Bacillus amyloliquefaciens* BAS23. *J. Microbiol. Biotechnol.* **2018**, *28*, 1527–1535. [CrossRef]

104. Sa, R.B.; An, X.; Sui, J.K.; Wang, X.H.; Ji, C.; Wang, C.Q.; Li, Q.; Hu, Y.R.; Liu, X. Purification and structural characterization of fengycin homologues produced by *Bacillus subtilis* from poplar wood bark. *Australas. Plant Pathol.* **2018**, *47*, 259–268. [CrossRef]

105. Zouari, I.; Jlaiel, L.; Tounsi, S.; Trigui, M. Biocontrol activity of the endophytic *Bacillus amyloliquefaciens* strain CEIZ-11 against *Pythium aphanidermatum* and purification of its bioactive compounds. *Biol. Control* **2016**, *100*, 54–62. [CrossRef]

106. Kaur, P.K.; Joshi, N.; Singh, I.P.; Saini, H.S. Identification of cyclic lipopeptides produced by *Bacillus vallismortis* R2 and their antifungal activity against *Alternaria alternata*. *J. Appl. Microbiol.* **2017**, *122*, 139–152. [CrossRef] [PubMed]

107. Ji, S.H.; Paul, N.C.; Deng, J.X.; Kim, Y.S.; Yun, B.S.; Yu, S.H. Biocontrol activity of *Bacillus amyloliquefaciens* CNU114001 against fungal plant diseases. *Mycobiology* **2013**, *41*, 234–242. [CrossRef]
108. Palyzová, A.; Svobodová, K.; Sokolová, L.; Novák, J.; Novotný, Č. Metabolic profiling of *Fusarium oxysporum* *f.* sp. *conglutinans* race 2 in dual cultures with biocontrol agents *Bacillus amyloliquefaciens*, *Pseudomonas aeruginosa*, and *Trichoderma harzianum*. *Folia Microbiol.* **2019**. [CrossRef]
109. Blom, J.; Rueckert, C.; Niu, B.; Wang, Q.; Borriss, R. The complete genome *of Bacillus amyloliquefaciens* subsp. *plantarum* CAU B946 contains a gene cluster for non-ribosomal synthesis of Iturin A. *J. Bacteriol.* **2012**, *194*, 1845–1846. [CrossRef] [PubMed]
110. Chen, X.H.; Koumoutsi, A.; Scholz, R.; Eisenreich, A.; Schneider, K.; Heinemeyer, I.; Morgenstern, B.; Voss, B.; Hess, W.R.; Reva, O.; et al. Comparative analysis of the complete genome sequence of the plant growth–promoting bacterium *Bacillus amyloliquefaciens* FZB42. *Nat. Biotechnol.* **2007**, *25*, 1007–1014. [CrossRef]
111. Chen, L.; Heng, J.; Qin, S.; Bian, K. A comprehensive understanding of the biocontrol potential of *Bacillus velezensis* LM2303 against *Fusarium* head blight. *PLoS ONE* **2018**, *13*, e0198560. [CrossRef]
112. Deng, Y.; Zhu, Y.; Wang, P.; Zhu, L.; Zheng, J.; Li, R.; Ruan, L.; Peng, D.; Sun, M. Complete genome sequence of *Bacillus subtilis* BSn5, an endophytic bacterium of *Amorphophallus konjac* with antimicrobial activity for the plant pathogen *Erwinia carotovora* subsp. *carotovora*. *J. Bacteriol.* **2011**, *193*, 2070–2071. [CrossRef]
113. Deng, Q.; Wang, R.; Sun, D.; Sun, L.; Wang, Y.; Pu, Y.; Fang, Z.; Xu, D.; Liu, Y.; Ye, R.; et al. Complete genome of *Bacillus velezensis* CMT-6 and comparative genome analysis reveals lipopeptide diversity. *Biochem. Genet.* **2019**. [CrossRef]
114. Reiss, A.; Jørgensen, L.N. Biological control of yellow rust of wheat (*Puccinia striiformis*) with Serenade® ASO (*Bacillus subtilis* strain QST713). *Crop Prot.* **2017**, *93*, 1–8. [CrossRef]
115. Cai, X.C.; Liu, C.H.; Wang, B.T.; Xue, Y.R. Genomic and metabolic traits endow *Bacillus velezensis* CC09 with a potential biocontrol agent in control of wheat powdery mildew disease. *Microbiol. Res.* **2017**, *196*, 89–94. [CrossRef] [PubMed]
116. Kang, X.; Zhang, W.; Cai, X.; Zhu, T.; Xue, Y.; Liu, C. *Bacillus velezensis* CC09: A potential 'vaccine' for controlling wheat diseases. *Mol. Plant Microbe Interact.* **2018**, *31*, 623–632. [CrossRef] [PubMed]
117. Schisler, D.A.; Khan, N.I.; Boehm, M.J. Ohio State University Research Foundation and US Department of Agriculture. *Bacillus* species NRRL B-30212 for reducing *Fusarium* head blight in cereals. U.S. Patent 7,001,755, 21 February 2006.
118. Da Luz, W.C. Empresa Brasileira de Pesquisa Agropecuaria EMBRAPA, Biocontrol of plant diseases caused by *Fusarium* species with novel isolates of *Pantoea agglomerans*. U.S. Patent 7,118,739, 10 October 2006.

 © 2019 by the authors. Licensee MDPI, Basel, Switzerland. This article is an open access article distributed under the terms and conditions of the Creative Commons Attribution (CC BY) license (http://creativecommons.org/licenses/by/4.0/).

Review

Biological Control of Citrus Postharvest Phytopathogens

Jaqueline Moraes Bazioli [1,2], João Raul Belinato [1], Jonas Henrique Costa [1], Daniel Yuri Akiyama [1], João Guilherme de Moraes Pontes [1], Katia Cristina Kupper [3], Fabio Augusto [1], João Ernesto de Carvalho [2] and Taícia Pacheco Fill [1,*]

1   Institute of Chemistry, Universidade Estadual de Campinas, CP 6154, 13083-970 Campinas, SP, Brazil
2   Faculty of Pharmaceutical Sciences, Universidade Estadual de Campinas, 13083-859 Campinas, SP, Brazil
3   Instituto Agronômico de Campinas (IAC), 13490-970 Cordeiropolis, SP, Brazil
*   Correspondence: taicia@iqm.unicamp.br; Tel.: +55-19-3521-3092

Received: 11 July 2019; Accepted: 25 July 2019; Published: 6 August 2019

Abstract: Citrus are vulnerable to the postharvest decay caused by *Penicillium digitatum*, *Penicillium italicum*, and *Geotrichum citri-aurantii*, which are responsible for the green mold, blue mold, and sour rot post-harvest disease, respectively. The widespread economic losses in citriculture caused by these phytopathogens are minimized with the use of synthetic fungicides such as imazalil, thiabendazole, pyrimethanil, and fludioxonil, which are mainly employed as control agents and may have harmful effects on human health and environment. To date, numerous non-chemical postharvest treatments have been investigated for the control of these pathogens. Several studies demonstrated that biological control using microbial antagonists and natural products can be effective in controlling postharvest diseases in citrus, as well as the most used commercial fungicides. Therefore, microbial agents represent a considerably safer and low toxicity alternative to synthetic fungicides. In the present review, these biological control strategies as alternative to the chemical fungicides are summarized here and new challenges regarding the development of shelf-stable formulated biocontrol products are also discussed.

Keywords: biological control; post-harvest phytopathogen; *Penicillium digitatum*; *Penicillium italicum*; *Geothrichum citri-aurantii*

Key Contribution: This review demonstrates the potential of alternative methods for the control of diseases that occur in the postharvest of citrus.

---

1. Introduction

Citrus is one of the most produced fruit genus. Grown in more than 100 countries, this group is composed by several species, including oranges, tangerines, mandarins, grapefruits, lemons, and limes [1]. The impact of citrus agroindustry in the international economy is huge. Besides their value as commodities, they also provide employment in many segments involved in its production cycle: harvesting, handling, transportation, and storage. In 2017, the global orange production reached 47.6 million metric tons (tons) and is expected to expand 4.2 million in 2018/2019 due to favorable weather in Brazil and United States, two of the most important orange producers in the world [2].

Over 20 different kinds of postharvest diseases have been reported in citrus and they are the main cause of fruit spoilage, resulting in massive economic losses [3]. Moreover, fungal growth in fruit may lead to production of mycotoxins, including potential carcinogenic agents such as citrinin and patulin [4], as well as tremorgenic compounds, for example tryptoquivalines [5], therefore representing a threat to human and animal life.

Green mold, blue mold, and sour rot, caused by *Penicillium digitatum*, *P. italicum*, and *Geotrichum citri-aurantii*, respectively, are the main citrus postharvest diseases [6]. *P. digitatum*, alone, is responsible for approximately 90% of total postharvest losses [7]. The fruits are contaminated through skin postharvest damage during their picking, packaging, storage, and transportation [3].

The interaction between citrus fruit and these phytopathogens is not fully understood, but some factors are known to affect this interaction in order to increase the fungus pathogenicity. *P. digitatum* and *P. italicum* are known to secrete organic acids during infection, leading to an optimal pH for its cell wall-degrading enzymes, such as polygalacturonases (PG) [8,9]. Moreover, *P. digitatum* also produce catalase during infection, an antioxidant enzyme that decomposes hydrogen peroxide, the main defense mechanism in citrus [10].

In order to deal with these fungi, chemical fungicides have been the main focus of research over the past decades. *P. digitatum* and *P. italicum* can be efficiently controlled by imazalil, thiabendazole, or pyrimethanil, but these fungicides are not effective against sour rot. For the control of *G. citri-aurantii*, guazatine, and propiconazole can be applied, although they are not allowed in some producing countries such as Brazil [11]. However, widespread use of chemical fungicides has caused the proliferation of resistant strains of these phytophatogens, compromising the effectiveness of these treatments [12]. Furthermore, concerns about environmental contamination and risks associated to human health have been raised around the accumulation of their residues in food.

Finding commercially viable, effective, alternative control methods has been a leading challenge for researchers, especially for controlling *G. citri-aurantii*, since there are fewer options available of acceptable chemical fungicides. Many alternatives have been proposed, including the application of antagonistic microorganisms and natural antimicrobial substances. Natural antimicrobial substances, especially plant extracts, are considered relatively safe, presenting low toxicity and high decomposability due to their natural origin, raising particular interest for use in these natural products [13].

Thus, the application of biocontrol agents has been an alternative for synthetic chemical fungicides. However, more research is necessary to understand their mechanism of action and effectiveness in different infection levels; this knowledge is crucial to implement their use as practical control agents. Therefore, the use of alternative postharvest biological control methods, both non-polluting and possessing low toxicity are reviewed here, highlighting advances presented in the literature in the recent years.

2. Alternative Control Methods

2.1. Microrganisms

Besides fungicides, other agricultural practices such as irradiation application (light emitting diode, gamma radiation, or UV radiation) [14–17], thermotherapy [18–20], biocontrol agents (BCA) [21–23], and salt solution [24,25] may be used to control postharvest diseases. In the past thirty years, there have been extensive research activities to explore and develop strategies based on microbial antagonists to biologically control postharvest pathogens [26–30]. This section focuses on the natural products and BCA described to be effective to control postharvest diseases.

BCA have been used in post-harvested fruits and its mechanism of action is poorly described in the literature for the majority of microorganisms (either bacteria or yeasts). However, it is supposed that more than one control mechanism could be acting simultaneously over the host-pathogen-antagonist and environment interactions [30]; others include: antibiosis [30,31], competition for nutrients or space [32], induction of resistance in citrus fruits [33,34], secretion of specific enzymes [35], stimulation of ROS in host tissues [36], mycoparasitism, and biofilm formation [37,38].

The use of yeasts as antagonists has been extensively studied, due to their high inhibitory capacity and the ability to colonize surfaces for a long period. The so-called "killer yeasts" yeast strains have the 'killer' phenotype (K+) and can produce the "killer proteins" that are potential antifungal agents; this feature is a biological advantage against others competing microbial [39–41]. Ferraz et al. reported that *Rhodotorula minuta, Candida azyma,* and *Aureobasidium pullulans* presented killer activity against the citrus pathogen *G. citri-aurantii,* deforming fungal hyphae and suppressing pathogen development [11]. *Saccharomyces cerevisiae* is another example of yeast that often presents killer activity [40,42–44].

Besides yeasts, bacteria are also promising BCA: *Bacillus* [45,46], *Lactobacillus* [47,48], and *Streptomyces sp.* [49,50] genus have been studied as BCA against citrus pathogens. Bacteria of *Bacillus* genus can act as antagonist through antibiotics or volatile organic compound (VOC) production that can induce the increase of plants resistance. Leelasuphakul et al. verified that strains of *Bacillus subtilis* found in soil were able to delay the spore germination of *Penicillium digitatum* by the action of water-soluble antibiotic secondary metabolites, proteins, enzymes, and VOC production [45]. As for *Lactobacillus,* metabolites such as 3-phenyllactic acid and allyl phenylacetate isolated from *L. plantarum* IMAU10014 had their antifungal activities against *Penicillium digitatum* observed *in vitro* by Wang et al. [47]. Finally, *Streptomyces sp.* had been tested *in vitro* and *in vivo* against *P. digitatum* and other pathogens [49,50]. Metabolites from *Streptomyces* RO3 cultures with molecular masses higher than 2000 Da showed fungicidal action and *in vivo* tests indicated that the green mold disease incidence decreased when treated with *Streptomyces* RO3 metabolites [49].

Among the alternatives reported in the literature, the induction of host resistance to pathogen by microorganisms with 'killer' activity has been pointed as a promising option for plant disease control, since they are active against a broad spectrum of pathogens and are safer than other alternatives [41,51]. For example, Parafati et al. related that the yeasts *Wickerhamomyces anomalus, Metschnikowia pulcherrima,* and *A. pullulans* increased the activities of peroxidase and superoxide dismutase in mandarins, reducing the incidence and severity of blue mold on these fruits [52]. Unfortunately, the activities and mechanisms of interaction of most 'killers' have not yet been well elucidated [11,41,44], being an open research field.

Another gap in the search for BCA against citrus pathogens is that the majority of the studies focuses only in *P. digitatum* and the other citrus pathogens such as *P. italicum* and *G. citri-aurantii* are less studied and also represent a problem for citriculture. As mentioned before, there are few fungicides active against sour rot and other approaches such as BCA could be explored to discover alternative methods to control this phytopathogen. However, few BCA are reported to control this disease. One possibility to solve this problem could be the evaluation of BCA already pointed as active against *P. digitatum* or other phytopathogens to control *G. citri-aurantii.*

Table 1 lists some known biocontrol agents and their mode of action against *P. digitatum, P. italicum,* and *G. citri-aurantii.*

Table 1. Biocontrol agents (BCA) used against *P. digitatum*, *P. italicum*, and *G. citri-aurantii*.

Antagonist	Agent	Mechanism	Target Pathogen	References
Yeast	*Wickerhamomyces anomalus* (or *Pichia anomala*)	Antibiosis, competition for nutrients, fruit resistance induction and 'killer' activity	*P. digitatum P. italicum*	[39–41,52]
	Saccharomyces cerevisiae	Competition for nutrients or space and 'killer' activity	*P. digitatum P. italicum*	[40,41,43,44]
	Candida oleophila	Resistance induction. Increase phenylalanine ammonia lyase activity and accumulation of the phytoalexins such as scoparone, scopoletin, and umbelliferone	*P. digitatum P. italicum*	[33,53]
	Saccharomycopsis crataegensis + sodium bicarbonate	Not specified	*P. digitatum*	[54]
	Kluyveromyces marxianus + sodium bicarbonate	Competition for nutrient and space. The salt stimulates *K. marxianus* growth and it inhibits fungal spore germination	*P. digitatum*	[55]
	Rhodosporidium paludigenum	Fruit resistance induction. Increase in ethylene production and expression of defensive genes	*P. digitatum*	[56]
	Pichia membranifaciens	Competition for nutrients or space	*P. digitatum*	[57]
	Metschnikowia pulcherrima, and *Aureobasidium pullulans*	Competition for nutrients and fruit resistance induction by influencing peroxidase and superoxide dismutase activities	*P. digitatum P. italicum*	[52]
	Candida stellimalicola	'Killer' activity, production of chitinase, and inhibition of conidial germination	*P. italicum*	[44]
	Cryptococcus laurentii associated with cinnamic acid	Different influence of cinnamic acid on the antagonistic yeast and the pathogen, leading to synergistic effect	*P. italicum*	[58]
	Metschnikowia citriensis	Biofilm formation, adhesion to mycelia, and iron depletion	*P. digitatum P. italicum*	[53]
	Pseudozyma antarctica	Direct parasitism	*P. digitatum P. italicum*	[53]
	Rhodotorula minuta, Candida azyma, and *Aureobasidium pullulans*	'Killer' activity and hydrolytic enzyme production	*G. citri-aurantii*	[11]
	Debaryomyces hansenii	Competition for space and nutrients	*P. digitatum P. italicum*	[32,59,60]
	Kazachstania exigua and *Pichia fermentans*	'Killer' activity	*P. digitatum P. italicum*	[41]
	Bacillus subtilis	Water soluble antibiotics, proteins, enzymes, and VOC production	*P. digitatum*	[43,45]

Table 1. *Cont.*

Antagonist	Agent	Mechanism	Target Pathogen	References
Bacteria	*Bacillus amyloliquefaciens*	Great amounts of antibiotics produced *in vitro*, however, still not effective for green mold control *in vivo*	*P. digitatum*	[46]
	Lactobacillus plantarum	Metabolites 3-phenyllactic acid and benzeneacetic acid, 2-propenyl ester with antifungal activity	*P. digitatum*	[47,48]
	Streptomyces sp.	Metabolites with higher mass than 2000 and fungicidal effect	*P. digitatum G. citri-aurantii*	[49,50]
	Streptomyces violascens	Extracellular antifungal compounds that inhibits fungal spore germination and antibiosis	*G. citri-aurantii*	[61]

2.2. Natural Products

Besides BCA, the use of chemicals isolated from natural sources may be an interesting approach to control *P. digitatum*; these include alkaloids [62], chitosan [26,63–66], carvacrol and thymol [67], citral [68], citronellal [69], and several other compounds isolated from essential oils (EOs) and plant extracts [22,70,71].

Olmedo et al. assessed the antifungal activity of six β-carboline alkaloids (harmine, harmol, norharmane, harmane, harmaline, and harmalol) against *P. digitatum* and *Botrytis cinerea*. They observed that harmol is more active than harmaline and harmalol, due to the differences in properties such as aromaticity, acidity, planarity, and polarity [62]. Table 2 shows some plant natural products that have been currently studied as control strategies against *P. digitatum*, *P. italicum*, and *G. citri-aurantii*; lists that are more extensive can be found in [22,71].

The natural products highlighted in Table 2 are summarized in Figure 1. The global Venn diagram (Figure 1) displays natural products activity distribution against *P. italicum*, *P. digitatum*, and *G. citri-aurantii*. The Venn diagram clearly indicates a significant number of natural products active against *P. digitatum* and *P. italicum*; however, few natural products have been studied to control *G. citri-aurantii*.

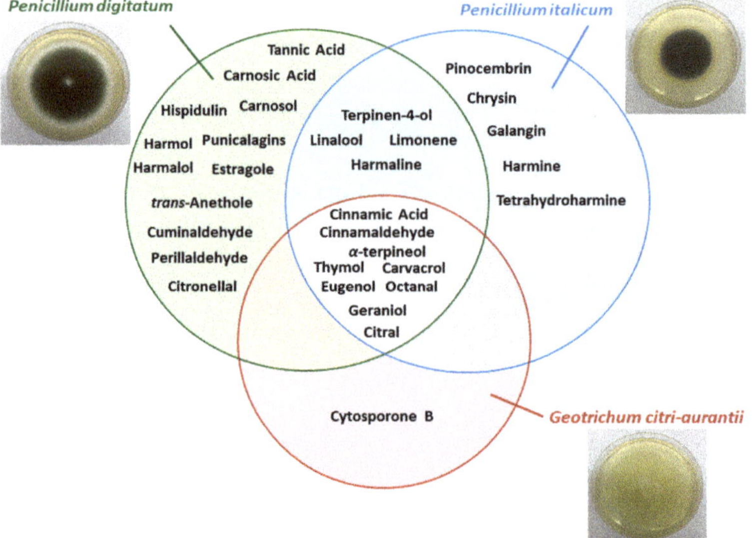

Figure 1. Venn diagram comparing the number of active natural products against the different postharvest citrus pathogens.

Table 2. Natural products extracted in plants as control strategies against *P. digitatum*, *P. italicum*, and *G. citri-aurantii*.

Plant/Fruit	Pathogen (s)	Extract/Method	Natural Products	Details	References
Chinese propolis	*P. italicum*	1) Ethyl acetate (3 times); 2) chloroform; 3) ethanol and water; 4) methanol	Pinocembrin	Pinocembrin acts against *P. italicum* through inhibition on respiration and interference of energy homeostasis	[72]
Citrus aurantium	*P. digitatum* *P. italicum*	Hydrodistillation (peels, leaves, and flowers)	α-terpineol, terpinen-4-ol, linalool, and limonene	Essential oils (EOs) of flowers and leaves reduced the growth of pathogen, while EO of peels was inactive	[73]
Citrus eticulate Blanco	*P. digitatum*	-	Citral	Antifungal activity of citral was tested *in vitro* and *in vivo* and combined with the wax showed potential for control applications	[68]
Citrus fruits	*P. italicum* *P. digitatum*	Commercial product	Octanal	Octanal inhibits the fungal mycelial growth	[74]
Citrus fruits	*P. italicum*	Commercial product	Citral	Citral inhibits the mycelial growth of *P. italicum* causing disruption of cell membrane integrity	[75]
Citrus paradise Macf. (Grapefruit fruit)	*P. digitatum*	-	Chitosan and salicylic acid	Chitosan combined with salicylic acid had better treatment of green mold than these isolated compounds, without compromising the quality of fruit.	[26]
Citrus sinensis Osbeck	*P. digitatum*	Commercial product	Citronellal	Citronellal was able to inhibit spores germination and mycelial growth. Just as citral, the compound combined with wax reduced the incidence rate	[69]
Laminaceae spp.	*P. digitatum* *P. italicum*	-	Carvacrol and thymol	The mechanisms that have been proposed for these compounds are: 1) morphological deformation and deterioration of the conidia and hyphae; 2) hydroxyl group and systems with delocalized electrons has important role for antimicrobial effect	[69]

Table 2. *Cont.*

Plant/Fruit	Pathogen (s)	Extract/Method	Natural Products	Details	References
Peganum harmala L. (harmal seeds)	*P. italicum*	Ethanol	Harmine, harmaline, and tetrahydroharmine (THH)	Harmal extracts showed strong antifungal activity against *P. italicum* and its activity is related to alkaloids harmine, harmaline e THH	[76]
Peganum harmala L.	*P. digitatum*	Commercial product	Harmol, harmaline, harmalol, harmane, and norharmane	It was tested the antifungal activity of β-carbolines against *P. digitatum* and *Botrytis cinerea*. Harmol showed highest antifungal activity after 24 h.	[63]
Pimpinella anisum and *Carum carvi*	*P. digitatum*	Hydrodistillation (seeds)	trans-anethole, estragole (anise oil), cuminaldehyde, and perillaldehyde (black caraway)	EO were able *in vitro* of reduce the germination, the mycelial growth of pathogen and the incidence of disease symptoms	[77]
Populus × *euramericana cv. 'Neva'* (poplar buds)	*P. italicum*	Dichloromethane	Flavonoids of pinocembrin, chrysin, and galangin	Antifungal compounds from poplar buds active fraction, identified by HPLC–MS, had antifungal effect in the fungal hyphae analyzed by scanning electron microscopy and transmission electron microscopy images	[78]
Punica Granatum	*P. digitatum*	Ethanol/water (4:1)	Phenolic compounds with a prevalence of punicalagins	Pomegranate peel extract has a broad range of antifungal activity	[13]
Ramulus cinnamomi	*P. digitatum P. italicum G. citri-aurantii*	Ethyl acetate and n-buthanol	Cinnamic acid and cinnamaldehyde	Through ^{1}H-NMR-based metabolomics it was identified the extracts related to antifungal activity of *Ramulus cinnamomi* after 4, 8, and 12 h. The antifungal mechanism of cinnamaldehyde it was also analyzed by ^{1}H-NMR	[79]
Rosmarinus officinalis L.	*P. digitatum*	Hydrodistillation (for EO) and methanol	Flavonoids, polyphenols, and essential oils	EO act in the fungal cells by disrupting the membrane permeability and the osmotic balance	[80]
Salvia fruticosa Mill.	*P. digitatum*	Ethyl acetate	Carnosic acid, carnosol, and hispidulin	Compounds that have antifungal properties, according to its compositions, structures/activity, and literature	[81]

Table 2. *Cont.*

Plant/Fruit	Pathogen (s)	Extract/Method	Natural Products	Details	References
Sapium baccatum	*P. digitatum*	Commercial product	Tannic acid	*In vitro* antifungal activity to *P. digitatum* was verified between 400 and 1000 µg mL^{-1} of tannic acid inoculated in Ponkan fruit was sufficient to inhibit the mycelial growth of 45% to 100%	[82]
Solanum nigrum	*P. digitatum*	Aqueous extract (leaves)	Alkaloids, flavonoids, saponins, steroids, glycosides, terpenoids, and tannins	Bioactive compounds that has pharmacological prospects for development of drugs	[83]
Thymus species (*T. leptobotris, T. riatarum, T. broussonnetii* subsp. *hannonis,* and *T. satureioides* subsp. *pseudomastichina*)	*P. digitatum P. italicum G. citri-aurantii*	Hydrodistillation	Thymol, carvacrol, geraniol, eugenol, octanal, and citral	EO of four *Thymus* species showed antifungal activity. Through GC–MS, MIC, and previous studies determined the principal active compounds	[84]
Thymus leptobotris	*P. digitatum P. italicum G. citri-aurantii*	Methanol, chloroform	Thymol and carvacrol	The antifungal screening from EO obtained from 21 plants showed that the EO from *Thymus leptobotris* had the highest fungistatic effect. The active compounds were identified in previous studies.	[85]
Thymus vulgaris L.	*P. italicum P. digitatum*	-	Thymol	EO of thyme inhibited the mycelium growth (MIC 0.13 µL mL^{-1}) and spore germination (MIC 0.50 µL mL^{-1}) *in vitro* and *in vivo*	[86]
Withania somnifera + *Acacia seyal*	*P. digitatum*	Methanol/acetone/water—7:7:1 *v/v* (dried plant powder—1:20 *w/v*)	Insoluble and soluble phenolic compounds	Application of plants extract (*W. somnifera* and *A. seyal*) in the sick host, induced plant resistance through change of phenolic concentration (phenylpropanoid pathway)	[87]

Garlic [88], neem [89], *Withania somnifera L.* and *Acacia seyal L.* [87], mustard, and radish [90] also have been reported as effective in controlling *P. digitatum*. Recently, Zhu et al. reported the antifungal activity of tannic acid on *P. digitatum*. *In vivo* tests showed significant decrease of disease signals of *P. digitatum* by inhibiting its mycelial growth and spore germination. Storage tests shown that tannins reduce the severity of green mold on citrus by 70%. The authors suggest that antifungal activity mechanism of tannic acid is related to the disruption of the cell walls and the plasmatic membrane, causing leakage of intracellular contents such as sugars [82].

Extracts from chili peppers and ginger were also proposed to control or inhibit postharvest diseases in citrus [91]. Singh et al. tested different concentrations of the plant extracts of the *Zingier officinale L.* (Ginger) and *Capsicum frutescence L.* (Chilly) against *P. digitatum*, *Aspergillus niger*, and *Fusarium* sp. isolated from naturally infected citrus. *P. digitatum*, specifically, had a reduction on colony development with inhibition zones of 51.5%, 69.2%, 74%, and 83.1% for *Zingier officinale L.* extract, and 56.4%, 64,1%, 76.6%, and 100% for *Capsicum frutescence L.* at concentration of 500, 1000, 2000, and 3000 ppm, respectively [91].

Besides the above-mentioned natural products used against *P. digitatum*, synergism between compounds have also been tested. Shi et al. studied the effects of chitosan and salicylic acid (SA), both isolated and mixed, on the control of green mold decay in grapefruit. The results showed that combination of chitosan with SA was effective to control green mold than either compound alone (significant efficacy of biocontrol agents against green mold decay induced by SA application in citrus fruits have previously been studied) [26]. Furthermore, significant reduction on lesion diameter and disease incidence was observed. Additionally, it was observed that treatment with chitosan/SA blends increased the content on ascorbic acid and total soluble solids in the fruits, providing a longer shelf life [26].

P. italicum causes the blue mold decay and represents one of the most problematic postharvest citrus infection, compromising fruit integrity during storage and transportation [92]. Currently, the blue mold is primarily controlled by the synthetic fungicide applications, such as thiabendazole and imazalil [44]. Regarding Imazalil-resistant biotypes *P. digitatum* is more common, whereas resistant *P. italicum* is rare [93].

Plant extracts studied as an alternative or complementary control agents to currently used fungicides may be attractive because of their potential antifungal activity, non-phytotoxicity and biodegradability [94–97]. Askarne et al. evaluated the antifungal activity of 50 species of plants collected in different regions of southern Morocco. *In vitro* antifungal activity showed that among them, *Anvillea radiata* and *Thymus leptobotrys* completely inhibited mycelial growth of *P. italicum* at concentrations of 10% m/v [98]. In addition, *Asteriscus graveolens*, *Bubonium odorum*, *Ighermia pinifolia*, *Inula viscosa*, *Halimium umbellatum*, *Hammada scoparia*, *Rubus ulmifolius*, *Sanguisorba minor*, and *Ceratonia siliqua* were also effective against *P. italicum* with inhibition of mycelial growth greater than 75%. The species on *in vitro* studies were also tested *in vivo* against the blue mold in citrus. The incidence of blue mold was significantly reduced to 5 and 25% when oranges were treated with aqueous extracts of *H. umbellatum* and *I. viscosa* (compared to 98% in the control), indicating the antifungal potential of these materials against *P. italicum*.

Kanan and Al-Najar reported the effective *in vitro* and *in vivo* antifungal activity of fenugreek (*Trigonella foenum-graecum L.*), harmal seeds (*Peganum harmala L.*), garlic cloves (*Allium sativum L.*), cinnamon bark (*Cinnamomum cassia L.*), sticky fleabane leaves (*Inula viscose L.*), nightshade leaves, and fruits (*Solanum nigrum L.*) against *P. italicum* isolates. Cinnamon, garlic, and sticky fleabane methanolic fractions resulted in complete inhibition of this pathogen [76].

The high antifungal activity against *P. italicum* of crude extracts cinnamon, as well as the corresponding methanolic, hexanic, and aqueous fractions was related to the high content of cinnamaldehyde, eugenol, cinnamic acid, flavonoids, alkaloids, tannins, anthraquinones, and phenolic compounds, some of them reported before as active antifungal agents. Among them, eugenol and cinnamaldehyde have been consistently reported to be the main antifungal components of cinnamon [79,99], representing potential, environmentally benign candidates for postharvest disease control. In addition, harmal extract was pointed also as highly effective extract against *P. italicum in vitro*. The activity of the harmal crude extracts may be related to the content of alkaloids such as harmine, harmaline, and tetrahydroharmine besides phenolic compounds that can alter the permeability of fungal cells [76]. The exact mechanism of action of phenols has not yet been determined; however, it is already known that they can inactivate essential enzymes and disrupt function of the genetic material [100].

Several studies have reported natural antifungal compounds against *P. italicum* produced by different sources. Essential oils, for instance, represent a low-toxic effects alternative and are reported to possess strong inhibitory effects on crop contaminated by *P. italicum* [74]. Among the major volatile constituents are limonene, β-linalool, α-terpineol, citral, and octanal, the latter reportedly exhibits antifungal activity against some postharvest pathogens such as *P. digitatum*, *P. italicum*, and *P. ulaiense* [101,102]. Regarding the mode of action, Tao et al. attributed its activity against mycelial growth of *P. italicum* and *P. digitatum* to the disruption of the cell membrane integrity and leakage of ions and other cell contents [74]. A similar mechanism of activity against postharvest citrus pathogens was attributed to citral, present in citrus EO that is able to alter the morphology of *P. italicum* hyphae [75].

Chinese propolis has been pointed as active against blue mold and the flavonoid pinocembrin (5,7-dihydroxyflavanone) was identified as one of its main active antifungal constituents. Peng et al. studied the inhibitory effect of this compound on *P. italicum* with particular attention to its response to the mycelial growth and energy metabolism by interfering in energy homeostasis and cell membrane damage of the pathogen. They observed that mycelial growth and spore germination were nearly completely inhibited with inhibitive percentage up to 93 and 97%, respectively, for pinocembrin concentrations of 400 mg/L [72].

Figure 2 shows the structures of the above mentioned antifungal compounds found in natural extracts and essential oils.

Sour rot of citrus, caused by *Geotrichum citri-aurantii*, represents another potentially devastating storage disease [103]. Pathogenicity of *Geotrichum citri-aurantii* on citrus fruit involves secretion of extracellular endo-polygalacturonases (PG) that aid in the rapid breakdown of infected tissues facilitating the disease [104]. The usual fungicides, with the partial exception of sodium o-phenylphenate (SOPP) and propiconazole [105], cannot actively controlled this disease. Yin et al. revealed the first time that cytosporone B—a compound isolated from the endophytic fungus *Cytospora sp.* and presenting a wide range of antitumor and antimicrobiotic activities—has a promising effect on the control of citrus decay caused by this pathogen, being comparable to that of fungicide prochloraz. Its mechanism of action is suggested to be related to the alteration of the morphology of pathogen cells, causing distortion of the mycelia and loss of membrane integrity [106].

Talibi et al. demonstrated that methanolic extracts *Cistus villosus*, *C. siliqua*, and *H. umbellatum* successfully reduced the disease incidence *in vitro* caused by *G. citri-aurantii* with no phytotoxic effects recorded on citrus; on the other hand, ethyl acetate extracts of *A. radiata*, *C. villosus*, and *C. siliqua* proved to be the best inhibitors of mycelial growth [107]. Other studies concerning the antifungal properties of organic extracts of plants include materials from *Cistus L.* species [108] and from extremophile plants from the Argentine Puna [109]. Incidence of sour rot was described to reduce significantly when fruits were treated with *Cistus populifolius* and *Cistus ladanifer* methanol extracts [108].

Zhou et al. also reported citral, octanal, and α-terpineol to have strong inhibition on *G. citri-aurantii*, being the former the most potent among them. They induced a decrease on the total lipid content of the cells, indicating the destruction of cellular membranes, disruption of cell membrane integrity, and leakage of cell components [110].

Liu et al. studied the antifungal activity of thyme (*Thymus sp.*) EO against postharvest sour rot on citrus fruit. It was shown that *G. citri-aurantii* cells treated with thyme EO showed morphology alteration (collapsed mycelia and arthroconidia structures); additionally, a marked enlargement of hypha wall thickness was observed [111]. Those effects can be assigned specially to the presence of thymol (a volatile terpenoid ubiquous in plants with strong, widespread antifungal activities).

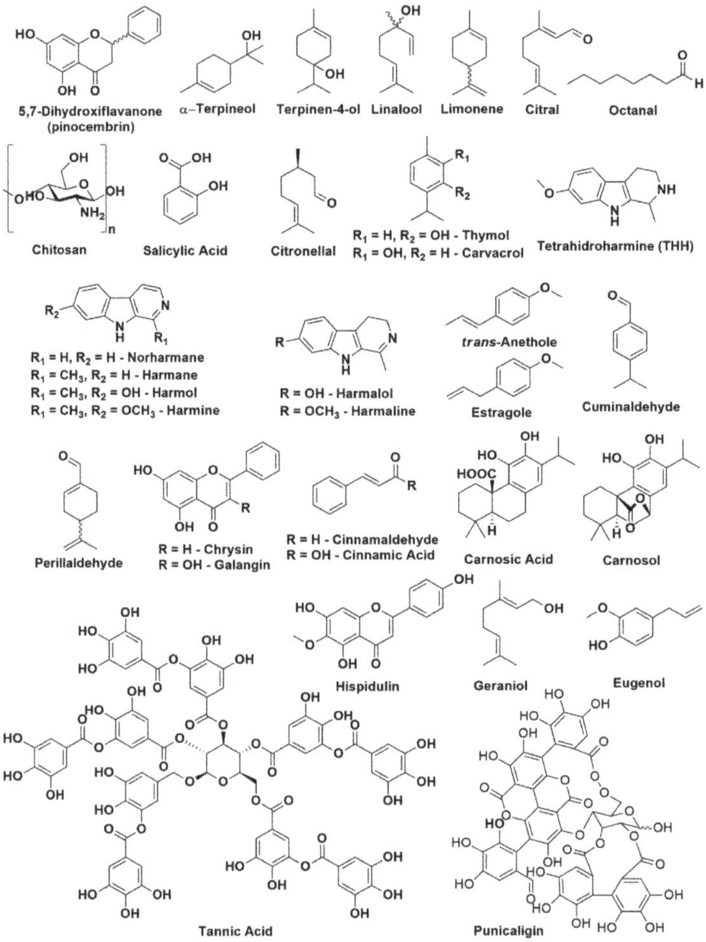

5,7-Dihydroxiflavanone (pinocembrin) α–Terpineol Terpinen-4-ol Linalool Limonene Citral Octanal

Chitosan Salicylic Acid Citronellal R₁ = H, R₂ = OH - Thymol R₁ = OH, R₂ = H - Carvacrol Tetrahidroharmine (THH)

R₁ = H, R₂ = H - Norharmane
R₁ = CH₃, R₂ = H - Harmane
R₁ = CH₃, R₂ = OH - Harmol
R₁ = CH₃, R₂ = OCH₃ - Harmine

R = OH - Harmalol
R = OCH₃ - Harmaline

trans-Anethole Estragole Cuminaldehyde

Perillaldehyde R = H - Chrysin R = OH - Galangin R = H - Cinnamaldehyde R = OH - Cinnamic Acid Carnosic Acid Carnosol

Hispidulin Geraniol Eugenol

Tannic Acid Punicaligin

Figure 2. Chemical structures of some antifungal compounds active against *P. digitatum* and *P. italicum* found in essential oils and natural extracts.

Regnier et al. screened 59 commercially available EO, as well as their major components, to determine their effects on mycelial growth of *G. citri-aurantii* [6]. Lemong (*Cymbopogon citratus*) was found to be the most cost-effective option; it was also found that a blend of the lemongrass and spearmint (*Mentha spicata*) EO could be an alternative for effective multi-target protection against *G. citri-aurantii*, *P. digitatum*, and *P. italicum* [6].

Xu et al. found that cassia (*Cassia sp.*) EOs have the ability to control postharvest pathogens and diseases, but their poor solubility in water might prevent its effective use. In order to circumvent this problem, they presented an aqueous microemulsion formulation of cassia EO with ethanol as co-surfactant and Tween 20 as surfactant as antifungal agent against *G. citri-aurantii*. Both *in vitro* and *in vivo* assays showed that cassia EO had stronger activity when encapsulated in the microemulsion [112].

The structure of some compounds above listed with activity against *G. citri-aurantii* are shown in Figure 3.

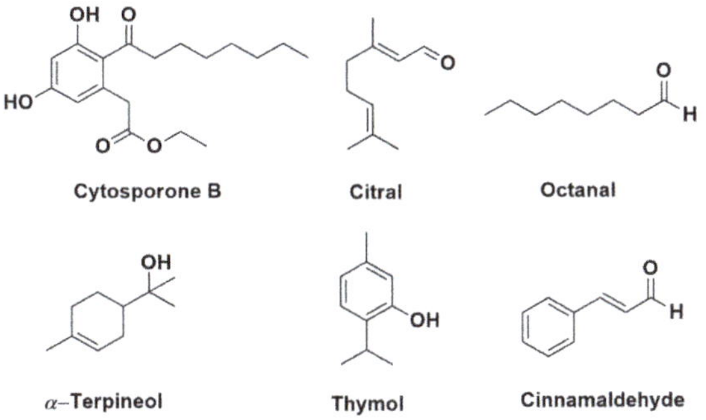

Figure 3. Chemical structures of compound found on plant extracts and essential oils with activity against *G. citri-aurantii*.

Considering that most plant-derived compounds are usually much less toxic to humans and fauna in general, as well as generally environmentally-friend when compared to fungicides, they should be considered safer and may represent a promising alternative to the existing chemical pesticides in the control of fungal diseases. However, it should be noted that there is a lack of studies concerning non-chemical postharvest treatment against *G. citri-aurantii* and very little research has been carried out to investigate the use of natural products to control citrus sour rot; therefore, special attention towards this pathogen is needed. Despite the promising antifungal activity of organic extracts of plants, the active antifungal compounds are frequently not identified; the knowledge of the substances accountable for the antifungal activity is fundamental to formulate viable commercial products and develop advanced biofungicides formulations as alternatives to synthetic fungicides to control the citrus postharvest diseases.

2.3. Commercial Biofungicides

Biofungicide is the general name given to microorganisms and naturally occurring compounds that possess the ability to control plant diseases [113]. Although the mechanisms of biocontrol in postharvest diseases have not been fully explained in many cases, effective colonization of wounds and competition for nutrients appear to be significant factors for many antagonists. Many microorganisms have shown good potential as basis for commercial biocontrol products due to their efficacy against fungal pathogens in field conditions. Although much research is being conducted in this area, a limited number of biofungicides are available commercially [114–116].

Several commercial biological control formulations based on *Trichoderma harzianum*, *A. pullulans*, *Bacillus subtilis*, *Streptomyces griseoviridis*, and *Gliocladium virens* [113,117] have been reported for application against different plant fungal pathogens. Most studies have focused on application of individual biocontrol agents without evaluating their combination with other microorganisms or even with chemical components. Nevertheless, combination BCA that are compatible with each other could offer a new and effective approach improving plant diseases control [114,118].

The major drawback in the commercialization of bioproducts based on BCA is the advancement in the production of shelf-stable formulated products that maintain biocontrol activity similar to that of the fresh cells. Although biofungicides have good action background against host pathogens, there are limitations to their use and effectiveness in the field. Growing demand and interest in bioproducts have led to many marketable brands but the absence of field application reliability of biofungicides has been a significant obstacle in the adoption of these approaches. Despite the remarkable results obtained with biofungicides in the laboratory experiments, some failed to provide consistent disease control in field conditions. Thus, these factors are potential contributors to the low dissemination of these products on the market [113,115,116].

Although a reasonable number of studies take into account the exploration of new bioactive compounds that may specifically act against citrus phytopathogens, few products come to commercialization and are widely marketed. In addition to the above problems, the workflow that starts with the discovery of the bioactive compound to the effective elaboration of the final product is still quite complex and is usually a long, iterative process that involves several steps. Figure 4 shows a usual workflow that could be involved in the development of a postharvest biofungicide.

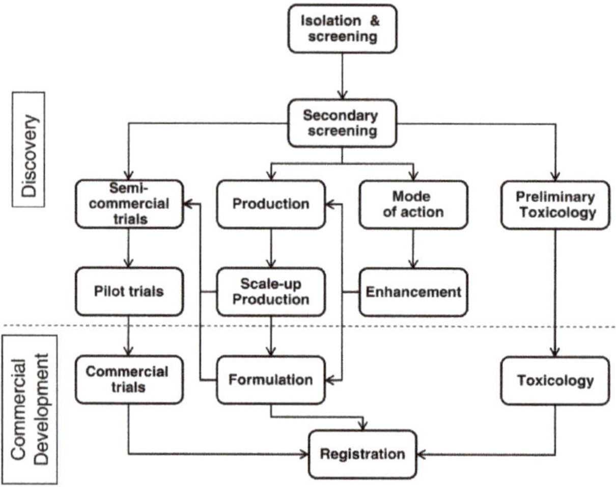

Figure 4. Outline flow of development of a postharvest biocontrol agent. Reprinted from [119]. Copyright 2012, Springer Nature.

According to Nunes, biocontrol agent development is comprised of two main steps: discovery and commercial development. Summarizing the required phases, the first one concerns the isolation and efficacy of the laboratory-based compounds to pilot tests. In this phase must be included evaluation of the action mechanism of the microorganisms, growth media required, improvements in biocontrol activity, and legal procedures involving the patent for the employment of the microorganism as a biopesticide. The commercial development of BCA involves steps including scale-up production, product formulation, biosafety of the microorganism, and registration [119]. The detailed workflow will not be discussed in detail but can be found in previous papers [116,119–121].

Indeed, the complex workflow is a weak point for the development of new products based on biomolecules. Moreover, the high cost of the production and the regulatory barriers to BCA registration in different countries do not encourage their dissemination and recurring issues that need to be overcome [119–122]. To date, four commercially biofungicides, based on microorganisms, were found for the control of postharvest citrus fruit. The yeast *Metschnikowia fructicola*, for instance, was reported as an efficient biological control agent of postharvest diseases of fruits and vegetables, and it is the bases of the commercial formulated product "Shemer" [122]. Its effect has been reported as a good control of decay in oranges being equivalent to oranges treated with the chemical fungicide, imazalil [123]. Nevertheless, just a few antagonists have achieved the commercial development stage as commercial products. Some of these biofungicides are represented in Table 3.

Table 3. Commercial biofungicides, based on microorganisms, for the control of postharvest citrus fruit.

Microorganism	Product	Targeted Pathogens	References
Candida oleophila	Aspire	*Botrytis, Penicillium*	[117]
Metschnikowia fructicola	Shemer	*Botrytis, Penicillium, Rhizopus, Aspergillus*	[124]
Pantoea agglomerans	Pantovital	*Penicillium, Botrytis, Monilinia*	[125]
Pseudomonas syringae	Biosave	*Penicillium, Botrytis, Mucor*	[126]

Aspire™, based on *Candida oleophila* [117], has been commercialized for some years but did not prevail due to low and inconsistent efficacy under commercial conditions, difficulties in market penetration, and perception of the customers and industry [113]. Other products, such as Shemer™ (based on the yeast *M. fructicola*), have been more successful [27], and are being used for both pre- and postharvest application on various fruits and vegetables, including citrus fruit, grapes, peaches, peppers, strawberries, and sweet potatoes.

Despite limited numbers of specific biofungicides, the demand for these products in agriculture as alternatives to synthetic pesticides has increased over the last few years. The adoption and widespread use of biofungicide will make it possible to produce food that are exempt from or have low values of chemical residues. This will contribute to the consumption of more natural, healthy, and safe foods with respect to fungicide usage [116,127]. Thus, both regulatory barriers and workflow-related procedures must be improved in order to overcome the challenges in the biofungicide market.

3. Conclusions

The information presented in this review reports the potential of alternative methods for the control of postharvest citrus diseases. Despite all the drawbacks regarding the development of new products, a considerable number of studies have been conducted concerning biocontrol strategies of citrus postharvest phytopathogens. Several studies have reported antifungal compounds, mostly against *P. digitatum*, which is responsible for the most important disease found in citrus fruits. Another interesting approach, with good results against green mold, is the application of the biocontrol agent in a mixture with low doses of chemical fungicides during citrus fruit processing. Nevertheless, few studies treat individually the particularities of each fungus and, therefore, many strategies of biocontrol still need to be studied taking into account the different metabolic and enzymatic fungi profile. In this context, there is a lack of studies on postharvest non-chemical treatment against *G. citri-aurantii*, which causes a decay not controlled by the conventional treatments, and therefore special attention toward this pathogen is necessary. It is very clear that in recent years the interest in biocontrol strategies that minimize the use of chemical pesticides is a worldwide trend, which has driven research in this field. Compounds from biological sources are usually much less toxic to humans as well as environmentally-friendly when compared to synthetic fungicides; for this reason they represent a promising alternative to the existing chemical pesticides in the treatment of fungal diseases. However, as discussed here, several challenges related to the workflow procedures and development of biofungicides still need to be overcome so that new technology can be employed during citrus fruit processing in order to lead to a commercially viable strategy that meets the needs of the producers.

Author Contributions: Writing—original draft preparation: J.M.B., J.R.B., J.H.C., D.Y.A., writing—review and editing: J.M.B., J.G.M.P., and K.C.K, F.A., J.E.C., T.P.F. supervised the conception and design, and final editing of the manuscript.

Funding: This work was supported by grants from Fundação de Amparo à Pesquisa do Estado de São Paulo (grant number FAPESP 2018/03670-0, 2016/20547-2, 2018/13027-8, and 2017/24462-4) and Coordenação de Aperfeiçoamento de Pessoal de Nível Superior, Brasil (CAPES), Finance Code 001.

Acknowledgments: The authors are grateful to the Brazilian institutions Fundação de Amparo à Pesquisa do Estado de São Paulo (grant number FAPESP 2018/03670-0, 2016/20547-2, 2018/13027-8, and 2017/24462-4) and

Coordenação de Aperfeiçoamento de Pessoal de Nível Superior, Brasil (CAPES), Finance Code 001 for the financial support.

Conflicts of Interest: The authors declare no conflict of interest.

References

1. Ismail, M.; Zhang, J. Post–harvest citrus diseases and their control. *Outlooks Pest Manag.* **2004**, *15*, 29–35. [CrossRef]
2. USDA/FAS. Citrus: World Markets and Trade. United States Department of Agriculture. Foreing Agricultural Service, February, p. 1–13, 2019. Available online: https://www.fas.usda.gov/data/citrus-world-markets-and-trade (accessed on 30 May 2019).
3. Chen, J.; Shen, Y.; Chen, C.; Wan, C. Inhibition of key citrus postharvest fungal strains by plant extracts in vitro and in vivo: A review. *Plants* **2019**, *8*, 26. [CrossRef]
4. Dukare, A.S.; Paul, S.; Nambi, V.E.; Gupta, R.K.; Singh, R.; Sharma, K.; Vishwakarma, R.K. Exploitation of microbial antagonists for the control of postharvest diseases of fruits: A review. *Crit. Rev. Food Sci. Nutr.* **2019**, *59*, 1498–1513. [CrossRef] [PubMed]
5. Ariza, M.R.; Larsen, T.O.; Duus, J.Ø.; Barrero, A.F. *Penicillium digitatum* metabolites on synthetic media and citrus fruits. *J. Agric. Food Chem.* **2002**, *50*, 6361–6365. [CrossRef] [PubMed]
6. Regnier, T.; Combrinck, S.; Veldman, W.; Du Plooy, W. Application of essential oils as multi-target fungicides for the control of *Geotrichum citri-aurantii* and other postharvest pathogens of citrus. *Ind. Crops Prod.* **2014**, *61*, 151–159. [CrossRef]
7. Zhu, C.; Sheng, D.; Wu, X.; Wang, M.; Hu, X.; Li, H.; Yu, D. Identification of secondary metabolite biosynthetic gene clusters associated with the infection of citrus fruit by *Penicillium digitatum*. *Postharvest Biol. Technol.* **2017**, *134*, 17–21. [CrossRef]
8. Barmore, C.R.; Brown, G.E. Polygalacturonase from citrus fruit infected with *Penicillium italicum*. *Phytopathol.* **1981**, *71*, 328–331. [CrossRef]
9. Prusky, D.; McEvoy, J.L.; Saftner, R.; Conway, W.S.; Jones, R. Relationship between host acidification and virulence of *Penicillium* spp. on apple and citrus fruit. *Phytopathology* **2004**, *94*, 44–51. [CrossRef] [PubMed]
10. Macarisin, D.; Cohen, L.; Eick, A.; Rafael, G.; Belausov, E.; Wisniewski, M.; Droby, S. *Penicillium digitatum* suppresses production of hydrogen peroxide in host tissue during infection of citrus fruit. *Phytopathology* **2007**, *97*, 1491–1500. [CrossRef] [PubMed]
11. Ferraz, L.P.; Cunha, T.; Silva, A.C.; Kupper, K.C. Biocontrol ability and putative mode of action of yeasts against *Geotrichum citri-aurantii* in citrus fruit. *Microbiol. Res.* **2016**, *188*, 72–79. [CrossRef]
12. Palou, L.; Smilanick, J.L.; Droby, S. Alternatives to conventional fungicides for the control of citrus postharvest green and blue moulds. *Stewart Postharvest Rev.* **2008**, *4*, 1–16. [CrossRef]
13. Li Destri Nicosia, M.G.; Pangallo, S.; Raphael, G.; Romeo, F.V.; Strano, M.C.; Rapisarda, P.; Droby, S.; Schena, L. Control of postharvest fungal rots on citrus fruit and sweet cherries using a pomegranate peel extract. *Postharvest Biol. Technol.* **2016**, *114*, 54–61. [CrossRef]
14. Ballester, A.R.; Lafuente, M.T. LED Blue light-induced changes in phenolics and ethylene in citrus fruit: Implication in elicited resistance against *Penicillium digitatum* infection. *Food Chem.* **2017**, *218*, 575–583. [CrossRef] [PubMed]
15. Kim, J.J.; Ben-Yehoshua, S.; Shapiro, B.; Henis, Y.; Carmeli, S. Accumulation of scoparone in heat-treated lemon fruit inoculated with *Penicillium digitatum* Sacc. *Plant Physiol.* **1991**, *97*, 880–885. [CrossRef] [PubMed]
16. Jeong, R.D.; Chu, E.H.; Lee, G.W.; Cho, C.; Park, H.J. Inhibitory effect of gamma irradiation and its application for control of postharvest green mold decay of Satsuma mandarins. *Int. J. Food Microbiol.* **2016**, *234*, 1–8. [CrossRef] [PubMed]
17. Olmedo, G.M.; Cerioni, L.; González, M.M.; Cabrerizo, F.M.; Volentini, S.I.; Rapisarda, V.A. UVA photoactivation of harmol enhances its antifungal activity against the phytopathogens *Penicillium digitatum* and *Botrytis cinerea*. *Front. Microbiol.* **2017**, *8*, 347. [CrossRef] [PubMed]
18. Nunes, C.; Usall, J.; Manso, T.; Torres, R.; Olmo, M.; García, J.M. Effect of high temperature treatments on growth of *Penicillium* spp. and their development on 'Valencia' oranges. *Food Sci. Technol. Int.* **2007**, *13*, 63–68. [CrossRef]

19. Fatemi, S.; Borji, H. The effect of physical treatments on control of *Penicillium digitatum* decay orange cv. Valencia during storage period. *Afr. J. Agric. Res.* **2011**, *6*, 5757–5760. [CrossRef]

20. Mulas, M. Combined effects of fungicides and thermotherapy on post-harvest quality of horticultural commodities. In *Fungicides—Beneficial and Harmful Aspects*; Thajuddin, N., Ed.; IntechOpen: London, UK, 2011; pp. 133–166.

21. Abraham, A.O.; Laing, M.D.; Bower, J.P. Isolation and in vivo screening of yeast and *Bacillus* antagonists for the control of *Penicillium digitatum* of citrus fruit. *Biol. Control* **2010**, *53*, 32–38. [CrossRef]

22. Talibi, I.; Boubaker, H.; Boudyach, E.H.; Aoumar, A.A.B. Alternative methods for the control of postharvest citrus diseases. *J. Appl. Microbiol.* **2014**, *117*, 1–17. [CrossRef]

23. Mohammadi, P.; Tozlu, E.; Kotan, R.; Şenol Kotan, M. Potential of some bacteria for biological control of postharvest citrus green mould caused by *Penicillium digitatum*. *Plant Protect. Sci.* **2017**, *53*, 134–143. [CrossRef]

24. Youssef, K.; Sanzani, S.M.; Ligorio, A.; Ippolito, A.; Terry, L.A. Sodium carbonate and bicarbonate treatments induce resistance to postharvest green mould on citrus fruit. *Postharvest Biol. Technol.* **2014**, *87*, 61–69. [CrossRef]

25. Fallanaj, F.; Ippolito, A.; Ligorio, A.; Garganese, F.; Zavanella, C.; Sanzani, S.M. Electrolyzed sodium bicarbonate inhibits *Penicillium digitatum* and induces defence responses against green mould in citrus fruit. *Postharvest Biol. Technol.* **2016**, *115*, 18–29. [CrossRef]

26. Shi, S.; Wang, F.; Lu, Y.; Deng, J. Combination of chitosan and salicylic acid to control postharvest green mold caused by *Penicillium digitatum* in grapefruit fruit. *Sci. Hortic.* **2018**, *233*, 54–60. [CrossRef]

27. Droby, S.; Wisniewski, M.; Macarisin, D.; Wilson, C. Twenty years of postharvest biocontrol research: Is it time for a new paradigm? *Postharvest Biol. Technol.* **2009**, *52*, 137–145. [CrossRef]

28. Sharma, R.R.; Singh, D.; Singh, R. Biological control of postharvest diseases of fruits and vegetables by microbial antagonists: A review. *Biol. Control* **2009**, *50*, 205–221. [CrossRef]

29. Spadaro, D.; Gullino, M.L. State of the art and future prospects of the biological control of postharvest fruit diseases. *Int. J. Food Microbiol.* **2004**, *91*, 185–194. [CrossRef]

30. Nunes, C.A.; Manso, T.; Lima-Costa, M.E. Postharvest biological control of citrus fruits. *Tree For. Sci. Biotechnol.* **2009**, *2*, 116–126.

31. Waewthongrak, W.; Pisuchpen, S.; Leelasuphakul, W. Effect of *Bacillus subtilis* and chitosan applications on green mold (*Penicilium digitatum* Sacc.) decay in citrus fruit. *Postharvest Biol. Technol.* **2015**, *99*, 44–49. [CrossRef]

32. Droby, S.; Chalutz, E.; Wilson, C.L.; Wisniewski, M. Characterization of the biocontrol activity of *Debaryomyces hansenii* in the control of *Penicillium digitatum* on grapefruit. *Can. J. Microbiol.* **1989**, *35*, 794–800. [CrossRef]

33. Droby, S.; Vinokur, V.; Weiss, B.; Cohen, L.; Daus, A.; Goldschmidt, E.E.; Porat, R. Induction of resistance to *Penicillium digitatum* in grapefruit by the yeast biocontrol agent *Candida oleophila*. *Phytopathology* **2002**, *92*, 393–399. [CrossRef] [PubMed]

34. Lu, L.; Lu, H.; Wu, C.; Fang, W.; Yu, C.; Ye, C.; Shi, Y.; Yu, T.; Zheng, X. *Rhodosporidium paludigenum* induces resistance and defense-related responses against *Penicillium digitatum* in citrus fruit. *Postharvest Biol. Technol.* **2013**, *85*, 196–202. [CrossRef]

35. Bar-Shimon, M.; Yehuda, H.; Cohen, L.; Weiss, B.; Kobeshnikov, A.; Daus, A. Characterization of extracellular lytic enzymes produced by the yeast biocontrol agent *Candida oleophila*. *Curr. Genet.* **2004**, *45*, 140–148. [CrossRef] [PubMed]

36. Macarisin, D.; Droby, S.; Bauchan, G.; Wisniewski, M. Superoxide anion and hydrogen peroxide in the yeast antagonist–fruit interaction: A new role for reactive oxygen species in postharvest biocontrol? *Postharvest Biol. Technol.* **2010**, *58*, 194–202. [CrossRef]

37. Castoria, R.; Caputo, L.; De Curtis, F.; De Cicco, V. Resistance of postharvest biocontrol yeasts to oxidative stress: A possible new mechanism of action. *Phytopathology* **2003**, *93*, 564–572. [CrossRef]

38. Benhamou, N. Potential of the mycoparasite, *Verticillium lecanii*, to protect citrus fruit against *Penicillium digitatum*, the causal agent of green mold: A comparison with the effect of chitosan. *Phytopathology* **2004**, *94*, 693–705. [CrossRef]

39. Aloui, H.; Licciardello, F.; Khwaldia, K.; Hamdi, M.; Restuccia, C. Physical properties and antifungal activity of bioactive films containing *Wickerhamomyces anomalus* killer yeast and their application for preservation of oranges and control of postharvest green mold caused by *Penicillium digitatum*. *Int. J. Food Microbiol.* **2015**, *200*, 22–30. [CrossRef]

40. Platania, C.; Restuccia, C.; Muccilli, S.; Cirvilleri, G. Efficacy of killer yeasts in the biological control of *Penicillium digitatum* on Tarocco orange fruits (*Citrus sinensis*). *Food Microbiol.* **2012**, *30*, 219–225. [CrossRef]

41. Comitini, F.; Mannazzu, I.; Ciani, M. *Tetrapisispora phaffii* killer toxin is a highly specific β-glucanase that disrupts the integrity of the yeast cell wall. *Microb. Cell Fact.* **2009**, *8*, 55. [CrossRef]

42. Perez, M.F.; Contreras, L.; Garnica, N.M.; Fernández-Zenoff, M.V.; Farías, M.E.; Sepulveda, M.; Ramallo, J.; Dib, J.R. Native killer yeasts as biocontrol agents of postharvest fungal diseases in lemons. *PLoS ONE* **2016**, *11*, e0165590. [CrossRef]

43. Kupper, K.C.; Cervantes, A.L.L.; Klein, M.N.; Silva, A.C. Avaliação de microrganismos antagônicos, *Saccharomyces cerevisiae* e *Bacillus subtilis* para o controle de *Penicillium digitatum*. *Rev. Bras. Frutic.* **2013**, *35*, 425–436. [CrossRef]

44. Cunha, T.; Ferraz, L.P.; Wehr, P.P.; Kupper, K.C. Antifungal activity and action mechanisms of yeasts isolates from citrus against *Penicillium italicum*. *Int. J. Food Microbiol.* **2018**, *276*, 20–27. [CrossRef] [PubMed]

45. Leelasuphakul, W.; Hemmanee, P.; Chuenchitt, S. Growth inhibitory properties of *Bacillus subtilis* strains and their metabolites against the green mold pathogen (*Penicillium digitatum* Sacc.) of citrus fruit. *Postharvest Biol. Technol.* **2008**, *48*, 113–121. [CrossRef]

46. Calvo, H.; Marco, P.; Blanco, D.; Oria, R.; Venturini, M.E. Potential of a new strain of *Bacillus amyloliquefaciens* BUZ-14 as a biocontrol agent of postharvest fruit diseases. *Food Microbiol.* **2017**, *63*, 101–110. [CrossRef] [PubMed]

47. Wang, H.K.; Yan, Y.H.; Wang, J.M.; Zhang, H.P.; Qi, W. Production and characterization of antifungal compounds produced by *Lactobacillus plantarum* IMAU10014. *PLoS ONE* **2012**, *7*, e29452. [CrossRef] [PubMed]

48. Matei, A.; Cornea, C.P.; Matei, S.; Matei, G.M.; Rodino, S. Comparative antifungal effect of lactic acid bacteria strains on *Penicillium digitatum*. *Bull. UASVM Food Sci. Technol.* **2015**, *72*, 226–230. [CrossRef]

49. Maldonado, M.C.; Orosco, C.E.; Gordillo, M.A.; Navarro, A.R. In vivo and in vitro antagonism of *Streptomyces* sp. RO3 against *Penicillium digitatum* and *Geotrichum candidum*. *Afr. J. Microbiol. Res.* **2010**, *4*, 2451–2456.

50. Najmeh, S.; Hosein, S.B.G.; Sareh, S.; Bonjar, L.S. Biological control of citrus green mould, *Penicillium digitatum*, by antifungal activities of *Streptomyces* isolates from agricultural soils. *Afr. J. Microbiol. Res.* **2014**, *8*, 1501–1509. [CrossRef]

51. Walling, L.L. Induced resistance: From the basic to the applied. *Trends Plant Sci.* **2001**, *6*, 445–447. [CrossRef]

52. Parafati, L.; Vitale, A.; Restuccia, C.; Cirvilleri, G. The effect of locust bean gum (LBG)-based edible coatings carrying biocontrol yeasts against *Penicillium digitatum* and *Penicillium italicum* causal agents of postharvest decay of mandarin fruit. *Food Microbiol.* **2016**, *58*, 87–94. [CrossRef] [PubMed]

53. Liu, Y.; Yao, S.; Deng, L.; Ming, J.; Zeng, K. Different mechanisms of action of isolated epiphytic yeasts against *Penicillium digitatum* and *Penicillium italicum* on citrus fruit. *Postharvest Biol. Technol.* **2019**, *152*, 100–110. [CrossRef]

54. Pimenta, R.S.; Silva, J.F.M.; Coelho, C.M.; Morais, P.B.; Rosa, C.A.; Corrêa, A., Jr. Integrated control of *Penicillium digitatum* by the predacious yeast *Saccharomycopsis crataegensis* and sodium bicarbonate on oranges. *Braz. J. Microbiol.* **2010**, *41*, 404–410. [CrossRef] [PubMed]

55. Geng, P.; Chen, S.; Hu, M.; Haq, M.R.; Lai, K.; Qu, F.; Zhang, Y. Combination of *Kluyveromyces marxianus* and sodium bicarbonate for controlling green mold of citrus fruit. *Int. J. Food Microbiol.* **2011**, *151*, 190–194. [CrossRef] [PubMed]

56. Lu, L.; Xu, S.; Zeng, L.; Zheng, X.; Yu, T. *Rhodosporidium paludigenum* induced resistance in Ponkan mandarin against *Penicillium digitatum* requires ethylene-dependent signaling pathway. *Postharvest Biol. Technol.* **2014**, *97*, 93–101. [CrossRef]

57. Spadaro, D.; Droby, S. Unraveling the mechanisms used by antagonistic yeast to control postharvest pathogens on fruit. *Acta Hortic.* **2016**, *1144*, 63–70. [CrossRef]

58. Li, J.; Li, H.; Ji, S.; Chen, T.; Tian, S.; Qin, G. Enhancement of biocontrol efficacy of *Cryptococcus laurentii* by cinnamic acid against *Penicillium italicum* in citrus fruit. *Postharvest Biol. Technol.* **2019**, *149*, 42–49. [CrossRef]

59. Chalutz, E. Postharvest biocontrol of green and blue mold and sour rot of citrus fruit by *Debaryomyces hansenii*. *Plant Dis.* **1990**, *74*, 134–137. [CrossRef]

60. Hernández-Montiel, L.G.; Ochoa, J.L.; Troyo-Diéguez, E.; Larralde-Corona, C.P. Biocontrol of postharvest blue mold (*Penicillium italicum* Wehmer) on Mexican lime by marine and citrus *Debaryomyces hansenii* isolates. *Postharvest Biol. Technol.* **2010**, *56*, 181–187. [CrossRef]

61. Choudhary, B.; Nagpure, A.; Gupta, R.K. Biological control of toxigenic citrus and papaya-rotting fungi by *Streptomyces violascens* MT7 and its extracellular metabolites. *J. Basic Microbiol.* **2015**, *55*, 1343–1356. [CrossRef]

62. Olmedo, G.M.; Cerioni, L.; González, M.M.; Cabrerizo, F.M.; Rapisarda, V.A.; Volentini, S.I. Antifungal activity of β-carbolines on *Penicillium digitatum* and *Botrytis cinerea*. *Food Microbiol.* **2017**, *62*, 9–14. [CrossRef]

63. Lu, L.; Yang, Y.L.J.; Azat, R.; Yu, T.; Zheng, X. Quaternary chitosan oligomers enhance resistance and biocontrol efficacy of *Rhodosporidium paludigenum* to green mold in satsuma orange. *Carbohydr. Polym.* **2014**, *113*, 174–181. [CrossRef] [PubMed]

64. Shao, X.; Cao, B.; Xu, F.; Xie, S.; Yu, D.; Wang, H. Effect of postharvest application of chitosan combined with clove oil against citrus green mold. *Postharvest Biol. Technol.* **2015**, *99*, 37–43. [CrossRef]

65. Tayel, A.A.; Moussa, S.H.; Salem, M.F.; Mazrou, K.E.; El-Tras, W.F. Control of citrus molds using bioactive coatings incorporated with fungal chitosan/plant extracts composite. *J. Sci. Food Agric.* **2016**, *96*, 1306–1312. [CrossRef] [PubMed]

66. El Guilli, M.; Hamza, A.; Clément, C.; Ibriz, M.; Barka, E.A. Effectiveness of postharvest treatment with chitosan to control citrus green mold. *Agriculture* **2016**, *6*, 12. [CrossRef]

67. Pérez-Alfonso, C.O.; Martínez-Romero, D.; Zapata, P.J.; Serrano, M.; Valero, D.; Castillo, S. The effects of essential oils carvacrol and thymol on growth of *Penicillium digitatum* and *P. italicum* involved in lemon decay. *Int. J. Food Microbiol.* **2012**, *158*, 101–106. [CrossRef] [PubMed]

68. Fan, F.; Tao, N.; Jia, L.; He, X. Use of citral incorporated in postharvest wax of citrus fruit as a botanical fungicide against *Penicillium digitatum*. *Postharvest Biol. Technol.* **2014**, *90*, 52–55. [CrossRef]

69. Wu, Y.; OuYang, Q.; Tao, N. Plasma membrane damage contributes to antifungal activity of citronellal against *Penicillium digitatum*. *J. Food Sci. Technol.* **2016**, *53*, 3853–3858. [CrossRef]

70. Sivakumar, D.; Bautista-Baños, S. A review on the use of essential oils for postharvest decay control and maintenance of fruit quality during storage. *Crop Prot.* **2014**, *64*, 27–37. [CrossRef]

71. Solgi, M.; Ghorbanpour, M. Application of essential oils and their biological effects on extending the shelf-life and quality of horticultural crops. *Trakia J. Sci.* **2014**, *12*, 198–210.

72. Peng, L.; Yang, S.; Cheng, Y.J.; Chen, F.; Pan, S.; Fan, G. Antifungal activity and action mode of pinocembrin from propolis against *Penicillium italicum*. *Food Sci. Biotechnol.* **2012**, *21*, 1533–1539. [CrossRef]

73. Trabelsi, D.; Hamdane, A.M.; Said, M.B.; Abdrrabba, M. Chemical composition and antifungal activity of essential oils from flowers, leaves and peels of Tunisian *Citrus aurantium* against *Penicillium digitatum* and *Penicillium italicum*. *J. Essent. Oil Bear Plants* **2016**, *19*, 1660–1674. [CrossRef]

74. Tao, N.; Jia, L.; Zhou, H.; He, X. Effect of octanal on the mycelial growth of *Penicillium italicum* and *P. digitatum*. *World J. Microbiol. Biotechnol.* **2014**, *30*, 1169–1175. [CrossRef] [PubMed]

75. Tao, N.; OuYang, Q.; Jia, L. Citral inhibits mycelial growth of *Penicillium italicum* by a membrane damage mechanism. *Food Control* **2014**, *41*, 116–121. [CrossRef]

76. Kanan, G.J.M.; Al-Najar, R.A.-W.K. In vitro and in vivo activity of selected plant crude extracts and fractions against *Penicillium italicum*. *J. Plant Prot. Res.* **2009**, *49*, 341–352. [CrossRef]

77. Aminifard, M.H.; Bayat, H. Antifungal activity of black caraway and anise essential oils against *Penicillium digitatum* on blood orange fruits. *Int. J. Fruit Sci.* **2018**, *18*, 307–319. [CrossRef]

78. Yang, S.; Liu, L.; Li, D.; Xia, H.; Su, X.; Peng, L.; Pan, S. Use of active extracts of poplar buds against *Penicillium italicum* and possible modes of action. *Food Chem.* **2016**, *196*, 610–618. [CrossRef] [PubMed]

79. Wan, C.; Li, P.; Chen, C.; Peng, X.; Li, M.; Chen, M.; Wang, J.; Chen, J. Antifungal activity of *Ramulus cinnamomi* explored by ^{1}H-NMR based metabolomics approach. *Molecules* **2017**, *22*, 2237. [CrossRef]

80. Hendel, N.; Larous, L.; Belbey, L. Antioxidant activity of rosemary (*Rosmarinus officinalis* L.) and its in vitro inhibitory effect on *Penicillium digitatum*. *Int. Food Res. J.* **2016**, *23*, 1725–1732.

81. Exarchou, V.; Kanetis, L.; Charalambous, Z.; Apers, S.; Pieters, L.; Gekas, V.; Goulas, V. HPLC-SPE-NMR Characterization of major metabolites in *Salvia fruticosa* Mill. extract with antifungal potential: Relevance of carnosic acid, carnosol, and hispidulin. *J. Agric. Food Chem.* **2015**, *63*, 457–463. [CrossRef]

82. Zhu, C.; Lei, M.; Andargie, M.; Zeng, J.; Li, J. Antifungal activity and mechanism of action of tannic acid against *Penicillium digitatum*. *Physiol. Mol. Plant Pathol.* **2019**, *107*, 46–50. [CrossRef]

83. Musto, M.; Potenza, G.; Cellini, F. Inhibition of *Penicillium digitatum* by a crude extract from *Solanum nigrum* leaves. *Biotechnol. Agron. Soc. Environ.* **2014**, *18*, 174–180.

84. Boubaker, H.; Karim, H.; El Hamdaoui, A.; Msanda, F.; Leach, D.; Bombarda, I.; Vanloot, P.; Abbad, A.; Boudyach, E.H.; Aoumar, A.A.B. Chemical characterization and antifungal activities of four *Thymus* species essential oils against postharvest fungal pathogens of citrus. *Ind. Crops Prod.* **2016**, *86*, 95–101. [CrossRef]

85. Ameziane, N.; Boubaker, H.; Boudyach, H.; Msanda, F.; Jilal, A.; Benaoumar, A.A. Antifungal activity of Moroccan plants against citrus fruit pathogens. *Agron. Sustain. Dev.* **2007**, *27*, 273–277. [CrossRef]

86. Vitoratos, A.; Bilalis, D.; Karkanis, A.; Efthimiadou, A. Antifungal activity of plant essential oils against *Botrytis cinerea, Penicillium italicum* and *Penicillium digitatum*. *Not. Bot. Horti Agrobot.* **2013**, *41*, 86–92. [CrossRef]

87. Mekbib, S.B.; Regnier, T.J.C.; Korsten, L. Control of *Penicillium digitatum* growth on citrus fruit using two plant extracts and their mode of action. *Phytoparasitica* **2007**, *35*, 264–276. [CrossRef]

88. Obagwu, J.; Korsten, L. Control of citrus green and blue molds with garlic extracts. *Eur. J. Plant Pathol.* **2003**, *109*, 221–225. [CrossRef]

89. Mossini, S.A.G.; Arrotéia, C.C.; Kemmelmeier, C. Effect of neem leaf extract and neem oil on *Penicillium* growth, sporulation, morphology and Ochratoxin A production. *Toxins* **2009**, *1*, 3–13. [CrossRef]

90. Baviskar, R.N. Anti-fungal activity of Launea pinnatifida and Argimone maxicana against post-harvest fungal pathogens in Apple fruits. *Int. J. of Life Sciences* **2014**, *2*, 346–349.

91. Singh, H.; Al-samarai, G.; Syarhabil, M. Exploitation of natural products as an alternative strategy to control postharvest fungal rotting of citrus. *IJSR* **2012**, *2*, 1–4. [CrossRef]

92. Palou, L. Penicillium digitatum, *Penicillium italicum* (Green Mold, Blue Mold). In *Postharvest Decay*; Academic Press: Cambridge, MA, USA, 2014; pp. 45–102.

93. Holmes, G.J.; Eckert, J.W. Sensitivity of *Penicillium digitatum* and *P. italicum* to postharvest citrus fungicides in California. *Phytopathol.* **1999**, *89*, 716–721. [CrossRef]

94. Du Plooy, W.; Regnier, T.; Combrinck, S. Essential oil amended coatings as alternatives to synthetic fungicides in citrus postharvest management. *Postharvest Biol. Technol.* **2009**, *53*, 117–122. [CrossRef]

95. Tripathi, P.; Dubey, N.K.; Shukla, A.K. Use of some essential oils as post-harvest botanical fungicides in the management of grey mould of grapes caused by *Botrytis cinerea*. *World J. Microbiol. Biotechnol.* **2008**, *24*, 39–46. [CrossRef]

96. Fawcett, C.H.; Spencer, D.M. Plant chemotherapy with natural products. *Annu. Rev. Phytopathol.* **1970**, *8*, 403–418. [CrossRef]

97. Tripathi, P.; Dubey, N.K. Exploitation of natural products as an alternative strategy to control postharvest fungal rotting of fruit and vegetables. *Postharvest Biol. Technol.* **2004**, *32*, 235–245. [CrossRef]

98. Askarne, L.; Talibi, I.; Boubaker, H.; Boudyach, E.H.; Msanda, F.; Saadi, B.; Serghini, M.A.; Aoumar, A.A.B. In vitro and in vivo antifungal activity of several Moroccan plants against *Penicillium italicum*, the causal agent of citrus blue mold. *Crop Prot.* **2012**, *40*, 53–58. [CrossRef]

99. Jham, G.N.; Dhingra, O.D.; Jardim, C.M.; Valente, V.M.M. Identification of the major fungitoxic component of cinnamon bark oil. *Fitopatol. Bras.* **2005**, *30*, 404–408. [CrossRef]

100. Telezhenetskaya, M.V.; D'yakonov, A.L. Alkaloids of *Peganum harmala*. Unusual reaction of peganine and vasicinone. *Chem. Nat. Compd.* **1991**, *27*, 471–474. [CrossRef]

101. Scora, K.M.; Scora, R.W. Effect of volatiles on mycelium growth of *Penicillium digitatum, P. italicum*, and *P. ulaiense*. *J. Basic Microbiol.* **1998**, *38*, 405–413. [CrossRef]

102. Droby, S.; Eick, A.; Macarisin, D.; Cohen, L.; Rafael, G.; Stange, R.; McColum, G.; Dudai, N.; Nasser, A.; Wisniewski, M.; et al. Role of citrus volatiles in host recognition, germination and growth of *Penicillium digitatum* and *Penicillium italicum*. *Postharvest Biol. Technol.* **2008**, *49*, 386–396. [CrossRef]

103. Mercier, J.; Smilanick, J.L. Control of green mold and sour rot of stored lemon by biofumigation with *Muscodor albus*. *Biol. Control* **2005**, *32*, 401–407. [CrossRef]

104. Nakamura, M.; Suprapta, D.N.; Iwai, H.; Arai, A.K. Comparison of endo-polygalacturonase activities of citrus and non-citrus races of *Geotrichum candidum*, and cloning and expression of the corresponding genes. *Mol. Plant Pathol.* **2001**, *2*, 265–274. [CrossRef] [PubMed]

105. McKay, A.H.; Förster, H.; Adaskaveg, J.E. Efficacy and application strategies for propiconazole as a new postharvest fungicide for managing sour rot and green mold of citrus fruit. *Plant Dis.* **2012**, *96*, 235–242. [CrossRef] [PubMed]

106. Yin, C.; Liu, H.; Shan, Y.; Gupta, V.K.; Jiang, Y.; Zhang, W.; Tan, H.; Gong, L. Cytosporone B as a biological preservative: Purification, fungicidal activity and mechanism of action against *Geotrichum citri-aurantii*. *Biomolecules* **2019**, *9*, 125. [CrossRef] [PubMed]

107. Talibi, I.; Askarne, L.; Boubaker, H.; Boudyach, E.H.; Msanda, F.; Saadi, B.; Aoumar, A.A.B. Antifungal activity of Moroccan medicinal plants against citrus sour rot agent *Geotrichum candidum*. *Lett. Appl. Microbiol.* **2012**, *55*, 155–161. [CrossRef] [PubMed]

108. Karim, H.; Boubaker, H.; Askarne, L.; Talibi, I.; Msanda, F.; Boudyach, E.H.; Saadi, B.; Aoumar, A.A.B. Antifungal properties of organic extracts of eight *Cistus* L. species against postharvest citrus sour rot. *Lett. Appl. Microbiol.* **2015**, *62*, 16–22. [CrossRef] [PubMed]

109. Sayago, J.E.; Ordoñez, R.M.; Kovacevich, L.N.; Torres, S.; Isla, M.I. Antifungal activity of extracts of extremophile plants from the Argentine Puna to control citrus postharvest pathogens and green mold. *Postharvest Biol. Technol.* **2012**, *67*, 19–24. [CrossRef]

110. Zhou, H.; Tao, N.; Jia, L. Antifungal activity of citral, octanal and α-terpineol against *Geotrichum citri-aurantii*. *Food Control* **2014**, *37*, 277–283. [CrossRef]

111. Liu, X.; Wang, L.P.; Li, Y.C.; Yu, T.; Zheng, X.D. Antifungal activity of thyme oil against *Geotrichum citri-aurantii* in vitro and in vivo. *J. Appl. Microbiol.* **2009**, *107*, 1450–1456. [CrossRef]

112. Xu, S.-X.; Li, Y.-C.; Liu, X.; Mao, L.-J.; Zhang, H.; Zheng, X.-D. In vitro and in vivo antifungal activity of a water-dilutable cassia oil microemulsion against *Geotrichum citri-aurantii*. *J. Sci. Food Agric.* **2012**, *92*, 2668–2671. [CrossRef]

113. Spadaro, D.; Droby, S. Development of biocontrol products for postharvest diseases of fruit: The importance of elucidating the mechanisms of action of yeast antagonists. *Trends Food Sci. Technol.* **2016**, *47*, 39–49. [CrossRef]

114. Alamri, S.A. The synergistic effect of two formulated biofungicides in the biocontrol of root and bottom rot of lettuce. *Biocontrol Sci.* **2014**, *19*, 189–197. [CrossRef] [PubMed]

115. Janisiewicz, W.J.; Jeffers, S.N. Efficacy of commercial formulation of two biofungicides for control of blue mold and gray mold of apples in cold storage. *Crop Prot.* **1997**, *16*, 629–633. [CrossRef]

116. Abbey, J.A.; Percival, D.; Abbey, L.; Asiedu, S.K.; Prithiviraj, B.; Schilder, A. Biofungicides as alternative to synthetic fungicide control of grey mould (*Botrytis cinerea*)—Prospects and challenges. *Biocontrol Sci. Technol.* **2019**, *29*, 241–262. [CrossRef]

117. Liu, J.; Sui, Y.; Wisniewski, M.; Droby, S.; Liu, Y. Review: Utilization of antagonistic yeasts tomanage postharvest fungal diseases of fruit. *Int. J. Food Microbiol.* **2013**, *167*, 153–160. [CrossRef] [PubMed]

118. Zamanizadeh, H.R.; Hatami, N.; Aminaee, M.M.; Rakhshandehroo, F. Application of biofungicides in control of damping disease off in greenhouse crops as a possible substitute to synthetic fungicides. *Int. J. Environ. Sci. Technol.* **2011**, *8*, 129–136. [CrossRef]

119. Nunes, C.A. Biological control of postharvest diseases of fruit. *Eur. J. Plant Pathol.* **2012**, *133*, 181–196. [CrossRef]

120. Mari, M.; Di Francesco, A.; Bertolini, P. Control of fruit postharvest diseases: Old issues and innovative approaches. *Stewart Postharvest Rev.* **2014**, *1*, 1–4. [CrossRef]

121. Anuagasi, C.L.; Okigbo, R.N.; Anukwuorji, C.A.; Okereke, C.N. The impact of biofungicides on agricultural yields and food security in Africa. *IJAT* **2017**, *13*, 953–978.

122. Piombo, E.; Sela, N.; Wisniewski, M.; Hoffmann, M.; Gullino, M.L.; Allard, M.W.; Levin, E.; Spadaro, D.; Droby, S. Genome sequence, assembly and characterization of two *Metschnikowia fructicola* strains used as biocontrol agents of postharvest diseases. *Front. Microbiol.* **2018**, *9*, 593. [CrossRef]

123. Wisniewski, M.; Macarisin, D.; Droby, S. Challenges and opportunities for the commercialization of postharvest biocontrol. In *Proceedings of the VI International Postharvest Symposium, Antalya, Turkey, 8–12 April 2009*; International Society for Horticultural Science: Leuven, Belgium, 2010; Volume 877, pp. 1577–1582. [CrossRef]

124. Kurtzman, C.P.; Droby, S. *Metschnikowia fructicola*, a new ascosporic yeast with potential for biocontrol of postharvest fruit rots. *Syst. Appl. Microbiol.* **2001**, *24*, 395–399. [CrossRef]

125. Viñas, I.; Usall, J.; Teixidó, N.; Sanchis, V. Biological control of major postharvest pathogens on apple with *Candida sake*. *Int. J. Food Microbiol.* **1998**, *40*, 9–16. [CrossRef]
126. Fravel, D.R.; Larkin, R.P. Availability and application of biocontrol products. In *Biological and Cultural Tests for Control of Plant Diseases*; Canaday, C.H., Ed.; APS Press: St. Paul, MN, USA, 1996; Volume 11, pp. 1–7.
127. Souza, J.R.B.; Kupper, K.C.; Augusto, F. In vivo investigation of the volatile metabolome of antiphytopathogenic yeast strains active against *Penicillium digitatum* using comprehensive two-dimensional gas chromatography and multivariate data analysis. *Microchem. J.* **2018**, *141*, 204–209. [CrossRef]

© 2019 by the authors. Licensee MDPI, Basel, Switzerland. This article is an open access article distributed under the terms and conditions of the Creative Commons Attribution (CC BY) license (http://creativecommons.org/licenses/by/4.0/).

Article

Application of Zearalenone (ZEN)-Detoxifying *Bacillus* in Animal Feed Decontamination through Fermentation

Shiau-Wei Chen [1], Han-Tsung Wang [1], Wei-Yuan Shih [2], Yan-An Ciou [1], Yu-Yi Chang [3], Laurensia Ananda [3], Shu-Yin Wang [3] and Jih-Tay Hsu [1,*]

[1] Department of Animal Science and Technology, National Taiwan University, No. 50, Lane 155, Sec 3, Keelung Rd, Taipei 10673, Taiwan; d98626002@ntu.edu.tw (S.-W.C.); rumen0808@gmail.com (H.-T.W.); g2299015@gmail.com (Y.-A.C.)

[2] Animal Resource Center, National Taiwan University, No. 118, Lane 155, Sec 3, Keelung Rd, Taipei 10673, Taiwan; weiyuanshih@ntu.edu.tw

[3] Graduate Institute of Biotechnology, Chinese Culture University, No. 55, Hwa-Kang Rd, Taipei 11114, Taiwan; yychang@ntu.edu.tw (Y.-Y.C.); lq_lin@ymail.com (L.A.); sywang@faculty.pccu.edu.tw (S.-Y.W.)

* Correspondence: jthsu@ntu.edu.tw; Tel.: +886-2-3366-4153

Received: 27 May 2019; Accepted: 6 June 2019; Published: 8 June 2019

Abstract: Zearalenone (ZEN) is an estrogenic mycotoxin which can cause loss in animal production. The aim of this study was to screen *Bacillus* strains for their ZEN detoxification capability and use a fermentation process to validate their potential application in the feed industry. In the high-level ZEN-contaminated maize ($5\ \mathrm{mg \cdot kg^{-1}}$) fermentation test, B2 strain exhibited the highest detoxification rate, removing 56% of the ZEN. However, B2 strain was not the strain with the highest ZEN detoxification in the culturing media. When B2 grew in TSB medium with ZEN, it had higher bacterial numbers, lactic acid, acetic acid, total volatile fatty acids, and ammonia nitrogen. The ZEN-contaminated maize fermented by B2 strain had better fermentation characteristics (lactic acid > 110 $\mathrm{mmol \cdot L^{-1}}$; acetic acid < 20 $\mathrm{mmol \cdot L^{-1}}$; pH < 4.5) than ZEN-free maize. Furthermore, B2 also had detoxification capabilities toward aflatoxins B1, deoxynivalenol, fumonisin B1, and T2 toxin. Our study demonstrated differences in screening outcome between bacterial culturing conditions and the maize fermentation process. This is important for the feed industry to consider when choosing a proper method to screen candidate isolates for the pretreatment of ZEN-contaminated maize. It appears that using the fermentation process to address the ZEN-contaminated maize problem in animal feed is a reliable choice.

Keywords: zearalenone; biological detoxification; *Bacillus*; fermentation

Key Contribution: Our study showed differences in screening outcome between the culture medium method and the maize fermentation process. In the high-level ZEN-contaminated maize fermentation test, the isolate B2 strain was stimulated by ZEN and exhibited the highest detoxification rate to ZEN (56%) with multiple detoxification capabilities toward other mycotoxins.

---

1. Introduction

Zearalenone (ZEN) is a nonsteroidal estrogenic mycotoxin produced by *Fusarium* species on cereal crops grown in warm, humid climates. Its contamination mostly occurs in preharvest periods rather than in storage periods [1]. Maize is susceptible to *Fusarium* infection and contamination by ZEN [2,3]. ZEN is classified as an endocrine disruptor, due to its estrogenic activity, which can disrupt the estrous cycles in animals [4]. Swine are more sensitive to ZEN effects than other species [2,5]. ZEN in diet as low as 1 ppm may lead to hyperestrogenic syndrome in gilts [5]. In addition, ZEN and its metabolites may affect humans through egg, milk, or meat. The potential threats of ZEN should not be ignored.

The mycotoxin's structure plays a key role in its toxicity [2,6]. When the lactone ring of ZEN is cleaved by esterase or lactonohydrolase, its ability to bind to estrogen receptors is compromised [7]. Various strategies have been developed to detoxify mycotoxins, but most of them have shown disadvantages in animal feed applications [6,8]. Several studies have suggested that biological detoxification is effective, specific, and safe in the decontamination of animal feed [1,6,9]. Biological detoxification depends on microorganism enzyme action to degrade mycotoxins and microbial cell walls to adsorb mycotoxins. In recent years, the mycotoxin detoxification ability of *Bacillus* spp. has been noticed. *Bacillus* spp. is Gram-positive, spore-forming bacteria. Many *Bacillus* spp. have been proposed for their qualified presumption of safety (QPS) status by the EFSA such as *B. amyloliquefaciens*, *B. licheniformis*, and *B. subtilis*. They are often utilized to produce enzymes or used as microbial feed additives [10]. With all these characteristics, *Bacillus* spp. are one of the safe and perfect candidates to act as a ZEN detoxification agent in animal feed.

Fermented liquid feed (FLF) is widely used in the animal industry and has many advantages. It can be produced by fermentation of the complete feed or by fermentation of cereals before being incorporated with other feed ingredients [11]. FLF has been shown to improve the growth performance of pigs and to decrease the incidence of enteric diseases in pigs by lowering enteric pathogen numbers [11,12]. The objective of this study was to screen ZEN-detoxifying *Bacillus* (ZDB) strains and put the candidate strains through fermentation of ZEN-contaminated maize to evaluate their field application potential for animal feed.

2. Results

2.1. Screening of Bacillus Strains for ZEN Detoxification Potential

In this study, 106 isolates were successfully isolated from fermented soybean products, soil, sewage, rumen fluid, and ruminant feces. All isolates were included in the ZEN detoxification capability test. The results showed that 14 isolates had a markedly greater ZEN detoxification ability than the others (Table 1). Out of these 14 isolates, seven isolates were from fermented soybean products, one from sewage, and the remaining from soil. The 16S rRNA gene sequence analysis indicated that these 14 isolates shared the highest identities with *Bacillus* spp. In the enterotoxin detection assays, nine out of the 14 strains were identified as enterotoxin producers. The five non-enterotoxin-producing ZEN-detoxifying *Bacillus* (ZDB) strains were kept for further study (Table 1).

Table 1. The characteristics of 14 isolates isolated from different sources with ZEN detoxification ability.

Strains	Source	Enterotoxins [b]			Emetic Toxin	Code [c]	16S rDNA Sequencing (Identify%)
		Nhe A	Nhe B	Hbl L2	Cereulide		
Isolate 1	Fermented soybean product	−	−	−	−	B1	*B. subtilis* (99)
Isolate 2	Fermented soybean product	+	−	−	−		
Isolate 3	Fermented soybean product	−	−	−	−	B2	*B. subtilis* (99)
Isolate 4	Fermented soybean product	+	−	−	−		
Isolate 5	Fermented soybean product	−	−	−	−	B3	*B. subtilis* (99)
Isolate 6	Fermented soybean product	−	−	−	−	B4	*Bacillus* sp. (100)
Isolate 7	Fermented soybean product	+	−	−	−		
Isolate 8	Sewage	+	−	−	−		
Isolate 9	Soil	+	−	−	−		
Isolate 10	Soil	+	−	−	−		
Isolate 11	Soil	+	−	−	−		
Isolate 12	Soil	+	+	−	−		
Isolate 13	Soil	−	−	−	−	B5	*Bacillus* sp. (100)
Isolate 14	Soil	+	−	−	−		

[a] Enterotoxins included nonhemolytic enterotoxin A (Nhe A), nonhemolytic enterotoxin B (Nhe B), and hemolysin BL (HBL). The "−" and "+" represent negative and positive response, respectively. [b] Five nontoxic ZEN-detoxifying *Bacillus* (ZDB) strains were selected for further study.

2.2. ZEN Detoxification Capability, Adsorption Ability, and Degradation Ability of ZDB Strains

The detoxification rate of strain B4 was 58%, followed by strain 17,441 (52%) in the culturing condition (Table 2). The detoxification rate of the other ZDB strains were between 28% and 43%. The adsorption ability of strain 17,441 (47%) was significantly higher than all of the other ZDB strains (Table 2) ($p < 0.05$). No significant differences were found between the degradation rate of all ZDB strains and type strain 17,441. The degradation rate of each ZDB strain (B1–B5) and strain 17,441 were 24, 35, 27, 31, 33, and 31%, respectively.

Table 2. The ZEN detoxification capability, absorption, and degradation ability of ZEN-detoxification *Bacillus* strains in TSB medium.

Bacillus Strains	Detoxification Rate (%) *	Absorption Rate (%) *	Degradation Rate (%) *
BCRC 17,441 [#]	51.9 ± 7.55 [a]	47.0 ± 9.26 [a]	31.0 ± 0.45
B1	41.8 ± 6.35 [b]	29.0 ± 3.43 [b]	24.0 ± 8.09
B2	41.4 ± 2.76 [b]	26.5 ± 6.11 [b]	35.0 ± 4.22
B3	28.1 ± 3.50 [c]	29.6 ± 3.79 [b]	27.2 ± 0.24
B4	58.1 ± 3.02 [a]	24.7 ± 9.31 [b]	31.0 ± 4.15
B5	43.1 ± 0.94 [b]	30.4 ± 2.86 [b]	32.8 ± 5.57

* The ZEN-detoxification *Bacillus* strains were inoculated in TSB containing 5 mg·L^{-1} ZEN and incubated at 37 °C for 24 h. The values represent the mean ± SD of triplicate experiments. Following the 24 h incubation, the cells and supernatants were separated by centrifugation. The separated cells were used for the ZEN adsorption ability test, and the collected supernatants were used for the ZEN degradation ability test. The values represent mean ± SD of triplicates. [#] *B. subtilis* (type strain BCRC 17,441) was used as standard strain for five ZDB strains to compare with. [a,b,c] Means in the same column with different superscript significantly differ ($p < 0.05$).

2.3. Enzymatic Profile of ZDB Strains

The API ZYM assay showed that most enzyme activities were positive (Table 3). All strains demonstrated esterase activity, which is linked to ZEN degradation ability. The B1 and B2 strains showed stronger esterase activity than the other strains, which was consistent with the high level of ZEN degradation observed with the B2 strain (Table 2).

Table 3. Enzyme activities of ZEN-detoxification *Bacillus* strains checked by the API ZYM system.

Enzyme	BCRC 17,441	B1	B2	B3	B4	B5
Alkaline phosphatase	+ [a]	+	+	++	+	++
Acid phosphatase	+	++	++	+	+	+
Esterase (C4)	+	++	++	+	+	+
Esterase lipase (C8)	+	+	+	++	+	+
Lipase (C14)	−	−	−	−	−	−
Leucine arylamidase	+	−	−	+	+	++
Valine arylamidase	+	−	−	+	+	+
Cystine arylamidase	−	−	−	+	−	+
Trypsin	−	−	−	−	−	−
α-chymotrypsin	+	−	−	+	+	+
Naphthol-AS-BI-phosphohydrolase	+	+	+	+	+	+
α-galactosidase	−	−	−	+	−	−
ß-galactosidase	+	+	−	+	+	+
ß-glucuronidase	−	−	−	−	−	−
α-glucosidase	+	−	−	++	+	+
ß-glucosidase	−	−	−	+	−	+
N-acetyl-ß-glucosaminidase	−	−	−	−	−	−
α-mannosidase	−	−	−	−	−	−
α-fucosidase	+	−	−	−	−	−

[a] The "−" and "+" represent negative and positive response, respectively. The "++" represents stronger enzyme activity response.

2.4. Detoxification of ZEN-Contaminated Maize by ZDB Strains

To ensure that the unsuccessful detoxification in ZEN-contaminated maize was not due to fermentation failure, the bacterial growth after 24 h and 48 h of fermentation were checked. At the beginning, the total number of bacteria was approximately 10^5 to 10^6 cfu·mL^{-1}. After 24 h, the bacterial number increased to 10^9 cfu·mL^{-1}, except for strain B1 (Table 4). After 48 h, the bacterial number of most strains slightly decreased and remained at 10^8 to 10^9 cfu·mL^{-1} (Table 4). All of the ZDB strains significantly reduced the ZEN content in maize after fermentation (Table 4). The B2 strain had the highest detoxification rate, removing 56% of the ZEN from the maize ($p < 0.05$). Conversely, strain 17,441 had the lowest detoxification rate. Among all ZDB strains, B2 strain showed the best detoxification capability, and it was selected for further research. The *gyrB* gene sequencing identified the B2 strain as *B. subtilis*.

Table 4. The bacterial count and detoxification rate of ZEN-detoxifying *Bacillus* strains in ZEN-contaminated (5 mg·kg^{-1}) maize after 24 h and 48 h fermentation.

Bacillus Strains	Bacterial Count (log CFU mL^{-1})		Detoxification of ZEN in Maize after 48 h (%)
	24 h	48 h	
BCRC 17,441	9.67 ± 0.561 [a]	8.69 ± 0.499 [a]	32.7 ± 10.42 [b,c]
B1	6.32 ± 0.922 [b]	7.09 ± 0.379 [b]	49.0 ± 9.43 [a,b]
B2	9.44 ± 0.175 [a]	9.20 ± 0.153 [a]	55.8 ± 6.20 [a]
B3	9.11 ± 0.494 [a]	8.81 ± 0.438 [a]	31.4 ± 10.96 [c]
B4	8.21 ± 1.861 [a]	8.79 ± 0.358 [a]	49.2 ± 8.46 [a,b]
B5	9.32 ± 0.210 [a]	9.19 ± 0.516 [a]	38.3 ± 3.57 [b,c]

The values represent mean ± SD of triplicate experiments. [a,b,c] Means in the same column with different superscript significantly differ ($p < 0.05$).

2.5. The Effect of ZEN on B2 Strain Growth in TSB Medium

The B2 strain was inoculated in TSB with or without ZEN to explore the influence of ZEN on its growth and metabolism. All parameters were increased significantly when B2 strain was inoculated in TSB with ZEN, including bacterial number, pH, NH$_3$-N, lactic acid, acetic acid, and total VFAs (Table 5). The results indicated that the growth rate of B2 strain was stimulated by ZEN, and the production of metabolites was enhanced.

Table 5. The characteristics of B2 strain in culture medium (TBS) without or with ZEN (5 mg·L^{-1}) after 24 h incubation.

Measurement	−ZEN	+ZEN
Bacterial count (log CFU mL^{-1}) (8 h)	10.4 ± 0.60	11.8 ± 0.58 *
Bacterial count (log CFU mL^{-1}) (24 h)	10.3 ± 0.05	11.3 ± 0.14 *
pH	4.77 ± 0.042	5.64 ± 0.056 *
NH$_3$-N (mmol L^{-1})	11.33 ± 0.891	27.57 ± 1.819 *
Lactic acid (mmol L$^{-1)}$)	208.2 ± 8.02	284.8 ± 20.02 *
Acetic acid (mmol L^{-1})	5.22 ± 0.269	11.55 ± 1.030 *
Total VFAs (mml L^{-1})	5.88 ± 0.325	12.41 ± 1.042 *

The values represent mean ± SD of quadruplicate experiments. * Significant difference ($p < 0.05$).

2.6. The Effect of ZEN on Fermentation Characteristics of B2 Strain in Maize

There was no statistically significant difference in the bacterial number of B2 strain between ZEN-free and ZEN-contaminated maize during 48 h and 72 h of fermentation (Table 6). After 48 h, the lactic acid concentration of fermented product from ZEN-contaminated maize was significantly higher than ZEN-free maize, resulting in a lower pH (Table 6) ($p < 0.05$). After 72 h, fermented ZEN-free maize product had significantly less NH$_3$-N and more acetic acid (Table 6). Overall, the fermented product of ZEN-contaminated maize had better fermentation characteristics (lactic acid > 110 mmol·L^{-1}; acetic acid < 20 mmol·L^{-1}; pH < 4.5) after 72 h of fermentation than ZEN-free maize.

Table 6. The fermented characteristics of B2 strain in ZEN-contaminated (5 mg·kg^{-1}) maize after 48 h and 72 h fermentation.

Measurement	−ZEN	+ZEN
Bacterial count (log CFU·mL^{-1}) (48 h)	8.65 ± 0.334	8.99 ± 0.289
Bacterial count (log CFU·mL^{-1}) (72 h)	8.84 ± 0.186	8.50 ± 0.327
pH (48 h)	4.58 ± 0.022 *	4.46 ± 0.070
pH (72 h)	4.35 ± 0.068	4.41 ± 0.079
NH$_3$-N (mmol·L^{-1}) (48 h)	6.03 ± 0.372	6.10 ± 0.259
NH$_3$-N (mmol·L^{-1}) (72 h)	5.97 ± 0.230	6.65 ± 0.269 *
Lactic acid (mmol·L^{-1}) (48 h)	67.8 ± 13.51	96.8 ± 8.88 *
Lactic acid (mmol·L^{-1}) (72 h)	88.8 ± 12.26	111.9 ± 15.33
Acetic acid (mmol·L^{-1}) (48 h)	21.3 ± 0.18	19.0 ± 1.02
Acetic acid (mmol·L^{-1}) (72 h)	21.6 ± 0.82 *	19.4 ± 0.62

The values represent mean ± SD of quadruplicate experiments. * Significant difference ($p < 0.05$).

2.7. Other Mycotoxins' (AFB1, DON, FB1, and T2 Toxin) Detoxification

Figure 1 illustrates that B2 strain significantly reduced the AFB1, DON, FB1, and T2 toxin content in TSB medium after 24 h cultivation ($p < 0.05$). The detoxification rates of AFB1, DON, FB1, and T2 toxin were 3.8, 25.0, 39.5, and 9.5%, respectively. The results showed that B2 strain had detoxification capability toward multiple mycotoxins.

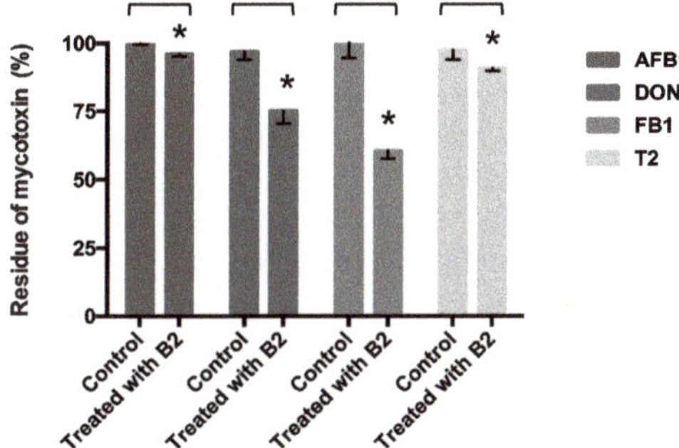

Figure 1. The mycotoxins detoxification of B2 strain in culturing condition. The values represent the mean ± SD of the quadruplicate experiments. Asterisks indicate that the residual percentage is significantly different among treatments ($p < 0.05$).

3. Discussion

The extreme weather conditions caused by global warming increase the risk of mycotoxin contamination of cereal crops. In some regions, the impact of ZEN on agriculture is the second highest impacting factor after aflatoxins [13]. It is virtually impossible to avoid the production of mycotoxins. Therefore, it is necessary to develop an effective detoxification strategy for mycotoxin-contaminated food or feed [3]. During the past decades, some microorganisms and their enzymes have been verified to detoxify ZEN, including fungi, yeast, and bacteria [14–23]. However, most of these microorganisms are not allowed for application in food and feedstuff. There are still some doubts surrounding the toxicity of microbial detoxification products and undesirable side effects from fermentation involving non-native microorganisms [24]. Based on this reason, the present study aimed to screen *Bacillus* spp. which are commonly applied in animal feed.

Bacillus spp. can form endospores which can resist unfavorable environment conditions such as heat, chemicals, and radiation [25]. In this study, all samples were cultured for 72 h in TSB with polymyxin B and isolated after heat treatment. Because of the endospore formation, *Bacillus* spp.

should survive after these treatments. Based on 16S rRNA gene sequence analysis, 14 isolates with better ZEN detoxification ability were identified as *Bacillus* spp. (Table 1).

Generally, there are two mechanisms through which microbes detoxify ZEN: degradation and adsorption. In the present study, all of the tested ZDB strains demonstrated esterase activity (Table 3), which is linked to their ZEN degradation ability. The B2 strain was one of stronger detoxifying strains (Table 2), and its detoxification rate was the highest out of all of the strains when tested with the ZEN-contaminated maize (Table 4). Previous research indicated that the efficiency of the ZEN degradation ability of *Bacillus* also depends on the initial ZEN concentration [26]. When the initial ZEN concentration was 0.02 and 5 mg L^{-1}, the degradation rate was found to be 100 and 18%, respectively [27]. In the present study, the initial ZEN concentration was 5 mg·L^{-1} for all experiments. The degradation rates of the ZDB strains were between 24 and 35 % which is greater than previously reported (Table 2). According to Reddy et al., the amounts of ZEN in grain ranges from a few µg·kg^{-1} to thousands of µg kg^{-1} worldwide [13]. The ZDB strains of the present study performed well at higher level ZEN contamination (mg·kg^{-1}).

All of the ZDB strains tested in this study were able to adsorb ZEN (Table 2). Previous research has suggested that the adsorption mechanism of *Bacillus* is similar to that of *Lactobacillus* because both are Gram-positive bacteria and have the same cell wall characteristics [27]. ZEN is mainly adsorbed by the surface hydrophobicity and the carbohydrate components of the *Lactobacillus* cell wall [28]. Therefore, ZEN adsorption by ZDB strains may rely on the same components. The high concentration of ZEN promotes the bacterial cell wall to contact with ZEN, which increases ZEN adsorption [29]. Although high concentration of ZEN was used in ZEN adsorption test, the ZEN adsorption rate of ZDB strains were still low. The data suggested that poor adsorption capacity of the ZDB strains may favor contact of ZEN with enzymes. In fact, adsorption of ZEN by microorganisms does not really remove the ZEN, which may be released back into the digestive tract when the digestive fluid continues to flush the bacterial surface [30]. It is worth noting that the ZEN detoxification capability of strain 17,441 in TSB was the second highest (52%) in the culturing condition (Table 2). Strain 17,441 also demonstrated the greatest adsorption capability (Table 2). However, strain 17,441 exhibited a lower detoxification rate than the other ZDB strains in the ZEN-contaminated maize detoxification experiment (Table 4). Therefore, the detoxification capability of a given strain in culturing condition is not necessarily the same as in the feedstuff fermentation process.

Suitable microorganisms must be easily applied in feed, and their detoxifying action must be fast enough in complex environments, e.g., the gastrointestinal tract or feed pretreatment. The pH has a significant impact on ZEN degradation [26], and the gastrointestinal tract environment may not be suitable for certain microorganisms and their enzymatic reactions. Moreover, previous research has found that acid-treated *Bacillus* adsorbed less ZEN than untreated cells [27], implying that low pH sites, such as the gastrointestinal tract, are unfavorable for the microbial degradation and adsorption of ZEN. ZEN is absorbed by intestinal epithelium within 30 min after entering the duodenum [31]. Because intestinal absorption is quick, ZEN detoxification must take place rapidly or be accomplished before feeding [6]. Therefore, application of the microbial fermentation process for animal feed detoxification could be a suitable strategy.

The fermentation process can cause loss of nutrients. Some studies suggest that FLF should be produced by fermenting the cereal ingredient instead of the complete feed [11,12]. Furthermore, fermentation of cereals often leads to a more rapid fermentation than compound feed. In the present study, maize contaminated with high level of ZEN (5 mg·kg^{-1}) was used as a substrate, and ZDB strains grew normally and retained their detoxification capacity (Table 4).

In order to control the growth of pathogenic bacteria, FLF should have a pH below 4.5 [32,33]. FLF should contain at least 75 mmol·L^{-1} lactic acid to avoid the growth of *Salmonella* spp. [33] and above 100 mmol·L^{-1} to decrease the number of enterobacteria [11]. FLF also has beneficial effect on daily gain, feed intake, and feed efficiency. However, a high concentration of acetic acid would make the FLF less palatable [33]. The acetic acid concentration of FLF should less than 40 mmol·L^{-1} [33].

In the present study, both the fermentation products of ZEN-free maize and ZEN-contaminated maize had good fermentation characteristics (lactic acid > 110 mmol·L^{-1}; acetic acid < 20 mmol·L^{-1}; pH < 4.5) after 72 h of fermentation. The results indicated that the B2 strain may be a suitable candidate for ZEN detoxification by fermentation.

Bacteria have developed complicated regulation systems to obtain nutrients from a wide range of sources. In *B. subtilis*, catabolite control protein A (ccpA) and codY are the major global regulators of transcription connected with carbon metabolism involving in the synthesis of lactic acid and acetic acid [34,35]. In the present study, the presence of ZEN significantly increased the lactic acid concentration in B2 strain (Tables 5 and 6). It is speculated that ZEN may have affected ccpA and codY of B2 strain, which needs further investigation. A significant increase in NH$_3$-N concentration (Tables 5 and 6) may imply that ZEN also affects protein metabolism in B2 strain. When B2 strain was inoculated in TSB with ZEN, the bacterial number was increased significantly (Table 5). It is worth noting that ZEN can be a potential growth promoter for B2 strain.

From a practical perspective, animal feed may not be only contaminated by one kind of mycotoxin [3]. The additive or synergistic interactions of co-occurring mycotoxins might lead to unpredictable toxicity [36]. An appropriate detoxification strategy should be able to detoxify multiple mycotoxins. The B2 strain has been confirmed to have the ability to detoxify AFB1, DON, FB1, and T2, which is very appealing.

4. Conclusions

Overall, the B2 strain may be a suitable candidate for ZEN detoxification by fermentation before feeding because it demonstrated strong esterase activity and exhibited the highest detoxification capability in maize with a high level of ZEN (5 mg·kg^{-1} maize).

5. Materials and Methods

5.1. Chemicals

HPLC grade methanol and acetonitrile were purchased from Sigma-Aldrich (St. Louis, MO, USA), chloroform and polymyxin B from Merck (Darmstadt, Germany), ZEN from Enzo Biochem (Farmingdale, NY, USA), aflatoxin B1 (AFB1), deoxynivalenol (DON), fumonisin B1 (FB1) and T2 toxin from Sigma-Aldrich (St. Louis, MO, USA), and tryptic soy broth (TSB) and tryptic soy agar (TSA) from Acumedia (Lansing, MI, USA).

5.2. Isolation of Bacillus Strains

In this study, fermented soybean products, soil, sewage, rumen fluid, and ruminant feces were collected for the isolation of ZEN-detoxification *Bacillus* (ZDB). One gram of each sample was suspended in 10 mL TSB containing polymyxin B (100,000 IU L^{-1}) and incubated for 72 h at 37 °C. All of the cultured samples were heated in a water bath for 15 min at 80 °C and then spread on TSA with polymyxin B and incubated at 37 °C for 24 h. Individual colonies from each plate were collected for further screening.

5.3. Screening of ZEN Detoxification Potential Strains

All isolates were inoculated at 1% (*v/v*) in TSB containing 5 mg L^{-1} ZEN for 24 h at 37 °C. Then, samples were centrifuged for 20 min at 8000× *g* at 4 °C; the supernatants were collected and extracted with an equal volume of chloroform and sonicated for 30 min. The organic phase was separated by centrifugation (500× *g* for 10 min at 25 °C) and dried with nitrogen gas at 63.5 °C. The residues were re-dissolved in 1 mL methanol and concentrated to 1/5 of the original volume by centrifugal vacuum concentrator (5301 VacuFuge, Eppendorf®, Hamburg, Germany) at 60 °C, then filtered through a 0.22 µm nylon syringe filter before loaded into HPLC (LC-2000Plus, JASCO, Tokyo, Japan) with a fluorescence detector (excitation and emission wavelengths were 274 and 440 nm) and the Luna®5 µm

C18(2) 100-Å, LC column (250 × 4.6 mm) (Phenomenex, Torrance, CA, USA) to detect residual ZEN. The mobile phase was acetonitrile solution (50:50, *v/v*).

5.4. Bacterial Strain Identification

The *Bacillus* isolates were identified through 16S rRNA gene sequencing. DNA was extracted from each isolate using the DNeasy plant mini kit (Qiagen, Valencia, CA, USA). The PCR products were sequenced using the BigDye terminator v3.1 cycle sequencing kit (Applied Biosystems, Foster City, CA, USA), and sequencing was performed on a DNA Analyzer (3730XL, Applied Biosystems, Foster City, CA, USA). The sequences (approximately 1500 bp) were compared with 16S rRNA gene sequences in the NCBI GenBank database using basic local alignment search tool (BLAST). The candidate strain was further identified by sequence analysis of *gyrB* gene sequence (approximately 1200 bp) [37].

5.5. Bacillus-Related Enterotoxin Detection

Nonhemolytic enterotoxin A (Nhe A), and nonhemolytic enterotoxin B (Nhe B) were detected by the *Bacillus* Diarrhoeal Enterotoxin Visual Immunoassay (BDE VIA™) (TECRA International Pty Ltd, Chatswood, Australia). Nhe B and hemolysin BL (HBL) were detected by the Duopath®Cereus Enterotoxins kit (EMD Millipore, Merck KGaA, Darmstadt, Germany). Cereulide was detected by the Singlepath®Emetic Tox. Mrk. Kit (EMD Millipore, Merck KGaA, Darmstadt, Germany).

5.6. ZEN Detoxification Capability Test in Culturing Condition (TSB Medium)

The ZDB strains were inoculated at 1% (*v/v*) in TSB containing 5 mg L^{-1} ZEN at 37 °C for 24 h. After that, samples were centrifuged (8000× *g* for 20 min at 4 °C) and supernatants were collected for the residual ZEN analysis. The ZEN detoxification of the ZDB strains was compared to a strain of *B. subtilis* (BCRC 17,441) from the Bioresource Collection and Research Center, Food Industry Research and Development Institute (Taiwan) to assess their relative detoxification capabilities [27].

5.7. ZEN Adsorption Ability and ZEN Degradation Ability

The ZDB strains were inoculated at 1% (*v/v*) in TSB at 37 °C for 24 h. Following the incubation, the cells and supernatants were separated by centrifugation (8000× *g* for 20 min at 4 °C). The cells were used for the ZEN adsorption ability test, and the supernatants were used for the ZEN degradation test. The separated cells were resuspended in 10 mL PBS containing ZEN (5 mg·L^{-1}) and incubated with constant shaking at 150 rpm. After 30 min, the culture was centrifuged (8000× *g* for 20 min at 4 °C), and the supernatants were collected and analyzed for residual ZEN. The ZEN degradation test was carried out by adding ZEN to the collected supernatants (at a final concentration of 5 mg L^{-1}) and incubated in a rotary shaking incubator (150 rpm) at 37 °C. After 24 h, the supernatant was analyzed for residual ZEN.

5.8. Enzymatic Profile of ZDB Strains

The API ZYM system (bioMérieux, Marcy l'Etoile, France) was used for the assay of enzymatic activities of the ZDB strains. After subcultured twice, the cells of each ZDB strain were collected through centrifugation (8000× *g* for 20 min) and resuspended in API suspension medium with turbidity adjusted to 5–6 McFarland. All detection tests were performed according to the manufacturer's instructions.

5.9. Detoxification of ZEN-Contaminated Maize by ZDB Strains

ZEN (5 mg·kg^{-1}) was added to 20 g of sterile ZEN-free maize, and then 60 mL of sterile distilled water was added. The inoculum of the candidate strain was added at 1% (*v/v*) and incubated at 37 °C for 48 h. After 24 and 48 h, samples were collected for monitoring bacterial number and ZEN detoxification activity. Bacterial counts were performed with TSA plating at 37 °C for 24 h. Finally, all fermentation residues were collected, freeze-dried, and analyzed for residual ZEN according to Ok et al. [38].

5.10. The Effect of ZEN on the Candidate Strain (B2 Strain) Growth in TSB Medium

The candidate strain (B2 strain) was inoculated at 1% (*v/v*) in TSB with or without ZEN (at a final concentration of 5 mg·L^{-1}), and incubated at 37 °C for 24 h. After 8 h and 24 h, samples were collected for checking bacterial numbers. At the end of incubation, the supernatants were collected and analyzed for the pH, NH_3-N, lactic acid, acetic acid, and total volatile fatty acids (VFAs). Bacterial count was done as previously. The pH value was measured with a pH meter (pH 22, Horiba, Kyoto, Japan). The NH_3-N concentration was determined as described by Chaney and Marbach [39]. The lactic acid concentration was determined by L-lactic acid assay kit (LC2653, Randox, Crumlin, UK). Acetic acid and total VFAs were analyzed using gas chromatography (GC7820A, Agilent, Santa Clara, CA, USA) with a flame ionization detector and the Nukol™ capillary GC column (size × I.D. 30 m × 0.25 mm, df 0.25 μm) (SUPELCO, Bellefonte, PA, USA). The carry gas was helium gas. The crotonic acid (25 g·L^{-1}) was used as an internal standard.

5.11. The Effect of ZEN on Fermentation Characteristics of the Candidate Strain (B2) in Maize

The candidate strain (B2 strain) was inoculated at 1% (*v/v*) in ZEN-free maize or ZEN-contaminated maize (at a final concentration of 5 mg·kg^{-1}) added 60 mL of sterile distilled water. The fermentation process was under aerobic conditions at 37°C for 72 h. After 48 h and 72 h, samples were collected for an analysis of bacterial count and pH value. After then, the supernatants were collected for analysis of NH_3-N, lactic acid, and acetic acid.

5.12. Other Mycotoxin (AFB1, DON, FB1 and T2 Toxin) Detoxification Test

The candidate strain (B2 strain) was inoculated at 1% (*v/v*) in individual TSB containing AFB1 (5 μg·L^{-1}), DON (400 μg·L^{-1}), FB1 (500 μg·L^{-1}), or T2 toxin (100 μg·L^{-1}), respectively. After incubated at 37 °C for 24 h, samples were centrifuged (8000× *g* for 20 min at 4 °C) and supernatants were collected for the residual mycotoxins analysis. Residual AFB1, DON, FB1, and T2 toxin were performed by using enzyme-linked immunosorbent assay kit (Vaccigen, New Taipei, Taiwan).

5.13. Statistical Analysis

The data of the effect of ZEN on the candidate strain (B2 strain) growth in TSB, fermentation characteristics in maize experiments, and mycotoxin (AFB1, DON, FB1, and T2 toxin) detoxification capability test were analyzed via *t*-test analysis. The data of ZDB strains' ZEN detoxification capability, adsorption ability, degradation ability, and detoxification of ZEN-contaminated maize experiment were analyzed by the general linear model procedure of SAS, Version 9.4 and expressed as the means ± standard deviation (SD) (SAS Institute Inc., Cary, NC, USA). Statistical differences were determined by Duncan's multiple range test, and significance was defined as $p < 0.05$.

Author Contributions: The authors' responsibilities were as follows: Conceptualization, S.-W.C. and J.-T.H.; Methodology, S.-W.C. and J.-T.H.; Software, S.-W.C. and J.-T.H.; Validation, S.-W.C. and J.-T.H.; Formal Analysis, S.-W.C., H.-T.W., W.-Y.S., Y.-A.C., Y.-Y.C. and J.-T.H.; Investigation: S.-W.C., H.-T.W., W.-Y.S., Y.-A.C., Y.-Y.C. and J.-T.H.; Resources, H.-T.W., S.-Y.W. and J.-T.H.; Data Curation, S.-W.C. and J.-T.H.; Writing-Original Draft Preparation, S.-W.C., L.A. and J.-T.H.; Writing-Review & Editing; J.-T.H.; Visualization, S.-W.C. and J.-T.H.; Supervision, J.-T.H.; Project Administration, J.-T.H.; Funding Acquisition, S.-Y.W. and J.-T.H

Funding: This research received no external funding.

Acknowledgments: The authors would like to thank all of their colleagues for their contributions to this research. We are grateful to Chong-Zhi Bai, Wei-Chen Chen, Yu-Chih, Huang, Zhi-Hong Liu, Chen Syu and Show-Show Yang for their help in the sample collection. We would also like to express our gratitude to Dr. Hsin Tsai for invaluable guidance.

Conflicts of Interest: The authors declare no conflict of interest.

References

1. Zinedine, A.; Soriano, J.M.; Moltó, J.C.; Mañes, J. Review on the toxicity, occurrence, metabolism, detoxification, regulations and intake of zearalenone: An oestrogenic mycotoxin. *Food Chem. Toxicol.* **2007**, *45*, 1–18. [CrossRef] [PubMed]
2. Chaytor, A.C.; Hansen, J.A.; van Heugten, E.; See, M.T.; Kim, S.W. Occurrence and decontamination of mycotoxins in swine feed. *Asian-Australas. J. Anim. Sci.* **2011**, *24*, 723–738. [CrossRef]
3. Schatzmayr, G.; Streit, E. Global occurrence of mycotoxins in the food and feed chain: Facts and figures. *World Mycotoxin J.* **2013**, *6*, 213–222. [CrossRef]
4. Kowalska, K.; Habrowska-Górczyńska, D.E.; Piastowska-Ciesielska, A.W. Zearalenone as an endocrine disruptor in humans. *Environ. Toxicol. Pharmacol.* **2016**, *48*, 141–149. [CrossRef] [PubMed]
5. Coulombe, R.A., Jr. Biological action of mycotoxins. *J. Dairy Sci.* **1993**, *6*, 880–891. [CrossRef]
6. Schatzmayr, G.; Zehner, F.; Täubel, M.; Schatzmayr, D.; Klimitsch, A.; Loibner, A.P.; Binder, E.M. Microbiologicals for deactivating mycotoxins. *Mol. Nutr. Food Res.* **2006**, *50*, 543–551. [CrossRef] [PubMed]
7. Vekiru, E.; Hametner, C.; Mitterbauer, R.; Rechthaler, J.; Adam, G.; Schatzmayr, G.; Krska, R.; Schuhmacher, R. Cleavage of zearalenone by *Trichosporon mycotoxinivorans* to a novel nonestrogenic metabolite. *Appl. Environ. Microbiol.* **2010**, *76*, 2353–2359. [CrossRef] [PubMed]
8. Huwig, A.; Freimund, S.; Käppeli, O.; Dutler, H. Mycotoxin detoxication of animal feed by different adsorbents. *Toxicol. Lett.* **2001**, *122*, 179–188. [CrossRef]
9. Upadhaya, S.D.; Park, M.A.; Ha, J.K. Mycotoxins and their biotransformation in the rumen: A review. *Asian-Australas. J. Anim. Sci.* **2010**, *23*, 1250–1260. [CrossRef]
10. EFSA. Introduction of a Qualified Presumption of Safety (QPS) approach for assessment of selected microorganisms referred to EFSA. *EFSA J.* **2007**, *587*, 1–16. [CrossRef]
11. Missotten, J.A.M.; Michiels, J.; Degroote, J.; De Smet, S. Fermented liquid feed for pigs: An ancient technique for the future. *J. Anim. Sci. Biotechnol.* **2015**, *6*, 4. [CrossRef] [PubMed]
12. Canibe, N.; Jensen, B.B. Fermented liquid feed-microbial and nutritional aspects and impact on enteric diseases in pigs. *Anim. Feed Sci. Technol.* **2012**, *173*, 17–40. [CrossRef]
13. Reddy, K.R.N.; Salleh, B.; Saad, B.; Abbas, H.K.; Abel, C.A.; Shier, W.T. An overview of mycotoxin contamination in foods and its implications for human health. *Toxin Rev.* **2010**, *29*, 3–26. [CrossRef]
14. El-Sharkawy, S.; Abul-Hajj, Y.J. Microbial cleavage of zearalenone. *Xenobiotica* **1988**, *18*, 365–371. [CrossRef] [PubMed]
15. Kakeya, H.; Takahashi-Ando, N.; Kimura, M.; Onose, R.; Yamaguchi, I.; Osada, H. Biotransformation of the mycotoxin, zearalenone, to a non-estrogenic compound by a fungal strain of *Clonostachys* sp. *Biosci. Biotechnol. Biochem.* **2002**, *66*, 2723–2726. [CrossRef] [PubMed]
16. Molnar, O.; Schatzmayr, G.; Fuchs, E.; Prillinger, H. *Trichosporon mycotoxinivorans* sp. nov., a new yeast species useful in biological detoxification of various mycotoxins. *Syst. Appl. Microbiol.* **2004**, *27*, 661–671. [CrossRef]
17. Varga, J.; Péteri, Z.; Tábori, K.; Téren, J.; Vágvölgyi, C. Degradation of ochratoxin A and other mycotoxins by *Rhizopus* isolates. *Int. J. Food Microbiol.* **2005**, *99*, 321–328. [CrossRef]
18. Altalhi, A.D.; El-Deeb, B. Localization of zearalenone detoxification gene(s) in pZEA-1 plasmid of *seudomonas putida* ZEA-1 and expressed in *Escherichia coli*. *J. Hazard Mater.* **2009**, *161*, 1166–1172. [CrossRef]
19. Yu, Y.; Qiu, L.; Wu, H.; Tang, Y.; Yu, Y.; Li, X.; Liu, D. Degradation of zearalenone by the extracellular extracts of *Acinetobacter* sp. SM04 liquid cultures. *Biodegradation* **2011**, *22*, 613–622. [CrossRef]
20. Cserháti, M.; Kriszt, B.; Krifaton, C.; Szoboszlay, S.; Háhn, J.; Tóth, S.; Nagy, I.; Kukolya, J. Mycotoxin-degradation profile of *Rhodococcus* strains. *Int. J. Food Microbiol.* **2013**, *166*, 176–185. [CrossRef]
21. Kosawang, C.; Karlsson, M.; Vélëz, H.; Rasmussen, P.H.; Collinge, D.B.; Jensen, B.; Jensen, D.F. Zearalenone detoxification by zearalenone hydrolase is important for the antagonistic ability of *Clonostachys rosea* against mycotoxigenic *Fusarium graminearum*. *Fungal Biol.* **2014**, *118*, 364–373. [CrossRef] [PubMed]
22. Sun, X.; He, X.; Xue, K.; Li, Y.; Xu, D.; Qian, H. Biological detoxification of zearalenone by *Aspergillus niger* strain FS10. *Food Chem. Toxicol.* **2014**, *72*, 76–82. [CrossRef] [PubMed]
23. Tan, H.; Hu, Y.; He, J.; Wu, L.; Liao, F.; Luo, B.; He, Y.; Zuo, Z.; Ren, Z.; Zhong, Z.; et al. Zearalenone degradation by two *Pseudomonas* strains from soil. *Mycotoxin Res.* **2014**, *30*, 191–196. [CrossRef] [PubMed]

24. Shetty, P.H.; Jespersen, L. *Saccharomyces cerevisiae* and lactic acid bacteria as potential mycotoxin decontaminating agents. *Trends Food Sci. Technol.* **2006**, *17*, 48–55. [CrossRef]

25. Setlow, P. Spores of *Bacillus subtilis*: Their resistance to and killing by radiation, heat and chemicals. *J. Appl. Microbiol.* **2006**, *101*, 514–525. [CrossRef] [PubMed]

26. Wang, G.; Yu, M.; Dong, F.; Shi, J.; Xu, J. Esterase activity inspired selection and characterization of zearalenone degrading bacteria *Bacillus pumilus* ES-21. *Food Control* **2017**, *77*, 57–64. [CrossRef]

27. Tinyiro, S.E.; Wokadala, C.; Xu, D.; Yao, W. Adsorption and degradation of zearalenone by *Bacillus* strains. *Folia Microbiol.* **2011**, *56*, 321–327. [CrossRef]

28. El-Nezami, H.; Polychronaki, N.; Lee, Y.K.; Haskard, C.; Juvonen, R.; Salminen, S.; Mykkänen, H. Chemical moieties and interactions involved in the binding of zearalenone to the surface of *Lactobacillus rhamnosus* strains GG. *J. Agric. Food Chem.* **2004**, *52*, 4577–4581. [CrossRef]

29. Sangsila, A.; Faucet-Marquis, V.; Pfohl-Leszkowicz, A.; Itsaranuwat, P. Detoxification of zearalenone by *Lactobacillus pentosus* strains. *Food Control* **2016**, *62*, 187–192. [CrossRef]

30. Dalié, D.K.D.; Deschamps, A.M.; Richard-Forget, F. Lactic acid bacteria-potential for control of mould growth and mycotoxins: A review. *Food Control* **2010**, *21*, 370–380. [CrossRef]

31. Olsen, M.; Malmlöf, K.; Pettersson, H.; Sandholm, K.; Kiessling, K.H. Plasma and urinary levels of zearalenone and alpha-zearalenol in a prepubertal gilt fed zearalenone. *Acta Pharmacol. Toxicol.* **1985**, *56*, 239–243. [CrossRef]

32. Van Winsen, R.L.; Lipman, L.J.A.; Biesterveld, S.; Urlings, B.A.P.; Snijders, J.M.A.; Van Knapen, F. Mechanism of Salmonella reduction in fermented pig feed. *J. Sci. Food Agric.* **2001**, *81*, 342–346. [CrossRef]

33. Missotten, J.A.M.; Goris, J.; Michiels, J.; Van Coillie, E.; Herman, L.; De Smet, S.; Dierick, N.A.; Heyndrickx, M. Screening of isolated lactic acid bacteria as potential beneficial strains for fermented liquid pig feed production. *Anim. Feed Sci. Tech.* **2009**, *150*, 122–138. [CrossRef]

34. Sonenshein, A.L. Control of key metabolic intersections in *Bacillus subtilis*. *Nat. Rev. Microbiol.* **2007**, *5*, 917–927. [CrossRef] [PubMed]

35. Shivers, R.P.; Dineen, S.S.; Sonenshein, A.L. Positive regulation of *Bacillus subtilis* ackA by CodY and CcpA: Establishing a potential hierarchy in carbon flow. *Mol. Microbiol.* **2006**, *62*, 811–822. [CrossRef]

36. Grenier, B.; Oswald, I. Mycotoxin co-contamination of food and feed: Meta-analysis of publications describing toxicological interactions. *World Mycotoxin J.* **2011**, *4*, 285–313. [CrossRef]

37. Wang, L.T.; Lee, F.L.; Tai, C.J.; Kasai, H. Comparison of *gyrB* gene sequences, 16S rRNA gene sequences and DNA-DNA hybridization in the *Bacillus subtilis* group. *Int. J. Syst. Evol. Microbiol.* **2007**, *57*, 1846–1850. [CrossRef]

38. Ok, H.E.; Choi, S.W.; Kim, M.; Chun, H.S. HPLC and UPLC methods for the determination of zearalenone in noodles, cereal snacks and infant formula. *Food Chem.* **2014**, *163*, 252–257. [CrossRef]

39. Chaney, A.L.; Marbach, E.P. Modified reagents for determination of urea and ammonia. *Clin. Chem.* **1962**, *8*, 130–132.

© 2019 by the authors. Licensee MDPI, Basel, Switzerland. This article is an open access article distributed under the terms and conditions of the Creative Commons Attribution (CC BY) license (http://creativecommons.org/licenses/by/4.0/).

 *toxins*

Article

Zearalenone Biodegradation by the Combination of Probiotics with Cell-Free Extracts of *Aspergillus oryzae* and Its Mycotoxin-Alleviating Effect on Pig Production Performance

Chaoqi Liu [1], Juan Chang [1], Ping Wang [1], Qingqiang Yin [1,*], Weiwei Huang [1], Xiaowei Dang [2], Fushan Lu [3] and Tianzeng Gao [4]

[1] College of Animal Science and Veterinary Medicine, Henan Agricultural University, Zhengzhou 450002, China; liuchaoqi2018@stu.henau.edu.cn (C.L.); changjuan2000@henau.edu.cn (J.C.); wangping@henau.edu.cn (P.W.); hww5501@stu.henau.edu.cn (W.H.)
[2] Henan Delin Biological Product Co. Ltd., Xinxiang 453000, China; hndlbio@hndlbio.com
[3] Henan Puai Feed Co. Ltd., Zhoukou 466000, China; lufushan@puaifeed.com
[4] Henan Guangan Biotechnology Co., Ltd., Zhengzhou 450001, China; gaotianzeng@groundgroup.com
*   Correspondence: qqy1964@henau.edu.cn

Received: 6 September 2019; Accepted: 18 September 2019; Published: 20 September 2019

Abstract: In order to remove zearalenone (ZEA) detriment—*Bacillus subtilis*, *Candida utilis*, and cell-free extracts from *Aspergillus oryzae* were used to degrade ZEA in this study. The orthogonal experiment in vitro showed that the ZEA degradation rate was 92.27% ($p < 0.05$) under the conditions that *Candida utilis*, *Bacillus subtilis* SP1, and *Bacillus subtilis* SP2 were mixed together at 0.5%, 1.0%, and 1.0%. When cell-free extracts from *Aspergillus oryzae* were combined with the above probiotics at a ratio of 2:1 to make mycotoxin-biodegradation preparation (MBP), the ZEA degradation rate reached 95.15% ($p < 0.05$). In order to further investigate the MBP effect on relieving the negative impact of ZEA for pig production performance, 120 young pigs were randomly divided into 5 groups, with 3 replicates in each group and 8 pigs for each replicate. Group A was given the basal diet with 86.19 µg/kg ZEA; group B contained 300 µg/kg ZEA without MBP addition; and groups C, D, and E contained 300 µg/kg ZEA added with 0.05%, 0.10%, and 0.15% MBP, respectively. The results showed that MBP addition was able to keep gut microbiota stable. ZEA concentrations in jejunal contents in groups A and D were 89.47% and 80.07% lower than that in group B ($p < 0.05$), indicating that MBP was effective in ZEA biodegradation. In addition, MBP had no significant effect on pig growth, nutrient digestibility, and the relative mRNA abundance of estrogen receptor alpha ($ER\alpha$) genes in ovaries and the uterus ($p > 0.05$).

Keywords: Zearalenone; biodegradation; probiotics; cell-free extracts of *Aspergillus oryzae*; pig production performance

Key Contribution: The combination of beneficial microbes and cell-free extracts from *A. oryzae* was prepared to degrade ZEA effectively in vitro. Its addition in pig diets could alleviate ZEA negative effect on pig production performances.

---

1. Introduction

Zearalenone (ZEA) is one of the nonsteroidal estrogenic mycotoxins produced by *Fusarium* species to cause reproduction disorders in female animals such as abortion, false estrus, and so on [1]. Some of ZEA derivatives competitively bind estrogen receptors in the uterus and ovaries for reducing the ability of estrogen to bind receptors [2,3]. Long-term exposure to low doses of ZEA could alter estrogen receptor beta genes and induce epigenetic modification to inhibit the development of the ovary [4]. However, the relative transactivation activity of ZEA for estrogen receptor alpha was higher than estrogen receptor beta [5]. In sorghum, maize, wheat, rice, barley, and other crops and their sideline products, ZEA can be detected in their natural state during storage, transportation, and processing [6]. It was reported that ZEA was present in 92% of maize, 88% of maize silage, and 97% of small grains samples with a range of 141.30–253.07 µg/Kg [7–9]. Another report showed that the incidences and maximal levels of ZEA in raw cereal grains were 46% and 3049 µg/kg respectively, according to the global occurrence data reported during the past 10 years [10]. Generally, pigs are more sensitive to ZEA than other animals. Therefore, it is necessary to take some methods to eliminate ZEA in the animal feeding process, especially for female pigs.

In order to reduce ZEA detriment, many countries and organizations in the world have limited the maximal ZEA contents in food, feed, and cereals. For example, Australia allows a maximal ZEA content of 50 µg/Kg in cereals. ZEA content in cereals and cereal products is not allowed to exceed 100 µg/Kg in Italy. The maximal ZEA content in vegetable oil and cereals is 200 µg/Kg in France [11]. The maximal ZEA content in swine diets is 250 µg/Kg in China and Europe. Even so, ZEA content often exceeds these thresholds because of the different environments and inadequate storage conditions.

In order to detoxify ZEA, some physical, chemical, and biological methods have been conducted [12–15]. Some reports have shown that ZEA biological degradation is more effective than other methods [16,17]. There were some reports about ZEA biodegradation by microorganisms including *Saccharomyces cerevisiae* [18], *Bacillus subtilis* (*B. subtilis*) ANSB01G [19], and *Lactobacillus plantarum* [20]. In this study, *B. subtilis* SP1, *B. subtilis* SP2, *Candida utilis* (*C. utilis*), and cell-free extracts of *Aspergillus oryzae* (*A. oryzae*) were selected and combined together to determine the effect on ZEA detoxification and pig production performance, which will supply a new feed additive for safe animal feeding and production.

2. Results

2.1. The Probiotics for ZEA Degradation

In order to obtain the optimal ratio of probiotics for ZEA degradation, an orthogonal design was used in this experiment. Table 1 indicates that the biggest ZEA degradation rate in the nine treatments was 92.27% ($p < 0.05$) under the conditions that *Candida utilis*, *Bacillus subtilis* SP1, and *Bacillus subtilis* SP2 were mixed together at a ratio of $A_1B_2C_2$ (i.e., 0.5%, 1.0%, and 1.0%, respectively). The analysis showed that the optimal addition ratio of *C. utilis*, *B. subtilis* SP1, and *B. subtilis* SP2 was $A_1B_1C_3$ (i.e., 0.50%, 0.50%, and 1.50%). However, further ZEA degradation experiments indicated that $A_1B_2C_2$ was better than $A_1B_1C_3$. Table 2 indicated that that the constructed model was accurate and acceptable ($p < 0.01$). Among the three strains of microbes, *C. utilis* had a significant effect on the model for ZEA degradation ($p < 0.01$), followed by *B. subtilis* SP1 *and B. subtilis* SP2 ($p > 0.05$).

Table 1. Zearalenone (ZEA) degradation rate by probiotics.

Number	*C. utilis* (%) A	*B. subtilis SP1* (%) B	*B. subtilis SP2* (%) C	ZEA Content (µg/L)	ZEA Degradation Rate (%)
1	0.5	0.5	0.5	79.14 ± 24.91e	85.00 ± 4.98ab
2	0.5	1	1	40.80 ± 6.68e	92.27 ± 1.30a
3	0.5	1.5	1.5	40.82 ± 12.25e	92.26 ± 2.45a
4	1	0.5	1	123.53 ± 14.87de	76.58 ± 2.41bc
5	1	1	1.5	163.59 ± 16.09cde	68.99 ± 0.78cd
6	1	1.5	0.5	247.85 ± 42.91c	53.02 ± 3.63e
7	1.5	0.5	1.5	238.45 ± 13.50cd	54.80 ± 7.96e
8	1.5	1	0.5	382.36 ± 102.93b	27.52 ± 13.24f
9	1.5	1.5	1	282.28 ± 20.81bc	46.49 ± 2.89e
Control	0	0	0	527.52 ± 63.55a	
T1	269.2	215.47	166		
T2	198.58	189.18	214.96		
T3	128.48	191.61	215.3		
X1	89.73	71.82	55.33		
X2	66.19	63.06	71.65		
X3	42.83	63.87	71.77		
R	46.9	8.76	16.44		
Impact order		A > C > B			
Optimal solution		$A_1B_1C_3$			

Note: T1, T2, and T3 mean the sums of all ZEA degradation rates at the levels of 0.50%, 1.00%, and 1.50% respectively; X1, X2, and X3 mean the averages of all ZEA degradation rates at the levels of 0.50%, 1.00%, and 1.50% respectively; *R* represents the D-value between the maximum and minimum averages of each factor at different levels, and a bigger *R* value indicates that the factor is more important for a higher ZEA degradation rate. Data with the same lowercase letters in the same columns are insignificantly different from each other ($p > 0.05$); while data with different lowercase letters in the same columns are significantly different from each other ($p < 0.05$); the same as below.

Table 2. Main effect analyses among different factors.

Sources	Sum of Squares	DF	Mean Square	*F*	*p*
Corrected Model	2808.72	6	468.12	73.68	0.0132 *
A	2545.95	2	1272.98	200.35	0.0051 *
B	83.97	2	41.99	6.61	0.1312
C	178.79	2	89.40	14.07	0.0661
Error	12.71	2	6.35		
Corrected total	2821.42	8			

Note: "*" shows significant differences.

2.2. ZEA Degradation by the Combined Probiotics and Cell-Free Extracts of A. oryzae

For measuring the effect of combined probiotics with cell-free extracts of *A. oryzae* on ZEA degradation, the effectiveness of extracts from *A. oryzae* was first determined. Figure 1 indicates that the ZEA degradation rate was 88.16% after 24 h enzymatic hydrolysis ($p < 0.05$); thereafter, there was no significant difference among the different groups ($p > 0.05$), even though ZEA degradation rate reached 98.62% after 48 h enzymatic reaction. Table 3 indicated that the best ZEA degradation rate was 95.15% when the ratio between probiotics and cell-free extracts of *A. oryzae* was 2:1 ($p < 0.05$). According to this ratio, mycotoxin-biodegradation preparation (MBP) was made for the further pig feeding experiment. ZEA degradation rate of individual probiotics was lower than both combinations of probiotics and cell-free extracts of *A. oryzae* ($p < 0.05$), indicating the effectiveness of combination; however, there was no significant difference between single cell-free extracts and both combinations for ZEA degradation ($p > 0.05$).

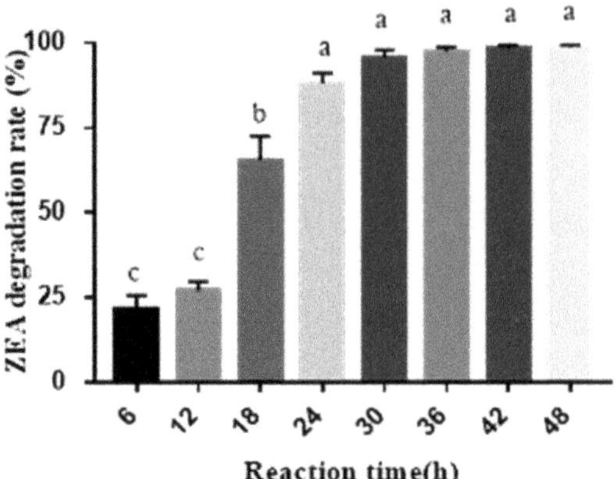

Figure 1. ZEA degradation by cell-free extracts of *A. oryzae* at different reaction times. Note: Data with the same lowercase letters in the bars are insignificantly different from each other ($p > 0.05$); while data with different lowercase letters in the bars are significantly different from each other ($p < 0.05$).

Table 3. ZEA degradation by the combined probiotics with cell-free extracts of *A. oryzae* for a 48 h reaction.

Groups	Probiotics: Cell-free extracts of *A. oryzae*	ZEA Content (µg/L)	ZEA Degradation Rate (%)
1	1:1	120.57 ± 21.56c	89.89 ± 1.81a
2	1:2	128.27 ± 79.74c	89.24 ± 6.69a
3	1:3	64.31 ± 83.73c	94.60 ± 7.02a
4	2:1	57.81 ± 12.83c	95.15 ± 1.08a
5	2:3	108.19 ± 28.55c	90.92 ± 2.39a
6	3:1	84.05 ± 73.25c	92.95 ± 6.14a
7	3:2	102.72 ± 72.17c	91.83 ± 6.05a
8	1:0	349.38 ± 34.13b	70.69 ± 2.86b
9	0:1	170.59 ± 64.49c	85.69 ± 5.41a
Control	0:0	1192.10 ± 86.55a	

Note: Data with the same lowercase letters in the same columns are insignificantly different from each other ($p > 0.05$); while data with different lowercase letters in the same columns are significantly different from each other ($p < 0.05$).

2.3. Effect of MBP on Pig Growth Performance and Nutrient Digestibility

In order to determine the effect of MBP on alleviating ZEA for pig production performance, feeding experiments were conducted. Table 4 indicates that there were no significant differences in average daily gain (ADG), average daily feed intake (ADFI), feed conversion rate (F/G), and digestibility of crude protein (CP), crude fat (CF), phosphorus (P), and calcium (Ca) among the 5 groups ($p > 0.05$). However, it showed that the above parameters could be improved by MBP addition in groups D and E, compared to group B.

Table 4. Effect of mycotoxin-biodegradation preparation (MBP) on growth performance and nutrient digestibility of pigs exposed to ZEA.

Items	Group A	Group B	Group C	Group D	Group E
Initial weight (Kg)	34.83 ± 1.06	34.92 ± 1.34	35.17 ± 2.35	34.88 ± 1.27	34.83 ± 1.89
Final weight (Kg)	90.00 ± 4.23	83.58 ± 4.47	82.63 ± 1.80	88.54 ± 2.47	88.33 ± 3.70
ADG (Kg)	0.9211 ± 0.0812	0.8088 ± 0.0811	0.7886 ± 0.0304	0.8913 ± 0.0412	0.8908 ± 0.0321
ADFI (Kg)	2.612 ± 0.151	2.402 ± 0.171	2.389 ± 0.080	2.614 ± 0.140	2.542 ± 0.079
F/G	2.842 ± 0.100	2.963 ± 0.121	3.019 ± 0.213	2.914 ± 0.020	2.771 ± 0.169
CP digestibility (%)	89.73 ± 0.16	91.97 ± 0.94	89.53 ± 1.67	89.23 ± 1.25	90.67 ± 1.93
CF digestibility (%)	76.99 ± 3.94	80.87 ± 5.56	82.63 ± 4.28	78.41 ± 3.11	75.32 ± 1.22
P digestibility (%)	88.19 ± 0.42	87.67 ± 0.48	86.38 ± 1.58	87.21 ± 0.20	88.98 ± 0.07
Ca digestibility (%)	78.83 ± 0.22	77.51 ± 0.75	77.44 ± 0.09	76.48 ± 1.19	77.21 ± 0.89

Note: Data without lowercase letters in the same rows are insignificantly different from each other ($p > 0.05$). Group A: control; Group B: 300.00 µg/kg ZEA; Groups C, D, and E: 300.00 µg/kg ZEA plus 0.05%, 0.10%, and 0.15% MBP, respectively. Note: average daily gain (ADG), average daily feed intake (ADFI), feed conversion rate (F/G), and digestibility of crude protein (CP) and crude fat (CF).

2.4. Pig Gut Microbiota Affected by MBP

The DGGE profile can reflect the structure of the gut bacterial community, and the number of bands indicates the richness of phylogenetic types. Figure 2 indicates that the numbers of discernible microbial bands in gilt large intestines were 18, 24, and 25 in group A; 18, 20, and 27 in group B; and 19, 23, and 26 in group D. The universal bands were 14, 13, and 13 in groups A, B, and D, respectively. The sizes of bands were about 220 bp. After statistical analysis of the bands, there was no significant difference in microbial richness among the three groups in Figure 3. The gut microbial similarity coefficient analysis in Figure 4 showed that similarity coefficients were 57.2%–72.9% in group A, 38.3%–72.8% in group B, and 69.1%–73.8% in group D, indicating that the gut microbiota was more stable by MBP addition in group D.

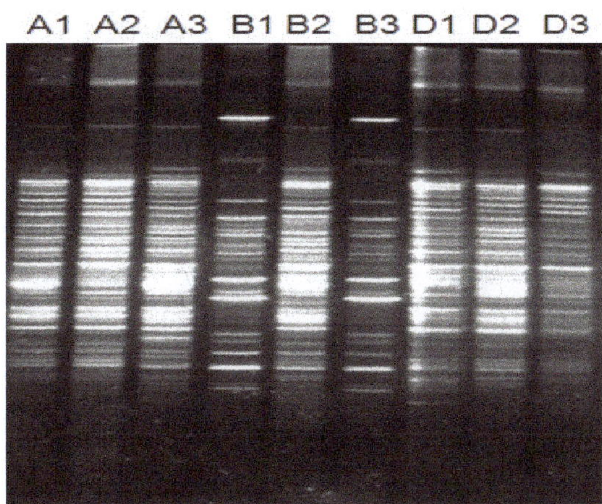

Figure 2. The electrophoresis diagram of DGGE in gilt large intestines. Note: Lanes A1–A3, contents of large intestine in group A; Lanes B1–B3, contents of large intestine in group B; Lanes D1–D3, contents of large intestine in group D.

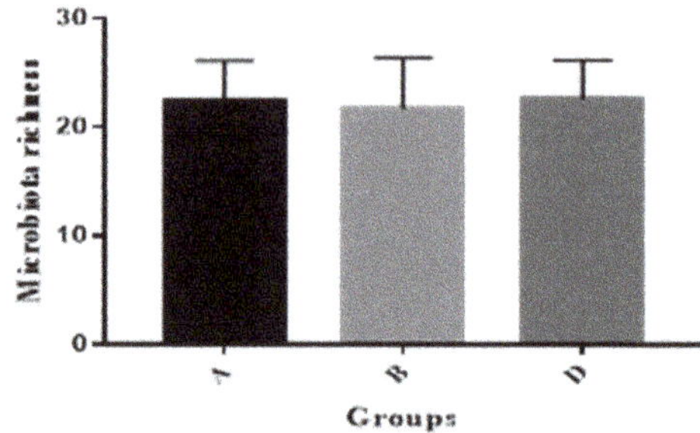

Figure 3. Microbial richness calculated by the number of bands in the electrophoresis diagram of DGGE.

Lane	A1	A2	A3	B1	B2	B3	D1	D2	D3
A1	100.0								
A2	57.2	100.0							
A3	58.8	72.9	100.0						
B1	23.2	31.8	38.3	100.0					
B2	60.0	64.7	72.8	44.7	100.0				
B3	27.5	26.3	35.8	55.6	39.8	100.0			
D1	47.5	64.7	68.9	45.1	68.0	34.9	100.0		
D2	45.2	65.5	72.7	38.7	66.9	25.4	73.8	100.0	
D3	52.6	63.0	65.8	32.0	62.6	19.1	69.2	69.1	100.0

Figure 4. The microbial similarity coefficients in different samples.

2.5. The Vulvar Area and Serum Parameters

Gilt vulvar area can reflect the effect of ZEA on female pig reproduction status. The effect of MBP on alleviating gilt vulvar area caused by ZEA in Figure 5 showed that, during the first 15 d, the vulvar area in group A increased slightly, while the other groups increased greatly, especially for group B. Fifteen days later, the vulvar area in each group increased slightly, and the order of vulvar area was: group B > group C > groups D and E > group A, indicating the effectiveness of MBP for weakening ZEA detriments.

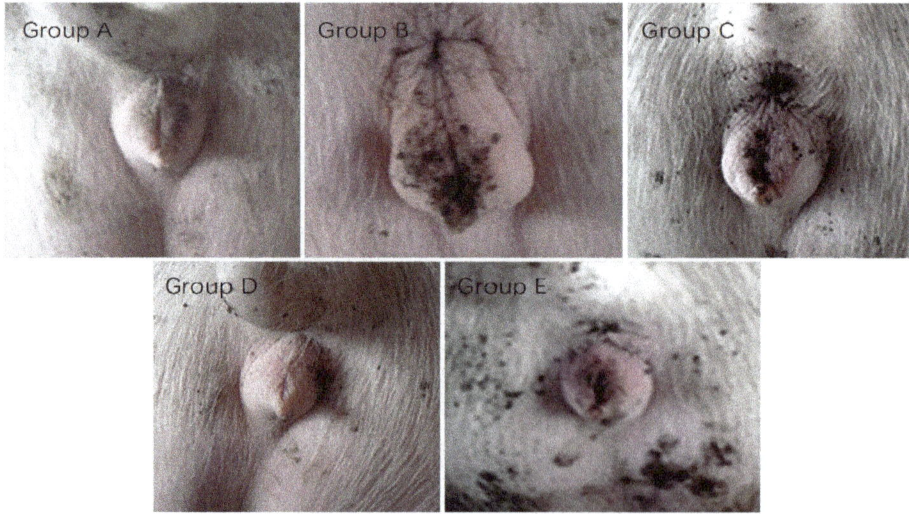

Figure 5. Effect of MBP on gilt vulvar representation.

In addition, ZEA or MBP had no significant effect on serum biochemical parameters such as urea nitrogen (UN), total cholesterol (TC), glucose (GLU), triglyceride, high-density lipoprotein (HDL), low-density lipoprotein (LDL), total protein (TP), albumin (ALB), globulin (GLO), albumin and globulin ratio (A/G), alanine aminotransferase (ALT), aspartate aminotransferase (AST), alkaline phosphatase (ALP), immunoglobulin G (IgG), immunoglobulin M (IgM), glutathione peroxidase (GSH-Px), and estradiol (E_2). It can be concluded that ZEA or MBP could not cause significant effects on the normal physiological and metabolic status of pigs in this study.

2.6. The Relative Organ Weight and ERα mRNA Abundance in Ovaries and the Uterus

ZEA may influence the relative organ weight and ERα mRNA abundance for female pigs. However, Table 5 shows that there was no significant difference for relative organ weight and ERα mRNA abundance in ovaries and the uterus among groups A, B, and D ($p > 0.05$), which showed that 300.00 µg/kg ZEA in pig diets was not enough to cause significant changes in relative organ weight and ERα mRNA abundance in ovaries and the uterus, even though MBP addition could alleviate ZEA hazards to some extent.

Table 5. Effect of MBP on serum E2 content, relative organ weight, and ERα mRNA in ovaries and the uterus.

Items	Group A	Group B	Group D
Heart (g/Kg)	3.241 ± 0.100	3.177 ± 0.311	3.616 ± 0.194
Liver (g/Kg)	15.97 ± 1.81	16.33 ± 0.80	18.67 ± 1.67
Spleen (g/Kg)	1.365 ± 0.249	1.605 ± 0.041	1.703 ± 0.164
Kidney (g/Kg)	2.960 ± 0.172	2.967 ± 0.221	3.432 ± 0.496
Uterus (g/Kg)	1.245 ± 0.752	1.932 ± 0.587	1.673 ± 0.318
Serum E2 (pg/ml)	109.03 ± 8.29	112.48 ± 15.75	103.88 ± 5.06
ERα mRNA abundance in ovaries	0.7899 ± 0.1021	1.171 ± 0.141	1.171 ± 0.122
ERα mRNA abundance in uterus	1.010 ± 0.086	1.111 ± 0.153	1.071 ± 0.191

Note: Data without lowercase letters in the same rows are insignificantly different from each other ($p > 0.05$).

2.7. ZEA Concentrations in Serum, Relative Tissues, and Gut

In order to realize the status of ZEA metabolism and deposit for pigs, the ZEA distribution was measured. Table 6 shows that residual ZEA in serum, longissimus dorsi, uterus, and liver in groups A, B, and D was not detected. ZEA concentrations in jejunal contents in groups A and D were 89.47% and 80.07% lower than that in group B ($p < 0.05$). ZEA concentrations in the large intestine in groups A and D were 68.57% ($p < 0.05$) and 12.79% ($p > 0.05$) lower than that in group B, respectively, indicating the effectiveness of MBP for ZEA degradation.

Table 6. Effect of MBP on ZEA concentrations in pig serum, tissues, and gut (µg/Kg).

Items	Group A	Group B	Group D
Serum	—	—	—
Longissimus dorsi	—	—	—
Uterus	—	—	—
Liver	—	—	—
Contents in jejunum	11.64 ± 0.27b	110.54 ± 16.19a	22.03 ± 8.20b
Contents in large intestine	43.45 ± 2.44b	138.23 ± 4.67a	120.55 ± 18.87a

Note: Data with the same lowercase letters in the same rows are insignificantly different from each other ($p > 0.05$); while data with different lowercase letters in the same rows are significantly different from each other ($p < 0.05$). "—" indicates no detection.

3. Discussion

As one of the most ubiquitous mycotoxins, ZEA has caused serious harm to animals and humans and has led to a great waste of food resources every year [21]. There are two ways to solve this problem. One way is to inhibit ZEA production from microbes, which is hard to conduct due to the difficult control of ZEA-excreting microbes in the environment; another way is to eliminate ZEA harm to animals and humans by chemical, physical, and biodegradable methods [12–15]. Although physical and chemical methods can eliminate ZEA harm, they also reduce the nutritional value of food and pollute the environment. Many researchers have indicated that ZEA biodegradation or absorption by microbes and mycotoxin-degrading enzymes is the most effective method [16,17,22,23].

It was found that the peroxiredoxin gene from *Acinetobacter* sp. SM04 was cloned in *Escherichia coli* BL21 to excrete one recombinant protein for ZEA detoxification [24]. *Saccharomyces cerevisiae* and *B. subtilis* have been found to be able to degrade ZEA around 90% [25,26]. *Broomyces pink* and *B. subtilis* were reported to open the ZEA lactone ring to form nontoxic compounds by hydrolyzing lactone bonds [27,28]. In order to increase the ZEA degradation rate, two strains of *B. subtilis* and one strain of *C. utilis* with good ZEA degradation abilities have been selected and combined with ZEA degradation enzymes from *A. oryzae* in this study.

A. oryzae is one kind of microbe that produces complex enzymes, which will help to increase nutrient availability and degrade mycotoxins. It was found that *A. oryzae* could convert ZEA into other metabolites and reduce the toxicity of toxins [29]. Another report showed that the deoxynivalenol degradation rate was over 92% after *A. oryzae* was inoculated with corn culture medium for 21 d [30]. In this research, cell-free extracts of *A. oryzae* could degrade ZEA by 98.62% after 48 h enzymatic reaction, which was better than the previous report. When cell-free extracts of *A. oryzae* were combined with the probiotics, the ZEA degradation rate was higher than the individual, indicating that there was a good cooperation between them. This result corresponds with the previous report for aflatoxin removal [31]. Generally, the probiotics are able to regulate gastrointestinal microbiota for animal health, except for its ZEA-removing ability. Therefore, the combination of beneficial bacteria and *A. oryzae* extracts will have great applications in the fields of animal production and food and feed processing.

ZEA is widely distributed in all kinds of grain crops and feedstuffs. The prevention and removal of ZEA pollution of grain has become a hotspot for researchers. It was found that the growth performance of weaned piglets fed with 1000 µg/Kg ZEA diet for 24 d and the young gilts fed with 1.5–2.0 mg/kg ZEA diet for 28 d had no significant changes [32,33]. In this study, after feeding with 300 µg/Kg ZEA diet for 60 d, ADG, ADFI, F/G, and nutrient digestibility of young pigs had no significant difference regardless of MBP addition or not. It is possible that this concentration of ZEA added in the diet was not enough to retard pig growth even though it may influence reproduction for female pigs.

This study showed that MBP addition was able to increase gut microbial similarity coefficients to help maintain gut microbial balance for pig health, in agreement with the former research, in which it was reported that *Saccharomyces cerevisiae* subsp. *boulardii* strain had the potential as feed additives to modulate bacterial populations associated with gut health for piglets [34].

After the diets contaminated with ZEA are fed to animals, ZEA will have three metabolic pathways: (1) ZEA is absorbed in the gut and remains in the animal's body; (2) it is degraded by the gut microbes and enzymes; and (3) some ZEA will be discharged with the feces. It was found that the residual amount of ZEA in gilt livers increased significantly when dietary ZEA contents were 500 or 2000 µg/Kg [35]. However, ZEA was not detected in serum, longissimus dorsi, the uterus, and liver in this study, inconsistent with the above. This may be due to the low ZEA content (300 µg/Kg) in pig diets. This research showed that ZEA concentrations in jejunal contents in group B was significantly higher than that in group A and group D, indicating that part of ZEA was eliminated by MBP in the gastrointestinal tract, which proved the effectiveness of MBP for degrading ZEA in pig guts.

Serum ALT, AST, and ALP are important indicators to measure the degree of liver lesions. Serum IgG and IgM are the main factors that participate in the humoral immune response. Their normal levels in this study indicated that liver cells and immune systems have not been damaged by ZEA at

300 µg/Kg dosage. The previous report showed that long-term exposure to a low dose of ZEA had no significant negative effect on serum ALT, AST, ALP, and E2 levels in pigs [36], which is in agreement with this study. The organ indexes of heart, liver, kidney, and spleen were not significantly changed by feeding the weaned piglets with 316 µg/Kg ZEA diets [37], which corresponds with this study. The reason may be due to the low ZEA dose in pig diets.

The chemical structure of ZEA is the same as estrogen; therefore, ZEA can competitively bind estrogen receptors to cause reproduction disorders, in which the red and swelling vulva of female pig is the common apparent symptom. This research showed that 300 µg/Kg ZEA could cause a red and swelling vulva; however, MBP addition was able to alleviate the symptoms, in agreement with the previous research with *Bacillus* addition [38]. It was found that MBP was better than only *Bacillus subtilis* for degrading ZEA and keeping the microbiota stable in the pig gut [38].

It was found that the relative expression of estrogen receptor gene ERα mRNA in the ZEA group was significantly higher than that in control group [39]. Another research study showed that the serum E2 of females was significantly decreased by the addition of 1.05 mg/kg ZEA in diets [32]. Generally, ZEA could cause DNA double-strand breaks and affect the proliferation of granulosa cells, and the addition of an estrogen receptor antagonist could improve this symptom, indicating that ZEA could damage granulosa cells through the estrogen receptor pathway [40]. It was reported that the relative expression levels of ERα genes in brain, liver, and gonads were not significantly affected by low doses of ZEA, but they were significantly affected by high doses of ZEA. This process changed the HPG axis by altering gene expression of the steroid hormone-encoding gene to affect reproductive function [41]. In this study, the relative expression of estrogen receptor gene ERα mRNA in ovaries and the uterus was not significantly different among the three groups, which does not correspond with the above results. The reason may be due to the low levels of ZEA in the diet in this study.

4. Conclusions

The combination of beneficial microbes (*B. subtilis* SP1, *B. subtilis* SP2, *C. utilis*) and cell-free extracts from *A. oryzae* could degrade ZEA effectively in vitro. MBP addition could alleviate ZEA negative effects in gilts by decreasing vulvar swelling, improving ZEA degradation in the jejunum, and keeping normal growth performance and gut microbiota stable.

5. Materials and Methods

5.1. Probiotics and Experimental Materials

B. subtilis SP1, *B. subtilis* SP2, *C. utilis*, and *A. oryzae* with high ZEA-degrading ability were purchased from China General Microbiological Culture Collection Center (CGMCC, Beijing, China). ZEA was purchased from Sigma-Aldrich (St. Louis, MO., USA) and diluted in methanol as stock solution (100 µg/mL). PBS buffer was prepared according to the previous protocol [42]. The compositions of media for *B. subtilis*, *C. utilis*, and *A. oryzae* incubations were prepared according to the published protocols [43,44]. Then, three kinds of probiotics were harvested and stored at 4 °C.

5.2. Degradation of ZEA by Probiotic Incubation

The visible counts of three kinds of microbes were adjusted to 1.0×10^8 CFU/mL, respectively. The orthogonal design (5 mL reaction system) with three factors (*B. subtilis* SP1, *B. subtilis* SP2, and *C. utilis*) and three levels (25, 50, and 75 µL) were used to investigate the effect of beneficial microbes on degrading ZEA. Each group contained 1 µg/mL ZEA. The contents of ZEA in the samples were determined by enzyme-linked immunosorbent assay (ELISA) according to Huang's report [44], which was highly consistent with high-performance liquid chromatography (HPLC). ZEA degradation rates were calculated according to the following formula: ZEA degradation rate = (1 − ZEA concentration in treatment/ZEA concentration in control) × 100%. All experiments were conducted in triplicate.

5.3. Preparation of Cell-Free Extracts from A. oryzae

A. oryzae was inoculated in solid medium at 30 °C for 3 d. After incubation, 15 g solid incubation was mixed with 300 mL physiological saline and put in a rotary shaker at 180 rpm for 2 h. The mixture was filtered with eight-layer gauze, centrifuged at 10,000 g for 10 min, passed through Whatman No.4 filter paper (20 to 25 μm pore diameters), and filtered with 0.22 μm Minisart High-flow filter (Sartorius Stedim Biotech Gmbh, Goettingen, Germany). Finally, it was stored at 4 °C for further use.

The final volumes of 60 mL filtrates with an initial ZEA concentration of 1 μg/mL in 250 mL conical flasks were incubated at 37 °C in a rotary shaker at 180 rpm, and the samples were collected after 0, 6, 12, 18, 24, 30, 36, 42, and 48 h incubation, respectively. The ZEA degradation rates were measured at different times. The ZEA degradation activity (31.0 U/L) from cell-free extracts of *A. oryzae* was determined with the previous protocol [42] and modified as the following: the amount of enzyme that could degrade 1 ng ZEA per min at pH 7.0 and 37 °C was defined as one unit.

5.4. ZEA Degradation by Probiotics Combined with Cell-Free Extracts of A. oryzae

The probiotics for degrading ZEA were added in YPD medium at the above ratio. The probiotics and cell-free extracts from *A. oryzae* were mixed at the volume ratios of 1:1, 1:2, 1:3, 2:1, 2:3, 3:1, 3:2, 1:0, and 0:1 as treatment groups. The final volume of the reaction system was adjusted to 3 mL with YPD medium. All the groups received 30 μL ZEA (100 μg/mL), were incubated at 37 °C in a rotary shaker at 180 rpm for 48 h, and were finally put in boiling water for 30 min to terminate the reaction.

5.5. Animals and Management

A total of 120 young pigs (Duroc × Landrace × Yorkshire) at the ages of 78–84 d were randomly divided into 5 groups, with 3 replicates in each group and 8 pigs (half male and half female) in each replicate. All animals used in this experiment were managed according to the guidelines of Animal Care and Use Ethics Committee in Henan Agricultural University (SKLAB-B-2010-003-01). The pigs were provided with a corn/soybean basal diet formulated to meet pig nutrient requirements according to the NRC (2012) [45]. The diets and water were provided ad libitum. The total experimental period was 60 d. After the feeding experiment, three gilts from groups A, B, and D were slaughtered for further analyses, respectively. Body weight and feed intake were recorded. ZEA used in animal feeding experiments was purchased from Wuhan 3B Scientific Corporation (Wuhan, China). The experimental designs were as follows:

Group A: Basal diet with 86.19 μg/kg ZEA
Group B: Basal diet containing 300.00 μg/kg ZEA
Group C: Basal diet containing 300.00 μg/kg ZEA and 0.05% MBP
Group D: Basal diet containing 300.00 μg/kg ZEA and 0.10% MBP
Group E: Basal diet containing 300.00 μg/kg ZEA and 0.15% MBP

5.6. Nutrient Digestibility Measurement

Fecal samples were taken without contamination from each of 5 pigs in each group for 3 d at the end of the experiment. The individual fecal sample was mixed, selected, and stored at −20 and 4 °C, respectively. The fecal samples stored at −20 °C were dried to determine nutrient digestibility. CP, CF, Ca, and P in diets and feces were measured with Kjeldahl, ether extract, potassium permanganate ($KMnO_4$), and ammonium molybdate (($NH_4)_6Mo_7O_{24}$) protocols, respectively [46]. The insoluble ash of hydrochloric acid in feed and feces was used as an indicator to calculate the nutrient digestibility with the following formula: Nutrient digestibility = (nutrient content in diets − nutrient content in feces)/nutrient content in diets × 100%.

5.7. Vulvar Area Measurement of Gilts

From the beginning of the experiment, vulvar areas of gilts were observed every day, and six gilts in each group were selected and measured at 15, 30, 45, and 60 d. The measurement method was based on a previous report, the vulvar area = $\pi \times$ vulvar width $\times$ vulvar length/4 [38].

5.8. Serum Parameter Determination

Blood samples were taken from the anterior vena cava of three pigs in each group. After the blood was tilted at room temperature for 3 h, the serum was collected by Transferpettor and stored in a centrifuge tube at −20 °C for further analysis. The serum biochemical parameters such as UN, TC, GLU, triglycerides, HDL, LDL, TP, ALB, GLO, A/G, ALT, AST, ALP, IgG, and IgM were measured with a 7600-020 Automatic Analyzer (Hitachi Ltd., Tokyo, Japan) in the Biochemical Laboratory of Zhengzhou University, Zhengzhou, China. The concentrations of GSH-Px and E_2 were respectively measured by ELISA quantification kits (Nanjing Jiancheng Bioengineering Institute, Nanjing, China).

5.9. Relative Organ Weight and ERα mRNA Abundance in Ovaries and the Uterus

After the feeding experiment, 3 gilts from 3 representative groups (groups A, B, and D) were selected and slaughtered, respectively. The gut contents were taken and frozen for further gut microbiota analysis, and the organs were taken and weighed. The formula for calculating the organ index was as follows: organ index = organ weight (g)/living weight (Kg). Some parts of ovaries and the uterus were stored in liquid nitrogen for mRNA abundance analyses.

In order to measure the expression levels of ERα in ovaries and the uterus, the total RNA was extracted by using RNAiso Plus kits (TaKaRa, Dalian, China). The result was checked with 1% agarose gel electrophoresis. The reaction system for cDNA consisted of 2 μL 4 × gDNA wiper Mix, 4 μL RNA template, and 2 μL RNase-free water. It was kept at 42 °C for 2 min, mixed with 2 μL 5×HiScript®IIqRT SuperMixII (Vazyme Biotech Co., Ltd., Nanjing, China), and then kept at 25 °C for 10 min, 50 °C for 30 min, and 85 °C for 5 min. The cDNA samples were stored at −20 °C for further use. The qPCR reaction consisted of 10.0 μL AceQ® qPCR SYBR® Green Master Mix, 0.4 μL ROX, 0.4 μL Primer F (10 μM), 0.4 μL Primer R (10 μM), 2.0 μL cDNA, and 6.8 μL RNase-free water (Vazyme Biotech Co., Ltd., Nanjing, China). The primers for ERα and glyceraldehyde-3-phosphate dehydrogenase (GAPDH) genes were as follows: Forward 5′-GACAGGAACCAGGGCAAGT-3′, Reverse 5′-ATGATGGATT TGAGGCACAC-3′ for ERα; Forward 5′-ATGGTGAAGGTCGGAGTGAA-3′, Reverse 5′-CGTGGGTGGAATCATACTGG-3′ for GAPDH. The thermal program of qPCR consisted of 1 cycle at 95 °C for 10 min, 40 cycles of 95 °C for 15 s, 57 °C for 30 s, and then stored at 37 °C for 30 s. Gene-specific amplification was determined by a melting curve analysis and agarose gel electrophoresis. The cycle threshold value was analyzed (iQ5 detection System) and transformed to a relative quantity using the $2^{-\Delta\Delta CT}$ method with the highest quantity scaled to 1 [47]. *GAPDH* gene was used as the reference gene for stability of expression to standardize the relative expression of the gene investigated.

5.10. DNA Extraction of Gut Bacteria and PCR-DGGE

About 1.0 g of gut sample from the large intestine was defrosted, put into a 10 mL centrifuge tube, washed with 5 mL PBS, and centrifuged at 500 g at 4 °C for 5 min to collect the supernatant according to the previous protocol [48]. The washing step was repeated 3 times. The supernatants from 3 washing steps were mixed together and centrifuged at 8000 g for 5 min to collect the pellet. The bacterial DNA was extracted with MiniBEST Bacterial Genomic DNA Extraction Kit (TaKaRa, Dalian, China), dissolved in 100 μL TE, and stored at −20 °C as the template for PCR amplification.

The primers of V3 regions of bacterial 16S rRNA genes [49] were F341: CGCCCGCCGCGCGCGGCGGGCGGGGCGGGGGCACGGGGGGCC TACGGGAGGCAGCAG, and R518: ATTACCGCGGCTGCTGG. The PCR system (50 μL) consisted of 25 μL Taq Master Mix, 5 μL extracted gut DNA, 1 μL primer F341-GC (10 μM), 1 μL R518 (10 μM), and 18 μL RNase-free water

(Beijing Comwin Biotech Co., Ltd. Beijing, China). The program consisted of initial DNA denaturation at 95 °C for 5 min, 35 cycles of 95 °C for 30 s, 55 °C for 30 s, 72 °C for 40 s, and a final extension at 72 °C for 10 min. The PCR products were detected by 1.5% agarose gel electrophoresis (Beijing Solarbio Science & Technology Co., Ltd., Beijing, China).

The PCR products were separated in 8% polyacrylamide gel containing a 35% to 60% gradient of urea and formamide increasing in the direction of electrophoresis. Electrophoresis was initiated by prerunning at a voltage of 200 V for 10 min and then at a constant voltage of 90 V for 14 h at 60 °C. After electrophoresis, the gel was stained with 0.5 μg/mL ethidium bromide for 30 min and then washed with deionized water for 10 min.

DGGE gels were analyzed using the software of Quantity One 4.6.6 (BioRad, California, USA, 2017). Microbial similarity coefficients (SC) between DGGE profiles were determined as follows: $SC = 2 \times J/(Nx + Ny)$, where Nx is the number of DGGE bands in lane x, Ny represents the number of DGGE bands in lane y, and J is the number of common DGGE bands [50].

5.11. Determination of ZEA Content in Serum and Tissues

Longissimus dorsi, liver, and uterus tissues were ground into powder with liquid nitrogen. A total of 1 mL serum or 5 g samples were added to 25 mL 70% methanol, shaken for 3 min, centrifuged at 10,000 g for 5 min, then 1 mL supernatant was mixed with 1 mL deionized water for further determination. The ZEA contents in all samples were measured by RIDASCREEN® FAST ZEA SC test kit (R-Biopharm, Darmstadt, Germany).

5.12. Statistical Analyses

The experimental data were analyzed as a single factor design by analysis of variance (ANOVA) using IBM SPSS-Statistics Program 20.0 (IBM, New York, NY, USA, 2012), and they are expressed as means and standard errors (SEs). The means were evaluated with Tukey's multiple range test, and differences were considered statistically significance at $p < 0.05$.

Author Contributions: C.L. and Q.Y. conceived and designed the experiments; W.H., C.L. and T.G. performed the experiments; J.C. and P.W. analyzed the data; X.D. and F.L. contributed reagents/materials/analysis tools; C.L. and Q.Y. wrote the paper.

Funding: This research was funded by the Natural Science Foundation of Henan Province (182300410029), and the Henan Key Scientific and Technological Project (171100110500).

Conflicts of Interest: The authors declare that there are no conflicts of interest associated with this research.

References

1. Caldwell, R.W.; Tuite, J.; Stob, M.; Baldwin, R. Zearalenone production by *Fusarium* species. *Appl. Microbiol.* **1970**, *20*, 31–34. [PubMed]
2. Takemura, H.; Shim, J.Y.; Sayama, K.; Tsubura, A.; Zhu, B.T.; Shimoi, K. Characterization of the estrogenic activities of zearalenone and zeranol in vivo and in vitro. *J. Steroid Biochem. Mol. Biol.* **2007**, *103*, 170–177. [CrossRef] [PubMed]
3. Gajecka, M. The effect of low-dose experimental zearalenone intoxication on the immunoexpression of estrogen receptors in the ovaries of pre-pubertal bitches. *Pol. J. Vet. Sci.* **2012**, *15*, 685–691. [CrossRef] [PubMed]
4. Benzoni, E.; Minervini, F.; Giannoccaro, A.; Fornelli, F.; Vigo, D.; Visconti, A. Influence of in vitro exposure to mycotoxin zearalenone and its derivatives on swine sperm quality. *Reprod. Toxicol.* **2008**, *25*, 461–467. [CrossRef] [PubMed]
5. Kuiper, G.G.J.M.; Lemmen, J.G.; Carlsson, B.; Corton, J.C.; Safe, S.H.; van der Saag, P.T.; van der Burg, B.; Gustafsson, J.A. Interaction of estrogenic chemicals and phytoestrogens with estrogen receptor β. *Endocrinology* **1998**, *139*, 4252–4263. [CrossRef]
6. Aiko, V.; Mehta, A. Occurrence, detection and detoxification of mycotoxins. *Toxicol. Appl. Pharm.* **2015**, *50*, 943–954. [CrossRef] [PubMed]

7. Kosicki, R.; Blajet-Kosicka, A.; Grajewski, J.; Twaruzek, M. Multiannual mycotoxin survey in feed materials and feedstuffs. *Anim. Feed Sci. Technol.* **2016**, *215*, 165–180. [CrossRef]

8. Phuong, N.H.; Thieu, N.Q.; Ogle, B.; Pettersson, H. Aflatoxins, fumonisins and zearalenone contamination of maize in the southeastern and central highlands provinces of Vietnam. *Agriculture* **2015**, *5*, 1195–1203. [CrossRef]

9. Oliveiraa, M.S.; Rochaa, A.; Sulyok, M.; Krskab, R.; Mallmann, C.A. Natural mycotoxin contamination of maize (*Zea mays* L.) in the south region of Brazil. *Food Control* **2017**, *73*, 127–132. [CrossRef]

10. Lee, H.J.; Ryu, D. Worldwide occurrence of mycotoxins in cereals and cereal-derived food products: Public health perspectives of their co-occurrence. *J. Agric. Food Chem.* **2017**, *65*, 7034–7051. [CrossRef]

11. FAO. *Worldwide Regulations for Mycotoxins in Food and Feed in 2003*; FAO Food and Nutrition Paper No. 81; Food and Agriculture Organization of the United Nations: Rome, Italy, 2004.

12. Zinedine, A.; Soriano, J.M.; Molto, J.C.; Manes, J. Review on the toxicity, occurrence, metabolism, detoxification, regulations and intake of zearalenone: An oestrogenic mycotoxin. *Food Chem. Toxicol.* **2007**, *45*, 1–18. [CrossRef] [PubMed]

13. Bordini, J.G.; Borsato, D.; Oliveira, A.S.; Ono, M.A.; Zaninelli, T.H.; Hirooka, E.Y.; Ono, E.Y.S. In vitro zearalenone adsorption by a mixture of organic and inorganic adsorbents: Application of the Box Behnken approach. *World Mycotoxin J.* **2015**, *8*, 291–299. [CrossRef]

14. Markovic, M.; Dakovic, A.; Rottinghaus, G.E.; Kragovic, M.; Petkovic, A.; Krajisnik, D.; Milic, J.; Mercurio, M.; de Gennaro, B. Adsorption of the mycotoxin zearalenone by clinoptilolite and phillipsite zeolites treated with cetylpyridinium surfactant. *Colloid Surf. B* **2017**, *151*, 324–332. [CrossRef] [PubMed]

15. Wang, L.; Wang, Y.; Shao, H.L.; Luo, X.H.; Wang, R.; Li, Y.F.; Li, Y.N.; Luo, Y.P.; Zhang, D.J.; Chen, Z.X. In vivo toxicity assessment of deoxynivalenol-contaminated wheat after ozone degradation. *Food Addit. Contam. A* **2017**, *34*, 103–112. [CrossRef] [PubMed]

16. Yu, Y.S.; Qiu, L.P.; Wu, H.; Tang, Y.Q.; Yu, Y.G.; Li, X.F.; Liu, D.M. Degradation of zearalenone by the extracellular extracts of *Acinetobacter* sp. SM04 liquid cultures. *Mycotoxin Biodegr.* **2011**, *22*, 613–622. [CrossRef] [PubMed]

17. Kriszt, R.; Krifaton, C.; Szoboszlay, S.; Cserhati, M.; Kriszt, B.; Kukolya, J.; Czeh, A.; Feher-Toth, S.; Torok, L.; Szoke, Z.; et al. A new zearalenone mycotoxin biodegradation strategy using non-pathogenic *Rhodococcus pyridinivorans* K408 strain. *PLoS ONE* **2012**, *7*, e43608. [CrossRef] [PubMed]

18. Tang, Y.Q.; Xiao, J.M.; Chen, Y.; Yu, Y.G.; Xiao, X.L.; Yu, Y.S.; Wu, H. Secretory expression and characterization of a novel peroxiredoxin for zearalenone detoxification in *Saccharomyces cerevisiae*. *Microbiol. Res.* **2013**, *168*, 6–11. [CrossRef]

19. Lei, Y.P.; Zhao, L.H.; Ma, Q.G.; Zhang, J.Y.; Zhou, T.; Gao, C.Q.; Ji, C. Degradation of zearalenone in swine feed and feed ingredients by *Bacillus subtilis* ANSB01G. *World Mycotoxin J.* **2014**, *7*, 143–151. [CrossRef]

20. Zhao, L.; Jin, H.T.; Lan, J.; Zhang, R.Y.; Ren, H.B.; Zhang, X.B.; Yu, G.P. Detoxification of zearalenone by three strains of *Lactobacillus plantarum* from fermented food in vitro. *Food Control* **2015**, *54*, 158–164. [CrossRef]

21. Guo, J.; Zhang, L.S.; Wang, Y.M.; Yan, C.H.; Huang, W.P.; Wu, J.; Yuan, H.T.; Lin, B.W.; Shen, J.L.; Peng, S.Q. Study of embryo toxicity of *Fusarium* mycotoxin butenolide using a whole rat embryo culture model. *Toxicol. In Vitro* **2011**, *25*, 1727–1732. [CrossRef]

22. Wang, M.; Yin, L.; Hu, H.; Selvaraj, J.N.; Zhou, Y.; Zhang, G. Expression, functional analysis and mutation of a novel neutral zearalenone-degrading enzyme. *Int. J. Biol. Macromol.* **2018**, *118*, 1284–1292. [CrossRef] [PubMed]

23. Pereyra, C.M.; Cavaglieri, L.R.; Chiacchiera, S.M.; Dalcero, A. The corn influence on the adsorption levels of aflatoxin B1 and zearalenone by yeast cell wall. *J. Appl. Microbiol.* **2013**, *114*, 655–662. [CrossRef] [PubMed]

24. Yu, Y.; Wu, H.; Tang, Y.; Qiu, L. Cloning, expression of a peroxiredoxin gene from *Acinetobacter* sp. SM04 and characterization of its recombinant protein for zearalenone detoxification. *Microbiol. Res.* **2012**, *167*, 121–126. [CrossRef] [PubMed]

25. Zhang, H.; Dong, M.; Yang, Q.; Apaliya, M.T.; Li, J.; Zhang, X. Biodegradation of zearalenone by *Saccharomyces cerevisiae*: Possible involvement of ZEN responsive proteins of the yeast. *J. Proteom.* **2016**, *143*, 416–423. [CrossRef]

26. Cho, K.J.; Kang, J.S.; Cho, W.T.; Lee, C.H.; Ha, J.K.; Bin, S.K. In vitro degradation of zearalenone by *Bacillus subtilis*. *Biotechnol. Lett.* **2010**, *32*, 1921–1924. [CrossRef]

27. El-Sharkawy, S.; Abul-Hajj, Y.J. Microbial cleavage of zearalenone. *Xenobiotica* **1988**, *18*, 365–371. [CrossRef] [PubMed]

28. Lei, Y.P. Mechanism of Degrading ZEA by ANSB01G Strain and Its Application Effect in Animal Production. Ph.D. Thesis, China Agricultural University, Beijing, China, 2014.

29. Brodehl, A.; Moller, A.; Kunte, H.J.; Koch, M.; Maul, R. Biotransformation of the mycotoxin zearalenone by fungi of the genera *Rhizopus* and *Aspergillus*. *FEMS Micobiol. Lett.* **2014**, *359*, 124–130. [CrossRef] [PubMed]

30. Tran, S.T.; Smith, T.K. Conjugation of deoxynivalenol by *Alternaria alternate* (54028 NRRL), *Rhizopusmicrosporus var. rhizopodiformis* (54029 NRRL) and *Aspergillus oryzae* (5509 NRRL). *Mycotoxin Res.* **2014**, *30*, 47–53. [CrossRef] [PubMed]

31. Zuo, R.Y.; Chang, J.; Yin, Q.Q.; Wang, P.; Yang, Y.R.; Wang, X.; Wang, G.Q.; Zheng, Q.H. Effect of the combined probiotics with aflatoxin B1-degrading enzyme on aflatoxin detoxification, broiler production performance and hepatic enzyme gene expression. *Food Chem. Toxicol.* **2013**, *59*, 470–475. [CrossRef]

32. Jiang, S.; Yang, Z.; Yang, W.; Gao, J.; Liu, F.; Chen, C.; Chi, F. Physiopathological effects of zearalenone in post-weaning female piglets with or without montmorillonite clay adsorbent. *Livest. Sci.* **2010**, *131*, 130–136. [CrossRef]

33. Oliver, W.T.; Miles, J.R.; Diaz, D.E.; Dibner, J.J.; Rottinghaus, G.E.; Harrell, R.J. Zearalenone enhances reproductive tract development, but does not alter skeletal muscle signaling in prepubertal gilts. *Anim. Feed Sci. Technol.* **2012**, *174*, 79–85. [CrossRef]

34. Brousseau, J.P.; Talbot, G.; Beaudoin, F.; Lauzon, K.; Roy, D.; Lessard, M. Effects of probiotics *Pediococcus acidilactici* strain MA18/5M and *Saccharomyces cerevisiae* subsp. *boulardii* strain SB-CNCM I-1079 on fecal and intestinal microbiota of nursing and weanling piglets. *J. Anim. Sci.* **2015**, *93*, 5313–5326. [CrossRef] [PubMed]

35. Wang, D.F.; Zhou, H.L.; Hou, G.Y.; Qi, D.S.; Zhang, N.Y. Soybean isoflavone reduces the residue of zearalenone in the muscle and liver of prepubertal gilts. *Animal* **2013**, *7*, 699–703. [CrossRef] [PubMed]

36. Gajecki, M. Evaluation of selected serum biochemical and haematological parameters in gilts exposed to 100 ppb of zearalenone. *Pol. J. Vet. Sci.* **2015**, *18*, 865–872.

37. Marin, D.E.; Pistol, G.C.; Neagoe, I.V.; Calin, L.; Taranu, I. Effects of zearalenone on oxidative stress and inflammation in weanling piglets. *Food Chem. Toxicol.* **2013**, *58*, 401–408. [CrossRef] [PubMed]

38. Zhao, L.H.; Lei, Y.P.; Bao, Y.H.; Jia, R.; Ma, Q.G.; Zhang, J.Y.; Chen, J.; Ji, C. Ameliorative effects of *Bacillus subtilis* ANSB01G on zearalenone toxicosis in pre-pubertal female gilts. *Food Addit. Contam. A* **2015**, *34*, 617–625. [CrossRef] [PubMed]

39. Niu, Q.S. Effect of *Fusarium* Toxins on Hematological Values, Development of Reproductive Organs, and Expression of Estrogen Receptor Genes in Post-Weaning Gilts. Ph.D. Thesis, Shandong Agricultural University, Taian, China, 2015.

40. Liu, X.L.; Wu, R.Y.; Sun, X.F.; Cheng, S.F.; Zhang, R.Q.; Zhang, T.Y.; Zhang, X.F.; Zhao, Y.; Shen, W.; Li, L. Mycotoxin zearalenone exposure impairs genomic stability of swine follicular granulosa cells in vitro. *Int. J. Biol. Sci.* **2018**, *14*, 294–305. [CrossRef]

41. Muthulakshmi, S.; Hamideh, P.F.; Habibi, H.R.; Maharajan, K.; Kadirvelu, K.; Mudili, V. Mycotoxin zearalenone induced gonadal impairment and altered gene expression in the hypothalamic-pituitary-gonadal axis of adult female zebrafish (*Danio rerio*). *J. Appl. Toxicol.* **2018**, *38*, 1388–1397. [CrossRef]

42. Cao, H.; Liu, D.L.; Mo, X.M.; Xie, C.F.; Yao, D.L. A fungal enzyme with the ability of aflatoxin B1 conversion: Purification and ESI-MS/MS identification. *Microbiol. Res.* **2011**, *166*, 475–483. [CrossRef]

43. Yin, Q.Q.; Fan, G.G.; Chang, J.; Zuo, R.Y.; Zheng, Q.H. Effect of the combined probiotics on inhibiting pathogenic *Escherichia coli* proliferation. *Adv. Mater. Res.* **2012**, *343*, 802–808.

44. Huang, W.W.; Chang, J.; Wang, P.; Liu, C.Q.; Yin, Q.Q.; Zhu, Q.; Lu, F.S.; Gao, T.Z. Effect of the combined compound probiotics with mycotoxin-degradation enzyme on detoxifying aflatoxin B1 and zearalenone. *J. Toxicol. Sci.* **2018**, *43*, 377–385. [CrossRef] [PubMed]

45. National Research Council (NRC). *Nutrient Requirements of Swine*, 11th rev. ed.; National Academic Press: Washington, DC, USA, 2012.

46. Jurgens, M.H. *Animal Feeding and Nutrition*, 8th ed.; Kendall/Hunt Publishing Company: Dubuque, IA, USA, 1997.

47. Livak, K.J.; Schmittgen, T.D. Analysis of relative gene expression data using real-time quantitative PCR and the 2 −ΔΔCT method. *Methods* **2001**, *25*, 402–408. [CrossRef] [PubMed]

48. Maibach, R.C.; Dutly, F.; Altwegg, M. Detection of Tropheryma whipplei DNA in feces by PCR using a target capture method. *J. Clin. Microbiol.* **2002**, *40*, 2466–2471. [CrossRef] [PubMed]

49. Liu, C.Q.; Zhu, Q.; Chang, J.; Yin, Q.Q.; Song, A.D.; Li, Z.T.; Wang, E.Z.; Lu, F.S. Effects of *Lactobacillus casei* and *Enterococcus faecalis* on growth performance, immune function and gut microbiota of suckling piglets. *Arch. Anim. Nutr.* **2017**, *71*, 120–133. [CrossRef] [PubMed]

50. Konstantinov, S.R.; Zhu, W.Y.; Williams, B.A.; Tamminga, S.; de Vos, W.M.; Akkermans, A.D.L. Effect of fermentable carbohydrates on piglet faecal bacterial communities as revealed by denaturing gradient gel electrophoresis analysis of 16S ribosomal DNA. *FEMS Microbiol. Ecol.* **2003**, *43*, 225–235. [CrossRef]

© 2019 by the authors. Licensee MDPI, Basel, Switzerland. This article is an open access article distributed under the terms and conditions of the Creative Commons Attribution (CC BY) license (http://creativecommons.org/licenses/by/4.0/).

Review

Assorted Methods for Decontamination of Aflatoxin M1 in Milk Using Microbial Adsorbents

Jean Claude Assaf [1,2,3,*], Sahar Nahle [1,2,3], Ali Chokr [2,3], Nicolas Louka [1], Ali Atoui [2] and André El Khoury [1]

[1] Centre d'Analyses et de Recherche (CAR), Unité de Recherche Technologies et Valorisation agro-Alimentaire (UR-TVA), Faculté des Sciences, Université Saint-Joseph de Beyrouth, Campus des sciences et technologies, Mar Roukos, Matn 1104-2020, Lebanon ; sahar.nahle@hotmail.com (S.N.); nicolas.louka@usj.edu.lb (N.L.); andre.khoury@usj.edu.lb (A.E.K.)

[2] Research Laboratory of Microbiology, Department of Life and Earth Sciences, Faculty of Sciences I, Lebanese University, Hadat Campus, Beirut P.O Box 5, Lebanon; alichokr@hotmail.com (A.C.); aatoui@ul.edu.lb (A.A.)

[3] Platform of Research and Analysis in Environmental Sciences (PRASE), Doctoral School of Sciences and Technologies, Lebanese University, Hadat Campus, Beirut P.O. Box 6573/14, Lebanon

* Correspondence: jeanclaude.assaf@net.usj.edu.lb; Tel.: +961-70-891-797

Received: 10 April 2019; Accepted: 15 May 2019; Published: 29 May 2019

Abstract: Aflatoxins (AF) are carcinogenic metabolites produced by different species of *Aspergillus* which readily colonize crops. AFM1 is secreted in the milk of lactating mammals through the ingestion of feedstuffs contaminated by aflatoxin B1 (AFB1). Therefore, its presence in milk, even in small amounts, presents a real concern for dairy industries and consumers of dairy products. Different strategies can lead to the reduction of AFM1 contamination levels in milk. They include adopting good agricultural practices, decreasing the AFB1 contamination of animal feeds, or using diverse types of adsorbent materials. One of the most effective types of adsorbents used for AFM1 decontamination are those of microbial origin. This review discusses current issues about AFM1 decontamination methods. These methods are based on the use of different bio-adsorbent agents such as bacteria and yeasts to complex AFM1 in milk. Moreover, this review answers some of the raised concerns about the binding stability of the formed AFM1-microbial complex. Thus, the efficiency of the decontamination methods was addressed, and plausible experimental variants were discussed.

Keywords: decontamination; mycotoxins; Aflatoxin M1; milk; binding; stability

Key Contribution: This review tackles current issues about AFM1 decontamination methods using different bio-adsorbents such as bacteria, yeasts or mixtures of both. The efficiency of these decontamination methods in addition to their plausible experimental variants, advantages, limitations and prospective applications were broadly discussed.

---

1. Introduction

Aflatoxins (AF) are secondary metabolites produced by several *Aspergillus* species, mainly by *Aspergillus flavus* and *A. parasiticus* [1–4]. The AF contamination of food and feed after mold colonization may occur at any stage extending from pre-harvest to consumption [5,6]. Thus, this can cause the direct or indirect contamination of different food commodities including cereals, corn, rice, and peanuts. Humid and warm environments are suitable for mold growth and AF production [7,8]. The group of AF includes more than 20 known metabolites; the most important are the naturally occurring ones such as B1, B2, G1, and G2 [9–11]. The toxicity of AF varies, but AFB1 remains the most toxic for humans and animals [12–14]. Briefly, after ingestion and absorption by an animal's gastrointestinal tract, AFB1 is then transformed in the liver into AFM1 and aflatoxin M2 (AFM2) [15–18]. It is noteworthy to mention that milk and its derivatives are widely consumed not only by adults but, more importantly, by infants [19]. Interestingly, Williams et al. reported that more than 4.5 billion people worldwide are at risk of exposure to foodstuffs contaminated with different levels of AF [20]. Upon investigation of its toxicity, the International Agency for Research on Cancer (IARC) has classified AFM1 as a group 1 human carcinogen [21–23]. The secretion of AFM1 in milk varies widely according to different factors including animal species, season and milking time, level of AFB1 intake, and volume of milk produced by the mammal in question [24–26]. Once in milk, AFM1 is not degraded and can resist different industrial treatments including milk sterilization or pasteurization, in addition to any other heat treatments [27–31]. For this reason, AFM1 contamination remains a serious problem, not only in produced milk but also in all its derived products including cheese, yoghurt, cream, and powdered milk [32,33]. Due to AF's harmful effects, several countries and international organizations have strictly regulated AF levels in feed and food [34,35]. Thus, the highest acceptable level of AFM1 in milk ranges between 0.05 µg/kg and 0.5 µg/kg, as established, respectively, by the European Union (EU) and the Food and Drug Administration (FDA) [36,37]. Therefore, the adopted AFM1 limit in milk (0.5 µg/kg) settled up in the USA, Brazil, Japan, and India is less restrictive compared to other countries such as France, Germany, Belgium, Australia, and Turkey (0.05 µg/kg) [38]. Importantly, the stricter the regulatory limits, the more food commodities are wasted, which results in a higher economic loss [39,40]. Accordingly, these regulations depend on several factors including the economic development level of each country, limits of consumption, and risk of exposure to AFM1 [41]. Moreover, the trade of any AF-contaminated products was also prohibited [40]. The implementation of Good Agricultural Practices (GAP) remains the best way to limit AF contamination in food and feedstuff but cannot guarantee their absolute prevention [42–44]. In addition, innovative technologies to cut pre- and post-harvest exposure to AF are strongly recommended. Some of these technologies include ozone fumigation [45], irradiation biological [46,47], and chemical agents [40,48,49]. Highly promising techniques such as using the biofilms of probiotic bacteria [50], chitin, and treated crustacean shells [51] are under rigorous investigation. Furthermore, the use of different biotransforming agents such as microorganisms and their purified enzymatic products can lead to the catabolization, cleaving, or transformation of the AF molecule to less or non-toxic metabolites [52]. Similarly, several clay materials including bentonite, hydrated sodium calcium aluminosilicate (HSCAS), zeolite, and activated charcoal have shown varying abilities to reduce AF in contaminated feedstuff [52]. Up until now, the most studied methods to mitigate AF contamination are mainly based on using biological adsorbents such as bacteria and yeasts [52–54].

This review aims to critically discuss different methods for AFM1 decontamination by microbial adsorption. Therefore, various treatments used for AFM1 decontamination by yeasts or bacteria will be broadly scrutinized, and some experimental variants will be highlighted in order to help researchers in improving the commonly used methods.

2. Toxic Effects of AFM1

Amongst all mycotoxins, AF present a high risk on the human health due to consumption of foods, including milk and dairy products, contaminated with their derivatives such as AFM1 [55,56]. Hence, the potential existence of AFM1 in milk, even in minor quantities, remains a worldwide alarming issue due to the consumption of wide range of contaminated dairy products [38]. Accordingly, the International Agency for Research on Cancer (IARC), following investigations on its toxicity, shifted AFM1 classification from group 2B to group 1 human carcinogens [22,57].

Briefly, AFB1 is initially absorbed by the gastrointestinal tract before being metabolized in the liver [58,59]. Within 15 min after ingestion, AFM1 could be detected in the blood of the lactating animal before being secreted in its milk and urine [15,58,60]. The biotransformation of AFB1 in the animal liver is carried out by cytochrome P450 enzymes, thus metabolizing AFB1 into hydroxylated AFM1 and AFB1 reactive epoxides [61,62]. While AFM1 is less toxic than AFB1, it is still highly harmful for humans and animal species [16,63]. As the major organ targeted by AFM1 is the liver, it is considered a hepatotoxic metabolite [64,65]. In addition, other damaging effects including immunity suppression, reduced milk production, and lower oxygen supply to body tissues may be caused by AFM1 [66–68]. The toxicity of AF, including AFM1, and its impact on its host is sex, age, species, and nutritional-behavior dependent [20,69]. It is important to highlight that breast-feeding is always encouraged for infants due to its nutritional qualities and is even recommended by the WHO for six months after birth [70]. Surprisingly, current studies on AFM1 in human breast milk conducted by Radonić et al. on samples from Serbia revealed alarming contamination levels [71]. Tests conducted on 60 samples showed that around 85% of colostrum and the totality of collected breast milk samples (four-to-eight months after delivery) were found highly contaminated with AFM1 in concentrations beyond tolerable levels [71]. Results of this study and other studies demonstrate the need to raise awareness about AFM1 presence in human milk [72].

3. Effective Strategies for AFM1 Reduction

3.1. Biological Control and Clay-Based Decontamination Methods

Strategies leading directly or indirectly to AFM1 reduction in milk vary from adopting good agricultural practices to using innovative detoxification methods [51,73,74]. Better management and monitoring of pre- and post-harvest conditions is an essential step to reduce AF contamination [5,75]. Several advanced techniques utilize biological methods such as bacteria, yeasts, and atoxigenic strains to reduce AF contamination in the field and during storage [76,77]. Thus, these "biocontrol" methods lead to the inhibition of fungal growth and AF production [52,78]. In addition, different types of mineral clays have been tested for their capability to bind AF in animal feeds [79,80]. These adsorbents, such as activated carbon (charcoal), zeolite, saponite-rich bentonite, and HSCAS, are able to bind AF, thus reducing AFB1 absorption in the gastrointestinal track and its carry-over as AFM1 in milk [52,81]. For example, the inclusion of HSCAS in dairy feed has resulted in the reduction of up to 50% of the concentration of AFM1 secreted in milk [82]. Furthermore, a recent study conducted by Carraro et al. revealed that bentonite was also effective in reducing AFM1 contamination in milk to levels below the European tolerable limits. Hence, the remaining residual bentonite amounts (0.4%) were in low quantities and showed no harmful effects on human health [83]. There are several limitations of using mineral adsorbents in beverages; they may affect its quality, color, texture, and various other physicochemical properties [54]. Therefore, due to their limitations, several mineral binders are kept for medical usage only [84]. In addition to their effect on food quality, many of these adsorbents are non-specific, non-environmental friendly, and even toxic at high concentration levels [54,85]. All these issues have led researchers to move toward more specific, non-toxic adsorbents, namely microbial ones such as Lactic Acid Bacteria (LAB) and yeasts [86–88].

3.2. Microbial Decontamination of AFM1

The use of probiotic yeasts and LAB to bind AF in contaminated liquid foods has been widely studied [89–93]. These biological adsorbents may be usually found in different foods including several dairy products such as milk. Their safe status, in addition to the high capability to bind mycotoxins, has lead researchers to test the ability of these adsorbents to bind AFM1 in milk and other liquids [86,90,94]. Consequently, AFM1 binding was reported to be rapid, and the binding percentage varied when changing different factors such as incubation time, temperature, pH, AFM1, and microbial concentrations [38,95]. The use of heat-killed cells is actually more favorable for milk decontamination than viable cells due to the contribution of the latter in product spoilage [86]. In order to assess the binding capability of these adsorbents without interference with the food matrix effect, AFM1 binding assays are initially conducted in buffer solutions such as phosphate-buffered saline (PBS) [74,96,97]. The efficiency of AFM1 binding by microbial adsorbents is detailed in this section.

Table 1. Summary of studies evaluating AFM1 binding by different bacterial strains.

Type	Strain	[AFM1]—[Cells]	Solution	Treatment	Incubation Time (37 °C)	Centrifugal Force (g/rpm)	Initial Binding (%)	Final Binding (%)	Reference
Bacteria	*L. rhamnosus GG*	[50 µg/L]—[10^{10}]	PBS	Viable	18 h	10 min—3000 g	55.62 ± 0.2 aλ	51.32 ± 0.3 a'*	[59]
	L. rhamnosus GG	[50 µg/L]—[10^{10}]	PBS	90 °C—1 h	18 h	10 min—3000 g	63.08 ± 0.3 a	59.67 ± 0.4 a'*	
	L. rhamnosus GG	[100 µg/L]—[5×10^8]	PBS	Viable	18 h	10 min—3000 g	1.38 ± 0.2 λ	0.51 ± 0.23*	
	L. acidophilus NCC 36	[5 µg/L]—[10^7]	PBS	Viable	0 h	15 min—3000 g	3.44 ± 3.04 β	-	
	L. acidophilus NCC 36	[5 µg/L]—[10^8]	PBS	Viable	0 h	15 min—3000 g	22.23 ± 10.76 β	-	
	L. acidophilus NCC 36	[5 µg/L]—[10^8]	PBS	Viable	24 h	15 min—3000 g	22.24 ± 4.67	-	
	L. acidophilus NCC 36	[20 µg/L]—[10^8]	PBS	Viable	0 h	15 min—3000 g	24.78 ± 1.39	-	
	L. acidophilus NCC 36	[20 µg/L]—[10^8]	PBS	Viable	24 h	15 min—3000 g	23.10 ± 5.19	-	
	L. acidophilus NCC 36	[5 µg/L]—[10^8]	PBS	90 °C—50 min	0 h	15 min—3000 g	26.38 ± 4.99	-	
	L. acidophilus NCC 36	[5 µg/L]—[10^8]	PBS	90 °C—50 min	24 h	15 min—3000 g	25.29 ± 5.03	-	
	L. acidophilus NCC 36	[20 µg/L]—[10^8]	PBS	90 °C—50 min	0 h	15 min—3000 g	26.22 ± 4.93	-	
	L. acidophilus NCC 36	[20 µg/L]—[10^8]	PBS	90 °C—50 min	24 h	15 min—3000 g	24.50 ± 4.40	-	[77]
	L. acidophilus NCC 36	[5 µg/L]—[10^8]	Reconstituted skim milk	Heat-killed	4 h	15 min—1800 g	23.73 ± 2.52	-	
	L. acidophilus NCC 36	[10 µg/L]—[10^8]	Reconstituted skim milk	Heat-killed	4 h	15 min—1800 g	24.13 ± 4.67	-	
	L. acidophilus NCC 36	[20 µg/L]—[10^8]	Reconstituted skim milk	Heat-killed	4 h	15 min—1800 g	25.07 ± 7.96	-	
	L. acidophilus NCC 36	[5 µg/L]—[10^8]	Reconstituted skim milk	Viable	4 h	15 min—1800 g	22.70 ± 3.36	-	
	L. rhamnosus	[5 µg/L]—[10^8]	PBS	Viable	0 h	15 min—3000 g	20.21 ± 6.16	-	
	L. rhamnosus	[5 µg/L]—[10^8]	PBS	Viable	24 h	15 min—3000 g	22.16 ± 7.14	-	
	L. rhamnosus	[20 µg/L]—[10^8]	PBS	Viable	0 h	15 min—3000 g	22.88 ± 7.11	-	
	L. rhamnosus	[20 µg/L]—[10^8]	PBS	Viable	24 h	15 min—3000 g	21.64 ± 1.66	-	
	L. rhamnosus	[5 µg/L]—[10^8]	PBS	90 °C—50 min	0 h	15 min—3000 g	23.37 ± 4.81	-	
	L. rhamnosus	[5 µg/L]—[10^8]	PBS	90 °C—50 min	24 h	15 min—3000 g	24.16 ± 3.33	-	
	L. rhamnosus	[20 µg/L]—[10^8]	PBS	90 °C—50 min	0 h	15 min—3000 g	27.78 ± 7.50	-	

Table 1. *Cont.*

Type	Strain	[AFM1]—[Cells]	Solution	Treatment	Incubation Time (37 °C)	Centrifugal Force (g/rpm)	Initial Binding (%)	Final Binding (%)	Reference
Bacteria	*L. rhamnosus*	[20 µg/L]—[10^8]	PBS	90 °C—50 min	24 h	15 min—3000 g	26.69 ± 5.48	-	[77]
	L. rhamnosus	[5 µg/L]—[10^8]	Reconstituted skim milk	Heat-killed	4 h	15 min—1800 g	25.13 ± 6.19	-	
Bacteria	*L. rhamnosus*	[10 µg/L]—[10^8]	Reconstituted skim milk	Heat-killed	4 h	15 min—1800 g	22.86 ± 9.33	-	[77]
	L. rhamnosus	[10 µg/L]—[10^8]	Reconstituted skim milk	Heat-killed	4 h	15 min—1800 g	22.86 ± 9.33	-	
	L. rhamnosus	[20 µg/L]—[10^8]	Reconstituted skim milk	Heat-killed	4 h	15 min—1800 g	26.27 ± 1.92	-	
	L. rhamnosus	[5 µg/L]—[10^8]	Reconstituted skim milk	viable	4 h	15 min—1800 g	21.74 ± 3.56	-	
	B. bifidum Bb13	[5 µg/L]—[10^8]	PBS	Viable	0 h	15 min—3000 g	23.48 ± 6.12	-	
	B. bifidum Bb13	[5 µg/L]—[10^8]	PBS	Viable	24 h	15 min—3000 g	26.65 ± 2.60	-	
	B. bifidum Bb13	[20 µg/L]—[10^8]	PBS	Viable	0 h	15 min—3000 g	24.77 ± 4.35	-	
	B. bifidum Bb13	[20 µg/L]—[10^8]	PBS	Viable	24 h	15 min—3000 g	26.33 ± 1.82	-	
	B. bifidum Bb13	[5 µg/L]—[10^8]	PBS	90 °C—50 min	0 h	15 min—3000 g	27.74 ± 2.97	-	
	B. bifidum Bb13	[5 µg/L]—[10^8]	PBS	90 °C—50 min	24 h	15 min—3000 g	25.12 ± 5.33	-	
	B. bifidum Bb13	[20 µg/L]—[10^8]	PBS	90 °C—50 min	0 h	15 min—3000 g	28.97 ± 3.49	-	
	B. bifidum Bb13	[20 µg/L]—[10^8]	PBS	90 °C—50 min	24 h	15 min—3000 g	27.31 ± 1.82	-	
	B. bifidum Bb13	[5 µg/L]—[10^8]	Reconstituted skim milk	Heat-killed	4 h	15 min—1800 g	25.41 ± 4.60	-	
	B. bifidum Bb13	[10 µg/L]—[10^8]	Reconstituted skim milk	Heat-killed	4 h	15 min—1800 g	25.64 ± 3.18	-	
	B. bifidum Bb13	[20 µg/L]—[10^8]	Reconstituted skim milk	Heat-killed	4 h	15 min—1800 g	27.31 ± 1.82	-	
	L. plantarum	[150 µg/L]—[10^{10}]	PBS	Viable	15 min	15 min—1800 g	5.60 ± 0.45 bA	3.71 ± 0.02 b'*	[69]
	L. plantarum	[150 µg/L]—[10^{10}]	PBS	100 °C—1 h	15 min	15 min—1800 g	13.11 ± 0.89 b	8.229 ± 0.03 b'*	

Table 1. *Cont.*

Type	Strain	[AFM1]—[Cells]	Solution	Treatment	Incubation Time (37 °C)	Centrifugal Force (g/rpm)	Initial Binding (%)	Final Binding (%)	Reference
Bacteria	*L. plantarum*	[150 µg/L]—[10^{10}]	PBS	viable	24 h	15 min—1800 g	8.09 ± 1.33 cA	5.571 ± 0.06 c′*	[69]
	L. plantarum	[150 µg/L]—[10^{10}]	PBS	100 °C—1 h	24 h	15 min—1800 g	14.14 ± 1.03 c	7.60 ± 0.03 c′*	
	E. avium	[150 µg/L]—[10^{10}]	PBS	Viable	15 min	15 min—1800 g	7.36 ± 1.10 d	5.19 ± 0.08 d′*	
	E. avium	[150 µg/L]—[10^{10}]	PBS	100 °C—1 h	15 min	15 min—1800 g	12.42 ± 2.20 d	7.070 ± 0.126 d′*	
	E. avium	[150 µg/L]—[10^{10}]	PBS	viable	24 h	15 min—1800 g	6.64 ± 1.40 e	2.69 ± 0.06 e′*	
	E. avium	[150 µg/L]—[10^{10}]	PBS	100 °C—1 h	24 h	15 min—1800 g	13.13 ± 2.14 e	7.446 ± 0.13 e′*	
Bacteria	*P. pentosaceus*	[150 µg/L]—[10^{10}]	PBS	Viable	15 min	15 min—1800 g	8.68 ± 1.24 f	5.36 ± 0.07 f′*	[69]
	P. pentosaceus	[150 µg/L]—[10^{10}]	PBS	100 °C—1 h	15 min	15 min—1800 g	15.16 ± 2.40 f	8.65 ± 0.14 f′*	
	P. pentosaceus	[150 µg/L]—[10^{10}]	PBS	viable	24 h	15 min—1800 g	7.76 ± 0.95 g	5.45 ± 0.079 g′*	
	P. pentosaceus	[150 µg/L]—[10^{10}]	PBS	100 °C—1 h	24 h	15 min—1800 g	13.86 ± 1.01 g	7.86 ± 0.07 g′*	
	L. gasseri	[150 µg/L]—[10^{10}]	PBS	Viable	15 min	15 min—1800 g	21.37 ± 2.76 h	16.91 ± 0.117 h′*	
	L. gasseri	[150 µg/L]—[10^{10}]	PBS	100 °C—1 h	15 min	15 min—1800 g	32.57 ± 1.96 h	20.6 ± 0.07 h′*	
	L. gasseri	[150 µg/L]—[10^{10}]	PBS	Viable	24 h	15 min—1800 g	22.77 ± 1.81 i	14.51 ± 0.017 i′*	
	L. gasseri	[150 µg/L]—[10^{10}]	PBS	100 °C—1 h	24 h	15 min—1800 g	32.30 ± 0.98 i	20.77 ± 0.012 i′*	
	L. bulgaricus	[150 µg/L]—[10^{10}]	PBS	Viable	15 min	15 min—1800 g	30.22 ± 1.43 kB	19.05 ± 0.05 k′*	
	L. bulgaricus	[150 µg/L]—[10^{10}]	PBS	100 °C—1 h	15 min	15 min—1800 g	36.32 ± 1.09 k	23.81 ± 0.05 k′*	
	L. bulgaricus	[150 µg/L]—[10^{10}]	PBS	Viable	24 h	15 min—1800 g	33.54 ± 1.56 B	18.02 ± 0.10 p′*	
	L. bulgaricus	[150 µg/L]—[10^{10}]	PBS	100 °C—1 h	24 h	15 min—1800 g	33.93 ± 1.91	23.5 ± 0.08 p′*	
	L. rhamnosus	[150 µg/L]—[10^{10}]	PBS	Viable	15 min	15 min—1800 g	17.13 ± 3.01 lC	14.96 ± 0.05 l′*	
	L. rhamnosus	[150 µg/L]—[10^{10}]	PBS	100 °C—1 h	15 min	15 min—1800 g	35.69 ± 3.13 l	23.02 ± 0.13 l′*	

Table 1. *Cont.*

Type	Strain	[AFM1]—[Cells]	Solution	Treatment	Incubation Time (37 °C)	Centrifugal Force (g/rpm)	Initial Binding (%)	Final Binding (%)	Reference
Bacteria	*L. rhamnosus*	[150 µg/L]—[10^{10}]	PBS	Viable	24 h	15 min—1800 g	27.79 ± 2.67 mC	16.51 ± 0.05 m'*	[69]
	L. rhamnosus	[150 µg/L]—[10^{10}]	PBS	100 °C—1 h	24 h	15 min—1800 g	45.67 ± 1.65 m	22.45 ± 0.063 m'*	
	B. lactis	[150 µg/L]—[10^{10}]	PBS	Viable	15 min	15 min—1800 g	16.89 ± 2.01 n	13.34 ± 0.115 n'*	
	B. lactis	[150 µg/L]—[10^{10}]	PBS	100 °C—1 h	15 min	15 min—1800 g	36.56 ± 2.46 n	23 ± 0.22 n'*	
	B. lactis	[150 µg/L]—[10^{10}]	PBS	Viable	24 h	15 min—1800 g	23.62 ± 4.13 o	13.71 ± 0.29 o'*	
	B. lactis	[150 µg/L]—[10^{10}]	PBS	100 °C—1 h	24 h	15 min—1800 g	35.84 ± 3.85 o	21.22 ± 0.316 o'*	
	L. rhamnosus strain GG	[150 µg/L]—[10^{10}]	skim milk	Viable	≈16 h	15 min—3500g	18.8 ± 1.9 pD	-	[78]
	L. rhamnosus strain GG	[150 µg/L]—[10^{10}]	skim milk	100 °C—1 h	≈16 h	15 min—3500 g	26.6 ± 3.2 pE	-	
	L. rhamnosus strain GG	[150 µg/L]—[10^{10}]	full cream milk	Viable	≈16 h	15 min—3500 g	26.0 ± 1.5 qD	-	
	L. rhamnosus strain GG	[150 µg/L]—[10^{10}]	full cream milk	100 °C—1 h	≈16 h	15 min—3500 g	36.6 ± 1.1 qE	-	
	LAB pool (*L.delbrueckii spp. Bulgaricus, L. rhamnosus and B.lactis*)	[0.5 µg/L]—[10^{10}]	UHT skim milk	100 °C—1 h	30 min	15 min—1800 g	11.5 ± 2.3	-	[66]
	LAB pool (*L.delbrueckii spp. Bulgaricus, L. rhamnosus and B.lactis*)	[0.5 µg/L]—[10^{10}]	UHT skim milk	100 °C—1 h	60 min	15 min—1800 g	11.7 ± 4.4	-	[66]

Results are the average ± SD for triplicates sample. Two-way ANOVA was conducted. Indicates a significant binding differences ($p < 0.05$) between: (*) Initial and final binding % of viable or heat-killed bacteria. (a, b, c, d, e, f, g, h, I, j, k, l, m, n, o, p, q, r, s, t, q) viable and heat-killed bacteria before washing. (a', b', c', d', e', f', g', h', i', j', k', l', m', n', o') viable and heat-killed bacteria after washing. (A, B, C, D, E) bacteria treated at different incubation time. (λ, β) bacteria treated at different AFM1 concentration.

Table 2. Summary of studies evaluating the AFM1 binding by different yeast strains.

Type	Strain	[AFM1]—[Cells]	Solution	Treatment	Incubation Time (37 °C)	Centrifugal Force (g/rpm)	Initial Binding (%)	Reference
Yeast	*S. cerevisiae*	[0.5 µg/L]—[10^9]	UHT skim milk	100 °C—1 h	30 min	15 min—1800 g	90.3 ± 0.3 A	[66]
	S. cerevisiae	[0.5 µg/L]—[10^9]	UHT skim milk	100 °C—1 h	60 min	15 min—1800 g	92.7 ± 0.7 A	
	Kluyveromyces lactis	[50 µg/L]—[10^9]	PBS	121 °C—10 min	72 h	15 min—6000 rpm	60.14 ± 2.5 λ	
	Kluyveromyces lactis	[50 µg/L]—[5 × 10^9]	PBS	121 °C—10 min	72 h	15 min—6000 rpm	69.14 ± 1.8 λ	[68]
	S. cerevisiae	[50 µg/L]—[10^9]	PBS	121 °C—10 min	72 h	15 min—6000 rpm	64.52 ± 1.83 β	
	S. cerevisiae	[50 µg/L]—[5 × 10^9]	PBS	121 °C—10 min	72 h	15 min—6000 rpm	78.74 ± 1.82 β	
	CYS-NV (*S. cerevisiae + k. lactis*)	[50 µg/L]—[5 × 10^9]	PBS	121 °C—10 min	72 h	15 min—6000 rpm	85.68 ± 1.84	

Results are the average ± SD for triplicates sample. Two-way ANOVA was conducted. Indicates a significant binding differences ($p < 0.05$) between: (A) Yeast treated at different incubation time. (λ, β) yeast treated at different AFM1 concentration.

Table 3. Summary of studies evaluating AFM1 binding by a mixture of yeasts and bacterial strains.

Type	Strain	[AFM1]—[Cells]	Solution	Treatment	Incubation Time (37 °C)	Centrifugal Force (g/rpm)	Initial Binding (%)	Reference
Mixture	LAB pool + *S. cerevisiae*	[0.5 µg/L]—[10^{10}] LAB pool + [10^9] *S. cerevisiae*	UHT skim milk	100 °C—1 h	30 min	15 min 1800 g	91.7 ± 0.5 A	[66]
	LAB pool + *S. cerevisiae*	[0.5 µg/L]—[10^{10}] LAB pool + [10^9] *S. cerevisiae*	UHT skim milk	100 °C—1 h	60 min	15 min 1800 g	100.0 ± 0.0 A	
	CPYS-NV (*B. bifidum* + *L. acidophilus* + *L. Plantarum* + *S. cerevisiae* + *k. lactis*)	[50 µg/L]—[5×10^9]	PBS	121 °C—10 min (b)—20 min (y)	72 h	-	87.92 ± 1.10	
	CPYS-NV (*B. bifidum* + *L. acidophilus* + *L. Plantarum* + *S. cerevisiae* + *k. lactis*)	[50 µg/L]—[5×10^9]	skim milk	121 °C—10 min (b) 121 °C—20 min (y)	12 h	-	80.56 ± 2.19 B	[68]
		[50 µg/L]—[5×10^9]	skim milk	121 °C—10 min (b) 121 °C—20 min (y)	24 h	-	86.64 ± 1.5 B	
		[50 µg/L]—[5×10^9]	skim milk	121 °C—10 min (b) 121 °C—20 min (y)	48 h	-	88.6 ± 1.3 C	
		[50 µg/L]—[5×10^9]	skim milk	121 °C—10 min (b) 121 °C—20 min (y)	72 h	-	90.88 ± 1.09 C	

Results are the average ± SD for triplicates sample. Two-way ANOVA was conducted. Indicates a significant binding differences ($p < 0.05$) between: (A, B, C) cells treated at different incubation time. (b): Bacterial strains (y): Yeast strains.

Tables 1–3 represent a summary of the literature using bacteria, yeasts, or a mixture of them for AFM1 decontamination. In addition, the stability of the formed complex in milk or in PBS before (Initial) and after (Final) washing is also highlighted.

3.2.1. Binding Efficiency of Bacterial Strains

AFM1 binding efficiency by different bacterial strains is shown in Table 1. Kabak et al. reported that the binding of AFM1 by viable *Lactobacillus* and *Bifidobacterium* strains in PBS depends on the AFM1 contamination level and incubation time. In addition, they indicated that heating the bacterial pellets did not improve their ability to remove AFM1 from PBS [98]. These findings were not consistent with Bovo et al. (2015), who reported that AFM1 bound by heat-killed *E. avium*, *L. plantarum*, *P. pentosaceus*, *B. lactis*, and *L. gasseri* was significantly greater than the amount bound by viable cells [91]. In fact, these results were in accordance with another study conducted by Assaf et al. in which the percentage of AFM1 bound by *L. rhamnosus* GG in PBS increased significantly, reaching up to 63.08% after heat treatment [74]. This binding increase was explained by the fact that AFM1 may adhere to bacterial cells via electrostatic bonding, thus suggesting that AFM1 is bound to cell wall components, namely polysaccharides and peptidoglycans [99,100]. Furthermore, during heat treatment, the cell wall components are affected by the denaturation of proteins, resulting in an increase in the hydrophobic nature of the cell's surface in addition to a possible formation of Maillard reaction products [101]. Hence, this denaturation allows AFM1 to bind to bacterial cell wall components that were not accessible when cells were intact [101]. A change in LAB concentration was sufficient to result in a variation of the amount of bound AFM1. According to Kabak et al., a reduction of the bacterial concentration resulted in a significant decrease of bound AFM1 in PBS [102]. This observation was congruent with an Assaf et al. finding, where the binding of AFM1 to *L. rhamnosus* GG greatly increased after increasing the bacterial concentration [74]. Besides, Kabak et al. mentioned that the amount of eliminated AFM1 was not affected by the contamination level of AFM1 in PBS [102].

However, similar findings in milk have been reported by Pierides et al., who demonstrated that heat-killed *L. rhamnosus* GG were able to more efficiently remove AFM1 than viable cells [103]. Accordingly, AFM1 removal in full cream milk (36.6%) was higher compared to skim milk (26.6%), though both were lesser than in PBS. In this regard, the same researchers justified the lower AFM1 removal in milk compared to PBS by the decrease in the availability of free AFM1 possibly associated with casein and other milk contents [103]. In addition, in 2013, Corassin et al. demonstrated that by using a pool of LAB in ultra-high temperature processing (UHT) skim milk, the bound amount of AFM1 has not significantly improved even after an increase in the incubation time [86]. These findings suggested that the binding process of AFM1 with LAB is completed in a fast manner.

3.2.2. Binding Efficiency of Yeast Strains

Upon using different type of yeast strains to bind AFM1 (Table 2), Corassin et al. reported that the binding of AFM1 by *Saccharomyces cerevisiae* in UHT skim milk was significantly higher (up to 92.7%) compared to the binding by LAB pool (up to 11.7%) [86]. Furthermore, in 2018, Abdelmotilib et al. stated that the combination of non-viable yeast strains (*Kluyveromyces lactis* and *S. cerevisiae*) had a higher AFM1 removal effect (85.68%) compared to separate yeast strains (up to 78.74%) [90]. Furthermore, the study showed that the removal of AFM1 by *Kluyveromyces lactis* increased significantly with an increasing yeast concentration [90].

3.2.3. Binding Efficiency of Bacteria and Yeasts Mixed Pools

Upon using a combination of bacterial strains (*L. Plantarum*, *L. acidophilus*, and *B. bifidum*) and yeast strains (*Kluyveromyces lactis* and *S. cerevisiae*) (Table 3), Abdelmotilib et al. demonstrated that this mixture showed the highest binding of AFM1 in a PBS medium (87.92%) [90]. Similarly Corassin et al. revealed that the amount of AFM1 bound by a mixture of LAB pool and *S. cerevisiae* in UHT skim milk was significantly higher, reaching up to 100% after 60 min of incubation [86]. In fact, this increase in the mixture's ability to remove AFM1 compared to bacteria or yeast strains was explained by the additive effect of both *S. cerevisiae* and LAB pool, making more accessible sites available for AFM1 fixation. After increasing the total microbial cell concentration in milk by forming a mixed pool of strains, the possible increase in AFM1 retention among microbial cells was not taken into consideration as a potential cause for this binding increase and should be further studied.

3.2.4. AFM1/Microbial Complex Binding Stability

Few studies assessed the stability of the AFM1/microbial complex after successive washing steps (Final binding). In this regard, in 2008, Kabak et al. reported that the binding of AFM1 to bacterial cells was partially reversible, and small amounts of AFM1 (up to 8.54%) were released back into the PBS solution [98]. Furthermore, Bovo et al. revealed that some AFM1 were released back into PBS after several washes but with greater percentages (up to 87%) [91]. Likewise, our observation indicated that after five successive washes, the percentage binding of AFM1 to viable and heat-treated *L. rhamnosus* GG decreased (up to 4.3%) [74]. Several investigators showed that the binding between AFM1 and bacteria is partially reversible, suggesting the implication of non-covalent bonds such as hydrogen bonds and van der Waals interactions [104,105]. A potential clarification of this variation in the released amounts of AFM1 can be explained by the difference in the binding sites between strains or to cross-linked interactions between AFM1 molecules and different microbes [106]. It is noteworthy to mention that the stability of the formed AFM1/microbial complex remains crucial; thus, a stable complex ensures a safe excretion of mycotoxins from the human body [38,107].

3.3. Plausible Experimental Variants

Depending on the performed experiments, the AFM1 binding assays using LAB or yeasts have shown some unexplained differences in the percentage of bound AFM1. In this section, we will try to explain the plausible causes of these differences and actions that could be taken in order to better clarify and analyze the given results.

3.3.1. Applied Heat Treatment

The heat treatment of bacteria, yeasts, or mixtures is not performed in a similar way (Tables 1–3). Hence, some tests are conducted by heating bacteria at 90–100 °C, while others are conducted by autoclaving at 121 °C. In both terms, the bacteria are heat-killed, but the effect of exerted heat on cell wall components (proteins, peptidoglycan, etc.) and their structures are not taken into consideration. However, depending on heating time, type, and temperature, reversible or irreversible denaturation events may take place [108]. Possibly, a reversible thermal denaturation of proteins or other cell wall components will cause their renaturation after heating [109]. Thus, this may cause an absence of significant changes in AFM1 binding after heat treatment, as shown by Kabak et al. (Table 1). Nevertheless, in another scenario, a variation of the heat rate may cause an irreversible denaturation of proteins and other cell wall components in addition to increasing the number of dissociated electrostatic bonds [74,109,110]. Consequently, it will affect the fixation of AFM1 on its binding sites in the cell wall. These results may explain some of the findings in which an increase in AFM1 binding was observed after heat treatment (Tables 1–3). In addition, heating may affect this binding through the formation of Maillard reaction products between proteins, polysaccharides, and peptides [101,111].

3.3.2. Working Temperature

In the conducted experiments for AFM1 binding by microbial adsorbents, the incubation temperature of AFM1-microbial suspension is clearly indicated (Tables 1–3). On the other hand, there is a lack of information regarding the working temperature of the carried out experiment that is usually not similar to the fixed incubation temperature of the suspended complex (e.g., 37 °C). Microbial pellets, PBS, and milk may have been stored cold before conducting the binding assay that may take place at room temperature. In addition, for heat-killed bacteria, the temperature of the bacterial pellets may not directly return to room temperature before being suspended in contaminated milk or PBS. Since we are dealing with an electrostatic type of bonding, a variation in the temperature may affect the binding of AFM1 to microbial cells [112]. Thus, for more accuracy, it may be better to indicate the working temperature for the performed assay.

3.3.3. Washing Steps

Testing the stability of the formed AFM1/microbial complex after successive washes was not always considered to be critical, and, for this reason, it was not conducted in all experiments (Tables 1–3). Due to the formation of electrostatic bonds, which are weak-to-intermediate in strength (hydrogen and van der Waals), it could be assumed that a certain amount of bound AFM1 may return to suspension (up to 87%) [74,91]. Nevertheless, electrostatic bonding may not be the only reason behind this decrease in bound AFM1 after washing steps. Accordingly, some AFM1 may be retained among bacteria or yeasts even without binding [74]. Therefore, conducting several washes until complete stabilization in the amount of bound AFM1 may be necessary to make appropriate assumptions regarding the actual binding percentage.

3.3.4. Filtration Step

The use of a filtration step to separate microbial adsorbents from AFM1 was not quite favorable due to the retention of some AFM1 that may take place in the filter even without the presence of any adsorbents [74]. Thus, the filtration of a suspension of AFM1 and bacteria may increase the retention of AFM1 in the filter due to membrane pore blockage by the bacteria and the formation of a cake layer. Furthermore, in order to avoid any malfunctions in High Performance Liquid Chromatography (HPLC), some manufacturer's instruction manuals [113] recommend filtering all samples before AF quantification [114–117]. In addition, it is worth mentioning that the use of a filtration step for supernatant samples after AFM1 binding may entail errors in the real amount of bound AFM1 [74]. For this reason, it might be better to keep controls for filtration steps at different AFM1 concentrations. In addition, these controls will help in sorting between retained AFM1 in filtration step and unrecovered AFM1 from milk following its clean-up by an immunoaffinity column (IAC). Furthermore, in order to reduce AFM1 retention in the filter, selecting the most appropriate filter membrane (membrane materials, pore size, etc.) may be crucial.

3.3.5. Centrifugation Step

In an AFM1 binding assay, the centrifugation steps are usually conducted to separate bacteria or yeasts from the containing medium (milk, PBS, etc.) [74,103]. For this reason, not much attention was given for the effect of centrifugal speed on AFM1 binding. Therefore, different speeds were used in the conducted experiments (Tables 1–3). We highlighted in our previous study that the centrifugation step is implicated in the binding of AFM1 to microbial adsorbents via increasing the contact among them [74]. For this reason, the centrifugation speed and time may affect the amount of bound AFM1, even without changes in other experimental conditions. Hence, the implication of the centrifugation step in AFM1 binding should not be ignored, and further studies may be needed.

3.3.6. Presence of an S-layer

The bacterial S-layer (surface layer) is a layer of thickness between 5 and 25 nm that forms the outermost cell envelope which covers the entire bacteria [118,119]. This layer is composed of proteins or glycoproteins arranged in different shapes in oblique, square, or hexagonal lattice symmetry [120,121]. In addition, the S-layer pore sizes range between 2 and 8 nm in diameter [118,122]. It is possible that this layer acts as a barrier against the entry of AFM1 and binding to the LAB cell wall peptidoglycans or polysaccharides. The potential role of the S-layer in the adsorption or retention of AFM1 should be elucidated after its extraction. It is important to mention that not all LAB have an S-layer [123]. A clear indication of its presence or absence gives a better understanding of the cell wall structure that may affect the binding of AFM1.

3.3.7. Detection and Quantification Techniques

To quantify AFM1 in milk, it is always essential to properly manipulate the samples and choose the appropriate detection method. The reversed-phase HPLC method is a widely used technique for its detection [114–117]. Other commonly known methods are the enzyme linked immune-sorbent assay (ELISA) and thin-layer chromatography (TLC) [49]. Both the ELISA and TLC methods are cost-effective and easy to handle. For these reasons, they are mainly used as AFM1 screening methods [124]. However, although the use of reverse phase HPLC is more expensive and requires skilled staff, it remains highly accurate with higher sensitivity and specificity [114]. The detection of AFM1 by HPLC is still of great importance due to its high sensitivity, accuracy, reliability, and possibility for column re-usage [114]. In contrast to ELISA, the HPLC method requires a clean-up of AFM1 from milk by using an IAC [116]. Along these lines, various experimental variants may occur at any stage, from sample handling to extraction, detection, or quantification of residual AFM1.

These inaccuracies are not only limited to the binding of AFM1 by microbial adsorbents but may extend to binding of different types of mycotoxins including aflatoxin B1, ochratoxin A, patulin, and other toxins following similar procedures [106,108,125,126].

4. Advantages and Limitations

The mechanism of AFM1 decontamination by LAB, yeasts, or mixtures present different advantages over other chemical, physical, or biological methods. This section highlights several advantages of the previously discussed methods in addition to different limitations which may act as a barrier toward their industrial commercialization and that need to be further investigated.

4.1. Advantages of Microbial Decontamination

4.1.1. Reduction of AFM1 Bioaccessibility

The use of microbial adsorbents to complex AFM1 may provide an additional strategy to reduce its bioavailability [38]. Hence, a decrease in the amount of free AFM1 for intestinal adsorption will occur. Serrano-Nino et al. revealed that the bioaccessibility of AFM1 in an in vitro digestive model was reduced after AFM1 binding by microbial probiotic strains [127]. Accordingly, *B. bifidum* NRRL B-41410 and *L. acidophilus* NRRL B-4495 were able to reduce the relative bioaccessibility of AFM1 by 45.17% and 32.20%, respectively. Moreover, tests conducted on mice revealed that the concurrent administration of *Lactobacillus plantarum* MON03 (LP) with AFM1 strongly reduced the adverse effects of AFM1 [128]. Therefore, there were no significant differences in tested parameters compared to the control mice [128]. In addition, several studies revealed that the binding of AF to LAB strains increased when simulated in a gastrointestinal environments. As a result of the exposure of LAB cells to bile, an alteration of proteins and phospholipids of the cell envelope may take place, resulting in increased binding [18,89,129,130].

4.1.2. Adsorption Specificity and Effectiveness

The use of either physical methods, such as heating and irradiation [131,132], or chemical methods, including solvent extraction, ammoniation, and ozone treatment [133–136], for AFM1 removal have many limitations. These detoxification methods are expensive, time consuming, and may cause significant nutritional losses compared to the microbial methods. Thus, microbial decontamination is found to be more effective and highly specific [137]. As observed in Tables 1–3, the binding of LAB with AFM1 varied not only between species but also within different strains of the same species, which is an additional indicator of the specificity of this type of binding. Hence, the additional confirmation of AFM1 binding specificity that is supposed to be exerted in specific sites of the microbial cell wall including polysaccharides and peptidoglycan, in addition to its binding mechanisms, needs to be further investigated.

4.1.3. Consumer Product Safety

The use of several probiotics that are "Generally Recognized as Safe" (GRAS) microorganisms for the milk detoxification of AFM1 has made this process safer. It is worth mentioning that probiotics can exert beneficial effects on the host, including consumers of milk and dairy products [138]. The above-discussed methods suggest that supplementing milk with probiotics may be a suitable solution for AFM1 removal in dairy products. In addition, the microbial control of AF production by LAB probiotic strains conferred better protection to milk and other contaminated dairy products during the storage period [52]. Different LAB such as *L. plantarum*, *L. fermentum*, and *L. rhamnosus* are known to be widely used as microbial inoculants for fermentation purposes including milk fermentation [139–141]. Therefore, the use of such microbial binders that are commonly found in dairy products and used in their processing is highly preferable. On the other hand, when using chemical or physical agents, some residues may be left in milk that will negatively affect the organoleptic quality of milk, putting the safety of the consumers at risk. Thus, using microbial adsorbents for the elimination of AFM1 from liquids such as milk remains a highly promising strategy.

4.2. Limitations of Microbial Decontamination

4.2.1. Microbial Supplementation Limits and Conditions

It is important to indicate that adding microbial agents to milk is acceptable to a certain limit. Current US standards require coliforms no greater than 10 cfu/mL in grade 'A' pasteurized fluid milk and a total plate counts of less than 20,000 cfu/mL [142]. Therefore, researchers must be aware of this issue that would be better if taken into consideration when performing the binding assays and fixing the needed microbial concentration. In addition, legislation concerning the total amount of dead and viable bacteria in milk and milk products varies regionally, and respecting these norms is a main concern for the safety of the consumers of dairy products. Adding different amounts of viable LAB, yeasts, or a mixture of both to milk is not quite favorable due to their uncontrolled proliferation. For example, yeasts such as *Kluyveromyces* sp. and *Saccharomyces* sp. usually cause milk spoilage by fermenting milk lactose [143].

4.2.2. Removal of Supplemented Microorganisms

If the concentrations of microbial agents necessary for AFM1 decontamination surpass the allowed limits, then they cannot remain in milk and a treatment for their removal is required. The removal of LAB or yeasts that were previously supplemented into milk using a filtration step is not easy to achieve due to several limiting factors. Within the required milk treatment for microbial removal, low membrane selectivity may take place in addition to its high operating costs [144]. Furthermore, size similarities between different milk components such as microbial adsorbents and fat can make this process harder to accomplish at low cost, and additional steps may be required. Thus, a proportion of the native milk fat globules which are similar in size to bacteria must be removed by centrifugal separation before conducting a microfiltration step [144]. It appears that this removal process is expensive, and important milk components may be lost, which means that they have to be re-supplemented later due to their significance to consumers.

4.2.3. Binding Reversibility

As previously reported in this review, the binding of AFM1 to microbial adsorbents is partially reversible. Hence, the non-covalent type of binding between microbial binders and AFM1 may be a main concern related to its industrial application. Since milk contamination by AFM1 and its maximum tolerable limits are not similar worldwide, the amount of supplemented yeasts, bacteria, or mixture of both in AFM1-contaminated milk needs to be regularly assessed. Thus, estimating the amounts of needed adsorbents is hard to achieve due to the partial reversibility of this type of bonding. Additionally, the stability of AF binding in milk or other liquids may differ according to different environmental condition including storage time, pH, milk temperature, and concentration of the used microbial adsorbents [145]. It remains tough to continuously monitor the amounts of free AFM1 and estimate the concentration of microbial agents in an unstable environment, especially when microbial additives bound to AFM1 are destined to be retained in milk.

5. Prospective Industrial Applications

Numerous microbial adsorbents have been tested in order to determine their potential ability to bind AF in milk and other dairy products, but, so far, researchers have not been able to commercially implement a fully reliable method. For this reason, several prospective methods for industrial applications are discussed in this section.

5.1. Microbial Fixation on Support or Membrane

The proposed method reported by Foroughi et al. in 2018 consists of immobilizing yeast such as *Saccharomyces cerevisiae* on perlite support to detoxify AFM1-contaminated milk [146]. The results showed a significant reduction in AFM1 concentration for all tested milk samples with various initial AFM1 contents. The highest reduction of AFM1 obtained was 81.3% after 80 min of milk circulation in the biofilter. This study revealed the high capability of immobilized yeast cells to detoxify AFM1 without any changes of its physicochemical properties. These promising results may be used for additional research such as fixing effective quantities of LAB, yeasts, or mixtures on a support or membrane that may be used to detoxify AFM1 by passing contaminated liquids through or over it. The formation of customized biofilters or cartridges containing these biological adsorbents may be more suitable for industrial application. Therefore, microbial cell immobilization is a remarkable method that may lead to different practical applications addressing not only AFM1 contamination in dairy products but also mycotoxins decontamination in beverages.

5.2. Microbial Biofilm Formation

A potential solution to the retention of microbial agents used for AFM1 decontamination in milk was shown in a study conducted by Assaf et al. in 2019 [50]. Tests were carried out to examine the ability of biofilm formed by probiotic LAB strains in tubes or in plates to eliminate AFM1. Hence, *L. rhamnosus* GG biofilm was able to significantly remove (up to 60.74%) AFM1 from contaminated whole milk. In addition, no significant difference in milk protein content was observed after AFM1 binding. Therefore, passing contaminated milk through or over the resultant biofilm for AFM1 decontamination could be a direct application of this method. It is important to mention that probiotic LAB and yeasts are able to form biofilms on different type of surfaces [147]. As such, their emerging applications in mycotoxin decontamination should be further elucidated.

5.3. Customized Rotating Mixer

As previously shown, an increase of AFM1 exposure to microbial adsorbents may affect the amount of AFM1 bound by LAB [74]. Thus, the binding of AFM1 to heat-killed *L. rhamnosus* GG increased when coupled to a mixing step such as pipetting [74]. For this reason, it may be suitable to use a customized rotating mixer that can increase the contact between AFM1 and microbial binders, thus decreasing the amount of microbial inoculum needed and the decontamination time. This procedure may be coupled with a filtration step to remove the supplemented microbial agents. Noting that even without any microbial supplementation, the increase of contact between AFM1 and milk components including LAB and proteins such as casein may result in an increase in the binding of free AFM1, thereby decreasing AFM1 bioavailability in contaminated milk [148–150].

6. Conclusions

Among all mycotoxins, the group of aflatoxins has received much attention due to their severe impact on human and animal health. In fact, numerous studies have investigated various microbial agents for their potential to bind AFM1. In this review, we aimed to investigate AFM1 decontamination methods by using microbial adsorbents and to emphasize the role of different experimental variants on complex binding and stability. Accordingly, this work highlights several experimental parameters that should be taken into consideration to optimize the binding of AFM1 in milk and other liquids. The decontamination of AFM1 using microbial adsorbents is still under vigorous investigation, and a better understanding of its binding mechanism and stability is needed. In addition, considerable testing of the physiochemical properties of milk after decontamination needs to be elucidated. Further studies on using these agents for AFM1 decontamination are still needed before the industrial implementation of the developed methods on milk products. This review highlights the use of different AFM1 decontamination methods and their plausible inaccuracies, thus answering some essential questions for a better understanding and improvement of these methods.

Author Contributions: Conceptualization, J.C.A.; Methodology, J.C.A., A.C., N.L., A.A. and A.E.K.; Software, S.N.; Validation, J.C.A., A.C., N.L., A.A. and A.E.K. Formal Analysis, J.C.A., A.C., N.L., A.A. and A.E.K.; Investigation, J.C.A., A.C., N.L., A.A. and A.E.K.; Resources, J.C.A.; Data Curation, J.C.A and S.N.; Writing-Original Draft Preparation, J.C.A.; Writing-Review & Editing, J.C.A, A.C., N.L., A.A. and A.E.K.; Visualization, J.C.A., S.N., A.C., N.L., A.A. and A.E.K; Supervision, A.C., N.L., A.A. and A.E.K.; Project Administration, J.C.A., A.C., N.L., A.A. and A.E.K; Funding Acquisition, A.A. and A.E.K.

Funding: This research and the APC were funded by the research council and the research and analysis center (CAR) at the Faculty of Sciences in Saint-Joseph University (USJ) and the Lebanese University through a grant number [5531/4].

Acknowledgments: This work was supported by the research council and the research and analysis center (CAR) at the Faculty of Sciences in Saint-Joseph University (USJ) and the Lebanese University.

Conflicts of Interest: The author declares no conflict of interest.

References

1. Bhatnagar, D.; Cary, J.W.; Ehrlich, K.; Yu, J.; Cleveland, T.E. Understanding the genetics of regulation of aflatoxin production and Aspergillus flavus development. *Mycopathologia* **2006**, *162*, 155–166. [CrossRef]
2. McLean, M.; Dutton, M.F. Cellular interactions and metabolism of aflatoxin: An update. *Pharmacol. Ther.* **1995**, *65*, 163–192. [CrossRef]
3. Dutton, M.F.; Healthcote, J. The structure, biochemical properties and origins of aflatoxin B2a nd G2a. *Chem. Ind.* **1968**, 418–421.
4. Marin, S.; Ramos, A.J.; Cano-Sancho, G.; Sanchis, V. Mycotoxins: Occurrence, toxicology, and exposure assessment. *Food Chem. Toxicol.* **2013**, *60*, 218–237. [CrossRef]
5. Torres, A.M.; Barros, G.G.; Palacios, S.A.; Chulze, S.N.; Battilani, P. Review on pre- and post-harvest management of peanuts to minimize aflatoxin contamination. *Food Res. Int.* **2014**, *62*, 11–19. [CrossRef]
6. Barkai-Golan, R.; Paster, N. Mouldy fruits and vegetables as a source of mycotoxins: Part 1. *World Mycotoxin J.* **2008**, *1*, 147–159. [CrossRef]
7. Rustom, I.Y.S. Aflatoxin in food and feed: Occurrence, legislation and inactivation by physical methods. *Food Chem.* **1997**, *59*, 57–67. [CrossRef]
8. Villers, P. Aflatoxins and safe storage. *Front. Microbiol.* **2014**, *5*, 1–6. [CrossRef] [PubMed]
9. Alshannaq, A.; Yu, J. Occurrence, Toxicity, and Analysis of Major Mycotoxins in Food. *Environ. Res. Public Health* **2017**, *14*, 632. [CrossRef]
10. Laciaková, A.; Cicoňová, P.; Mate, D.; Laciak, V. Aflatoxins and possibilities for their biological detoxification. *Med. Weter.* **2008**, *64*, 276–279.
11. Mahmood Fashandi, H.; Abbasi, R.; Mousavi Khaneghah, A. The detoxification of aflatoxin M1 by Lactobacillus acidophilus and Bifidobacterium spp.: A review. *J. Food Process. Preserv.* **2018**, *42*, 1–10. [CrossRef]

12. Yu, J.; Payne, G.A.; Campbell, B.C.; Guo, B.; Cleveland, T.E.; Robens, J.F.; Keller, N.P.; Bennett, J.W.; Nierman, W.C. Mycotoxin production and prevention of aflatoxin contamination in food and feed. *Aspergilli Genom. Med. Asp. Biotechnol. Res. Methods* **2007**, 457–472. [CrossRef]

13. Sklan, D.; Klipper, E.; Friedman, A.; Shelly, M.; Makovsky, B. The Effect of Chronic Feeding of Diacetoxyscirpenol, T-2 Toxin, and Aflatoxin on Performance, Health, and Antibody Production in Chicks. *J. Appl. Poult. Res.* **2001**, *10*, 79–85. [CrossRef]

14. Dhanasekaran, D.; Shanmugapriya, S. Aflatoxins and Aflatoxicosis in Human and Animals. In *Aflatoxins— Biochemistry and Molecular Biology*; InTechOpen: London, UK, 2011; pp. 222–254.

15. Battacone, G.; Nudda, A.; Cannas, A.; Borlino, A.C.; Bomboi, G.; Pulina, G. Excretion of Aflatoxin M1 in Milk of Dairy Ewes Treated with Different Doses of Aflatoxin B1. *J. Dairy Sci.* **2003**, *86*, 2667–2675. [CrossRef]

16. Kumar, P.; Mahato, D.K.; Kamle, M.; Mohanta, T.K. Aflatoxins: A Global Concern for Food Safety, Human Health and Their Management. *Front. Microbiol.* **2017**, *7*, 1–10. [CrossRef] [PubMed]

17. Oatley, J.T.; Rarick, M.D.; Ji, G.E.; Linz, J.E. Binding of Aflatoxin B 1 to Bifidobacteria In Vitro. *J. Food Prot.* **2016**, *63*, 1133–1136. [CrossRef]

18. Peltonen, K.D.; El-Nezami, H.; Haskard, C.; Ahokas, J.; Salminen, S. Aflatoxin B1 Binding by Dairy Strains of Lactic Acid Bacteria and Bifidobacteria. *Dairy Sci.* **2001**, *84*, 2152–2156. [CrossRef]

19. Adejumo, O.; Atanda, O.; Raiola, A.; Somorin, Y.; Bandyopadhyay, R.A. Correlation between aflatoxin M1 content of breast milk, dietary exposure to aflatoxin B1 and socioeconomic status of lactating mothers in Ogun State, Nigeria. *Food Chem. Toxicol.* **2013**, *56*, 171–177. [CrossRef] [PubMed]

20. Williams, J.H.; Phillips, T.D.; Jolly, P.E.; Stiles, J.K.; Jolly, C.M.; Aggarwal, D. Human aflatoxicosis in developing countries: A review of toxicology, exposure, potential health consequences, and interventions. *Am. J. Clin. Nutr.* **2018**, *80*, 1106–1122. [CrossRef]

21. Darwish, W.S.; Ikenaka, Y.; Nakayama, S.M.M.; Ishizuka, M. An Overview on Mycotoxin Contamination of Foods in Africa. *Toxicology* **2014**, *76*, 789–797. [CrossRef]

22. Angelis, I. De Aflatoxin M1 absorption and cytotoxicity on human intestinal in vitro model. *Toxicon* **2006**, *47*, 409–415.

23. Marchese, S.; Polo, A.; Ariano, A.; Velotto, S.; Costantini, S.; Severino, L. Aflatoxin B1 and M1: Biological properties and their involvement in cancer development. *Toxins (Basel)* **2018**, *10*, 1–19. [CrossRef]

24. Quinn, G.; Gary, P.J.; Damiano, C.; Teehan, G. Treatment of Resistant Hypertension: An Update in Device Therapy. In *ERRATUM*; InTechOpen: London, UK, 2018; pp. 1–11.

25. Britzi, M.; Friedman, S.; Miron, J.; Solomon, R.; Cuneah, O.; Shimshoni, A.J.; Soback, S.; Ashkenazi, R.; Armer, S.; Shlosberg, A. Carry-Over of Aflatoxin M1 in Cows in Mid- and Late-Lactation. *Toxins* **2013**, *5*, 173–183. [CrossRef] [PubMed]

26. Jalili, M.; Scotter, M. A review of aflatoxin M1 in liquid milk. *Iran. J. Health Saf. Environ.* **2015**, *2*, 283–295.

27. Sweeney, M.J.; Dobson, A.D.W. Mycotoxin production by Aspergillus, Fusarium and Penicillium species. *Food Microbiol.* **1998**, *43*, 141–158. [CrossRef]

28. Creppy, E.E. Update of survey, regulation and toxic effects of mycotoxins in Europe. *Toxicology* **2002**, *127*, 19–28. [CrossRef]

29. Park, D.L. Effect of Processing on Aflatoxin. *Adv. Exp. Med. Biol.* **2002**, *504*, 173–179. [PubMed]

30. Motawee, M. Reduction of Aflatoxin M1 Content during Manufacture and Storage of Egyptian Domaiti Cheese. *Vet. Med.* **2013**, *2013*, 11. [CrossRef]

31. Galvano, F.; Galofaro, V.; Galvano, G. Occurrence and stability of aflatoxin M1 in milk and milk products a worldwide review. *Food Prot.* **1996**, *59*, 1079–1090. [CrossRef]

32. Bovo, F.; Ganev, K.C.; Mousavi, A.; Portela, B.; Cruz, A.G.; Granato, D.; Corassin, C.H.; Augusto, C.; Oliveira, F.; Sant, A.S. The occurrence and effect of unit operations for dairy products processing on the fate of aflatoxin M1: A review. *Food Control* **2016**, *68*, 310–329.

33. Motawee, M.; Meyr, K.; Bauer, J. Incidence of Aflatoxins M1 and B1 in raw milk ans some dairy products In Damietta. *Agric. Sci.* **2001**, *29*, 719–725.

34. Jonker, M.A.; Schothorst, R.C.; Egmond, H.P. van Regulations relating to mycotoxins in food Perspectives in a global and European context. *Anal. Bioanal. Chem.* **2007**, *389*, 147–157.

35. Juan, A.; Ritieni, J.M. Determination of trichothecenes and zearalenones in grain cereal, flour and bread by liquid chromatography tandem mass spectroscopy. *Food Chem.* **2012**, *134*, 2389–2397. [CrossRef] [PubMed]

36. Food and Drug Administration (FDA). *Compliance Program Guidance Manual*; FDA: Silver Spring, MD, USA, 2005.

37. European Commission. Commission Regulation (EC) No. 1881/2006 of 19 December 2006. *Off. J. Eur. Union* **2006**, *364*, 5.

38. Ismail, A.; Akhtar, S.; Levin, R.E.; Ismail, T.; Riaz, M.; Amir, M. Aflatoxin M1: Prevalence and decontamination strategies in milk and milk products. *Crit. Rev. Microbiol.* **2015**, *42*, 418–427. [CrossRef]

39. Hm, M.; Raghava, S.; Umesha, S. *Mycotoxins in Food and Agriculture-Challenges and Opportunities*; MedCrave Group LLC: Oklahoma, OK, USA, 2017.

40. Udomkun, P.; Nimo, A.; Nagle, M.; Müller, J.; Vanlauwe, B.; Bandyopadhyay, R. Innovative technologies to manage aflatoxins in foods and feeds and the profitability of applicatione—A review. *Food Control* **2017**, *76*, 127–138. [CrossRef]

41. Kendra, D.F.; Dyer, R.B. Opportunities for biotechnology and policy regarding mycotoxin issues in international trade. *Food Microbol.* **2007**, *119*, 147–151. [CrossRef]

42. Karlovsky, P.; Suman, M.; Berthiller, F.; Meester, J. De; Eisenbrand, G.; Perrin, I.; Oswald, I.P.; Speijers, G. Impact of food processing and detoxification treatments on mycotoxin contamination. *Mycotoxin Res.* **2016**, *32*, 179–205. [CrossRef]

43. Kabak, B.; Dobson, A.D.W. Strategies to Prevent Mycotoxin Contamination of Food and Animal Feed: A Review. *Food Sci. Nutr.* **2006**, *46*, 593–619. [CrossRef]

44. Waliyar, F.; Osiru, M.; Ntare, B.R.; Kumar, K.V.K.; Sudini, H.; Traore, A.; Diarra, B. Post-harvest management of aflatoxin contamination in groundnut. *World Mycotoxin J.* **2014**, *8*, 245–252. [CrossRef]

45. Palou, L.; Crisosto, C.H.; Smilanick, J.L.; Adaskaveg, J.E.; Zoffoli, J.P. Effects of continuous 0.3 ppm ozone exposure on decay development and physiological responses of peaches and table grapes in cold storage. *Postharvest Biol. Technol.* **2002**, *24*, 39–48. [CrossRef]

46. Kanapitsas, A.; Batrinou, A.; Aravantinos, A.; Markaki, P. Effect of γ-radiation on the production of aflatoxin B1 by Aspergillus parasiticus in raisins (*Vitis vinifera* L.). *Radiat. Phys. Chem.* **2015**, *106*, 327–332. [CrossRef]

47. Markov, K.; Mihaljević, B.; Domijan, A.M.; Pleadin, J.; Delaš, F.; Frece, J. Inactivation of aflatoxigenic fungi and the reduction of aflatoxin B1 invitro and in situ using gamma irradiation. *Food Control* **2015**, *54*, 79–85. [CrossRef]

48. Barra, P.; Etcheverry, M.; Nesci, A. Efficacy of 2,6-di (t-butyl)-p-cresol (BHT) and the entomopathogenic fungus Purpureocillium lilacinum, to control Tribolium confusum and to reduce aflatoxin B1 in stored maize. *Stored Prod. Res.* **2015**, *64*, 72–79. [CrossRef]

49. Magan, N.; Olsen, M. Mycotoxins in Food: Detection and Control. *Food Sci. Technol.* **2005**, *40*, 337–342.

50. Assaf, J.C.; EL Khoury, A.; Chokr, A.; Louka, N.; Atoui, A. A novel method for elimination of aflatoxin M1 in milk using Lactobacillus rhamnosus GG biofilm. *Dairy Technol.* **2019**, *70*, 1–9. [CrossRef]

51. Assaf, J.C.; El Khoury, A.; Atoui, A.; Louka, N.; Chokr, A. A novel technique for aflatoxin M1 detoxification using chitin or treated shrimp shells: In vitro effect of physical and kinetic parameters on the binding stability. *Appl. Microbiol. Biotechnol.* **2018**, *102*, 6687–6697. [CrossRef]

52. Giovati, L.; Magliani, W.; Ciociola, T.; Santinoli, C.; Conti, S.; Polonelli, L. AFM1 in Milk: Physical, Biological, and Prophylactic Methods to Mitigate Contamination. *Toxins (Basel)* **2015**, *7*, 4330–4349. [CrossRef]

53. Wu, Q.; Jezkova, A.; Yuan, Z.; Pavlikova, L.; Dohnal, V.; Kuca, K. Biological degradation of aflatoxins. *Drug Metab.* **2009**, *41*, 1–7. [CrossRef]

54. Jard, G.; Liboz, T.; Mathieu, F.; Guyonvarch, A.; Lebrihi, A. Review of mycotoxin reduction in food and feed: From prevention in the field to detoxification by adsorption or transformation. *Food Addit. Contam. Part A* **2011**, *28*, 1590–1609. [CrossRef]

55. Stoloff, L.; Van Egmond, H.P.; Park, D.L. Rationales for the establishment of limits and regulations for mycotoxins. *Food Addit. Contam. ISSN* **2015**, *8*, 213–221. [CrossRef] [PubMed]

56. Wood, G.E. Mycotoxins in Foods and Feeds in the United States. *Food Drug Adm.* **2018**, *70*, 3941–3949. [CrossRef] [PubMed]

57. Vouk, V.B. *Methods for Assessing the Effects of Mixtures of Chemicals Scope 30 SGOMSEC 3*; Wiley: New York, NY, USA, 1989; Volume 27, ISBN 0471911232.

58. Fallah, A.A. Assessment of aflatoxin M1 contamination in pasteurized and UHT milk marketed in central part of Iran. *Food Chem. Toxicol.* **2010**, *48*, 988–991. [CrossRef] [PubMed]

59. Liew, W.P.P.; Nurul-Adilah, Z.; Than, L.T.L.; Mohd-Redzwan, S. The binding efficiency and interaction of Lactobacillus casei Shirota toward aflatoxin B1. *Front. Microbiol.* **2018**, *9*, 1–12. [CrossRef]

60. Urine, H.; Zhang, L.; Hu, X.; Xiao, Y.; Chen, J.; Xu, Y.; Fremy, J.; Chu, F.S. Correlation of Dietary Aflatoxin BI Levels with Excretion of Aflatoxin M. *Cancer Res.* **1987**, *47*, 1848–1852.

61. Kuilman, M.E.M.; Maas, R.F.M. Cytochrome P450-mediated metabolism and cytotoxicity of aflatoxin B1 in bovine hepatocytes. *Toxicol. Vitr.* **2000**, *14*, 321–327. [CrossRef]

62. Murphy, P.A.; Hendrich, S.; Landgren, C.; Bryant, C.M. Food mycotoxins: An update. *J. Food Sci.* **2006**, *71*, 51–65. [CrossRef]

63. Li, H.; Xing, L.; Zhang, M.; Wang, J.; Zheng, N. The Toxic Effects of Aflatoxin B1 and Aflatoxin M1 on Kidney through Regulating L-Proline and Downstream Apoptosis. *Biomed. Res. Int.* **2018**, *2018*, 1–11. [CrossRef]

64. Sun, Z.; Lu, P.; Gail, M.H.; Pee, D.; Zhang, Q.; Ming, L.; Wang, J.; Wu, Y.; Liu, G.; Wu, Y.; et al. Increased Risk of Hepatocellular Carcinoma in Male Hepatitis B Surface Antigen Carriers with Chronic Hepatitis Who Have Detectable Urinary Aflatoxin Metabolite M1. *Hepatology* **1999**, *30*, 379–383. [CrossRef] [PubMed]

65. Lu, H.; Li, Y. Effects of bicyclol on aflatoxin B1 metabolism and hepatotoxicity in rats. *Acta Pharmacol. Sin.* **2002**, *23*, 942–945. [PubMed]

66. Aydin, A. Determination of Aflatoxin B1 levels in powdered red pepper. *Food Control* **2007**, *18*, 1015–1018. [CrossRef]

67. Aydin, A.; Gunsen, U. Total Aflatoxin, Aflatoxin B1 and Ochratoxin A Levels in Turkish Wheat Flour. *Food Drug Anal.* **2008**, *16*, 48–53.

68. Ajani, J.; Chakravarthy, D.V.S.; Tanuja, P.; Vali Pasha, K. Aflatoxins. *Indian J. Adv. Chem. Sci.* **2014**, *3*, 49–60.

69. Zheng, Z.; Zhang, T. Recent Trends in Microbiological Decontamination of Aflatoxins in Foodstuffs. In *Aflatoxins-Recent Advances and Future Prospects*; InTechOpen: London, UK, 2013; pp. 59–92.

70. Brundtland, H.; Bellamy, C. Global strategy for infant and young child feeding. In *WHO*; WHO Library: Geneva, Switzerland, 2003; pp. 1–30.

71. Radonić, J.R.; Tanackov, S.D.K.; Mihajlović, I.J.; Zorica, S.; Miloradov, M.B.V.; Škrinjar, M.M.; Turk, M.M. Occurrence of aflatoxin M1 in human milk samples in Vojvodina, Serbia: Estimation of average daily intake by babies. *J. Environ. Sci. Health Part B* **2016**, *52*, 59–63. [CrossRef] [PubMed]

72. Mohammed, S.; Munissi, J.E.; Nyandoro, S. Aflatoxin M1 in raw milk and aflatoxin B1 in feed from household cows in Singida, Tanzania. *Food Addit. Contam.* **2016**, *9*, 85–90. [CrossRef] [PubMed]

73. Bata, Á.; Lásztity, R. Detoxification of mycotoxin-contaminated food and feed by microorganisms. *Trends Food Sci. Technol.* **1999**, *10*, 223–228. [CrossRef]

74. Assaf, J.C.; Atoui, A.; El Khoury, A.; Chokr, A.; Louka, N. A comparative study of procedures for binding of aflatoxin M1 to Lactobacillus rhamnosus GG. *Brazilian J. Microbiol.* **2017**, *49*, 120–127. [CrossRef]

75. Choudhary, A.K.; Kumari, P. Management of Mycotoxin Contamination in Preharvest and Post Harvest Crops: Present Status and Future Prospects. *J. Phytol.* **2010**, *2*, 37–52.

76. Atehnkeng, J.; Ojiambo, P.S.; Cotty, P.J.; Bandyopadhyay, R. Field efficacy of a mixture of atoxigenic Aspergillus flavus Link: FR vegetative compatibility groups in preventing aflatoxin contamination in maize (*Zea mays* L.). *Biol. Control* **2014**, *72*, 62–70. [CrossRef]

77. Dorner, J.W.; Cole, R.J. Effect of application of nontoxigenic strains of Aspergillus flavus and A. parasiticus on subsequent aflatoxin contamination of peanuts in storage. *J. Stored Prod. Res.* **2002**, *38*, 329–339. [CrossRef]

78. Abdallah, M.; Ameye, M.; De Saeger, S.; Audenaert, K.; Haesaert, G. Biological Control of Mycotoxigenic Fungi and Their Toxins: An Update for the Pre-Harvest Approach. In *Mycotoxins—Socio-Economic and Health Impact as Well as Pre- and Postharvest Management Strategies*; InTechOpen: London, UK, 2018.

79. Di Gregorio, M.C.; De Neeff, D.V.; Jager, A.V.; Corassin, C.H.; Carão, Á.C.D.P.; De Albuquerque, R.; De Azevedo, A.C.; Oliveira, C.A.F. Mineral adsorbents for prevention of mycotoxins in animal feeds. *Toxin Rev.* **2014**, *33*, 125–135. [CrossRef]

80. Grenier, B.; Applegate, T.J. *Reducing the Impact of Aflatoxins in Livestock and Poultry*; Perdue University Education Store: Lafayette, IN, USA, 2013; pp. 1–7.

81. Phillips, T.D.; Clement, B.A.; Park, D.L. Approaches to Reduction of Aflatoxin in Foods and Feeds. In *The Toxicology of Aflatoxins: Human Health, Veterinary, and Agricultural Significance*; David, L., Eaton, J.D.G., Eds.; Academic Press: Cambridge, MA, USA, 1994; pp. 383–406.

82. Galvano, F.; Pietri, A.; Bertuzzi, T.; Fusconi, G.; Galvano, M.; Piva, A.; Piva, G. Reduction of Carryover of Aflatoxin from Cow Feed to Milk by Addition of Activated Carbons. *J. Food Prot.* **1996**, *59*, 551–554. [CrossRef]

83. Carraro, A.; De Giacomo, A.; Giannossi, M.L.; Medici, L.; Muscarella, M.; Palazzo, L.; Quaranta, V.; Summa, V.; Tateo, F. Clay minerals as adsorbents of aflatoxin M1 from contaminated milk and effects on milk quality. *Appl. Clay Sci.* **2014**, *88–89*, 92–99. [CrossRef]

84. Devreese, M.; Antonissen, G.; De Backer, P.; Croubels, S. Efficacy of active carbon towards the absorption of deoxynivalenol in pigs. *Toxins (Basel)* **2014**, *6*, 2998–3004. [CrossRef] [PubMed]

85. Maxim, L.D.; Niebo, R.; McConnell, E.E. Bentonite toxicology and epidemiology—A review. *Inhal. Toxicol.* **2016**, *28*, 591–617. [CrossRef]

86. Corassin, C.H.; Bovo, F.; Rosim, R.E.; Oliveira, C.A.F. Efficiency of Saccharomyces cerevisiae and lactic acid bacteria strains to bind aflatoxin M1 in UHT skim milk C.H. *Food Control* **2013**, *31*, 80–83. [CrossRef]

87. Färber, P.; Brost, I.; Adam, R.; Holzapfel, W. HPLC based method for the measurement of the reduction of aflatoxin B1 by bacterial cultures isolated from different African foods. *Mycotoxin Res.* **2000**, *16*, 141. [CrossRef] [PubMed]

88. Khaneghah, A.M.; Chaves, R.D.; Akbarirad, H. Detoxification of Aflatoxin M1 (AFM1) in Dairy Base Beverages (Acidophilus Milk) by Using Different Types of Lactic Acid Bacteria-Mini Review. *Curr. Nutr. Food Sci.* **2017**, *13*, 78–81. [CrossRef]

89. El-Nezami, H.; Kankaanpää, P.; Salminen, A.J. Physicochemical alterations enhance the ability of dairy strains of lactic acid bacteria to remove aflatoxin from contaminated media. *Food Prot.* **1998**, *61*, 466–468. [CrossRef]

90. Abdelmotilib, N.M.; Hamad, G.; Salem, E.G. Aflatoxin M1 Reduction in Milk by a Novel Combination of Probiotic Bacterial and Yeast Strains. *Nutr. Food Saf.* **2018**, *8*, 83–99. [CrossRef]

91. Bovo, F.; Corassin, C.H.; Rosim, R.E. Efficiency of Lactic Acid Bacteria Strains for Decontamination of Aflatoxin M1 in Phosphate Buffer Saline Solution and Skimmed milk. *Food Bioprocess. Technol.* **2015**, *6*, 2230–2234. [CrossRef]

92. Var, I.; Kabak, B.; Brandon, E.F.A.; Blokland, M.H. Effects of probiotic bacteria on the bioaccessibility of aflatoxin B1 and Ochratoxin A using an in vitro digestion model under fed conditions. *Environ. Sci. Health B-Pestic.* **2009**, *44*, 472–480.

93. Patel, A.; Shah, N.; Shukla, D. Biological Control of Mycotoxins by Probiotic Lactic Acid Bacteria. In Proceedings of the Dynamism in Dairy Industry and Consumer Demands, Gujarat, India, 4–5 February 2017.

94. Patel, A.; Sv, A.; Shah, N.; Verma, K.D. Lactic acid bacteria as metal quenchers to improve food safety and quality. *AgroLife Sci.* **2017**, *6*, 146–154.

95. El-Nezami, H.; Mykkänen, H.; Haskard, C.; Salminen, S.; Salminen, E. Lactic acid bacteria as a tool for enhancing food safety by removal of dietary toxin. In *Lactic Acid Bacteria: Microbiological and Functional Aspects*; Lahtinen, S., Ouwehand, A.C., Salminen, S., von Wright, A., Eds.; CRC Press: New York, NY, USA, 2004; pp. 397–406.

96. Perczak, A.; Goliński, P.; Bryła, M.; Waśkiewicz, A. The efficiency of lactic acid bacteria against pathogenic fungi and mycotoxins. *Arh. Hig. Rada Toksikol.* **2018**, *69*, 32–45. [CrossRef] [PubMed]

97. Piotrowska, M. The adsorption of ochratoxin a by lactobacillus species. *Toxins (Basel)* **2014**, *6*, 2826–2839. [CrossRef] [PubMed]

98. Taylor, P.; Kabak, B.; Var, I. Factors affecting the removal of aflatoxin M 1 from food model by Lactobacillus and Bifidobacterium strains Factors affecting the removal of aflatoxin M 1 from food model by Lactobacillus and Bifidoba. *Environ. Sci. Health Part B* **2008**, *3*, 617–624.

99. Hosono, A.; Yoshimura, A.; Otani, H. Desmutagenic property of cell walls of Streptococcus faecalis on the mutagenicities induced by amino acid pyrolysates. *Milchwissenschaft* **1988**, *43*, 168–170.

100. Rajendran, R.; Ohta, Y. Binding of heterocyclic amines by lactic acid bacteria from miso, a fermented Japanese food. *Microbiology* **1998**, *44*, 109–115. [CrossRef]

101. Haskard, C.A.; El-nezami, H.S.; Kankaanpa, P.E.; Salminen, S.; Ahokas, J.T. Surface Binding of Aflatoxin B1 by Lactic Acid Bacteria. *Appl. Environ. Microbiol.* **2001**, *67*, 3086–3091. [CrossRef]

102. Taylor, P.; Kabak, B.; Var, I. Factors affecting the removal of aflatoxin M1 from food model by Lactobacillus and Bifidobacterium strains. *J. Environ. Sci. Health Part B* **2008**, *43*, 37–41.

103. Pierides, M.; El-nezami, H.; Peltonen, K.; Salminen, S. Ability of Dairy Strains of Lactic Acid Bacteria to Bind Aflatoxin M 1 in a Food Model. *Food Prot.* **2000**, *63*, 645–650. [CrossRef]

104. Yiannikouris, A.; Poughon, L.; Franc, J.; Dussap, C.; Jeminet, G. Chemical and Conformational Study of the Interactions Involved in Mycotoxin Complexation with B–D-Glucans. *Biomacaromolecules* **2006**, *7*, 1147–1155. [CrossRef]

105. Shetty, P.H.; Jespersen, L. Saccharomyces cerevisiae and lactic acid bacteria as potential mycotoxin decontaminating agents. *Trends Food Sci. Technol.* **2006**, *17*, 48–55. [CrossRef]

106. Sajid, M.; Mehmood, S.; Niu, C.; Yuan, Y.; Yue, T. Effective Adsorption of Patulin from Apple Juice by Using Non-Cytotoxic Heat-Inactivated Cells and Spores of Alicyclobacillus Strains. *Toxins (Basel)* **2018**, *10*, 344. [CrossRef]

107. Ismail, A.; Levin, R.E.; Riaz, M.; Akhtar, S.; Gong, Y.Y.; de Oliveira, C.A.F. Effect of different microbial concentrations on binding of aflatoxin M1 and stability testing. *Food Control* **2017**, *73*, 492–496. [CrossRef]

108. Zhong, L.; Carere, J.; Lu, Z.; Lu, F.; Zhou, T. Patulin in Apples and Apple-Based Food Products: The Burdens and the Mitigation Strategies. *Toxins (Basel)* **2018**, *10*, 475. [CrossRef]

109. Tsong, T.Y.; Gross, C.J. Reversibility of Thermally Induced Denaturation of Cellular Proteins. *Ann. N. Y. Acad. Sci.* **1994**, *720*, 65–78. [CrossRef] [PubMed]

110. Schön, A.; Clarkson, B.R.; Jaime, M.; Freire, E. Temperature Stability of Proteins: Analysis of Irreversible Denaturation Using Isothermal Calorimetry. *Proteins* **2018**, *85*, 2009–2016. [CrossRef]

111. Teodorowicz, M.; Van Neerven, J.; Savelkoul, H. Food processing: The influence of the maillard reaction on immunogenicity and allergenicity of food proteins. *Nutrients* **2017**, *9*, 835. [CrossRef]

112. Zinedine, A.; Faid, M.; Benlemlih, M. In Vitro Reduction of Aflatoxin B1 by Strains of Lactic Acid Bacteria Isolated from Moroccan Sourdough Bread. *Int. J. Agric. Biol.* **2005**, *7*, 67–70.

113. Application note 102. *Separation and Quantitation of Aflatoxins M1 and M2 Using TFA Derivatization and HPLC*; Sigma-Aldrich Co.: Missouri, USA, 1996.

114. Turnera, W.; Nicholas, S.; Piletskyb, S.A. Analytical methods for determination of mycotoxins: A review. *Anal. Chim. Acta* **2009**, *632*, 168–180. [CrossRef] [PubMed]

115. Espinosa-calderón, A.; Contreras-medina, L.M.; Muñoz-huerta, R.F.; Millán-almaraz, J.R.; Gerardo, R.; González, G.; Torres-pacheco, I. Methods for Detection and Quantification of Aflatoxins. In *Aflatoxins—Detection, Measurement and Control*; InTechOpen: London, UK, 2011; pp. 110–128.

116. Chun, S.; Yuan, Y.; Eremin, S.A.; Jong, W. Detection of aflatoxin M1 in milk products from China by ELISA using monoclonal antibodies. *Food Control* **2009**, *20*, 1080–1085.

117. Adibpour, N.; Soleimanian-Zad, S.; Sarabi-Jamab, M.; Tajalli, F. Effect of storage time and concentration of aflatoxin m1 on toxin binding capacity of L. acidophilus in fermented milk product. *J. Agric. Sci. Technol.* **2016**, *18*, 1209–1220.

118. Schuster, B.; Pum, D.; Sleytr, U.B. S-layer stabilized lipid membranes (Review). *Eur. PMC Funders Gr.* **2010**, *3*, 1–19. [CrossRef]

119. Sleytr, U.B.; Messner, P.; Pum, D.; Sára, M. Crystalline bacterial cell surface layers (S layers): From supramolecular cell structure to biomimetics and nanotechnology. *Angew. Chem. Int. Ed.* **1999**, *38*, 1034–1054. [CrossRef]

120. Sleytr, U.B.; Schuster, B.; Egelseer, E.; Pum, D. S-layers: Principles and applications. *FEMS Microbiol.* **2014**, *38*, 823–864. [CrossRef] [PubMed]

121. Rodrigues-Oliveira, T.; Belmok, A.; Vasconcellos, D.; Schuster, B.; Kyaw, C.M. Archaeal S-layers: Overview and current state of the art. *Front. Microbiol.* **2017**, *8*, 1–17. [CrossRef] [PubMed]

122. Pum, D.; Sleytr, U.B. S-layer Proteins for Assmebling Ordered Naanoparticle Arrays. In *Nanobioelectronics—for Electronics, Biology, and Medicine*; Springer: New York, NY, USA, 2007.

123. Fagan, R.P.; Fairweather, N.F. Biogenesis and functions of bacterial S-layers. *Nat. Rev. Microbiol.* **2014**, *12*, 211–222. [CrossRef]

124. Manetta, A.C. Aflatoxins: Their Measure and Analysis. In *Aflatoxins—Detection, Measurement and Control*; InTechOpen: London, UK, 2011; pp. 93–106.

125. Fuchs, S.; Sontag, G.; Stidl, R.; Ehrlich, V.; Kundi, M.; Knasmüller, S. Detoxification of patulin and ochratoxin A, two abundant mycotoxins, by lactic acid bacteria. *Food Chem. Toxicol.* **2008**, *46*, 1398–1407. [CrossRef]

126. Rushing, B.R.; Selim, M.I. Aflatoxin B1: A review on metabolism, toxicity, occurrence in food, occupational exposure, and detoxification methods. *Food Chem. Toxicol.* **2018**, *124*, 81–100. [CrossRef]

127. Serrano-niño, J.C.; Cavazos-garduño, A.; Hernandez-mendoza, A.; Applegate, B.; Ferruzzi, M.G. Assessment of probiotic strains ability to reduce the bioaccessibility of aflatoxin M1 in artificially contaminated milk using an in vitro digestive model. *Food Control* **2013**, *31*, 202–207. [CrossRef]

128. Salah-Abbès, J.B.; Abbès, S.; Jebali, R.; Haous, Z.; Oueslati, R. Potential preventive role of lactic acid bacteria against Aflatoxin M1 immunotoxicity and genotoxicity in mice. *J. Immunotoxicol.* **2015**, *12*, 107–114. [CrossRef] [PubMed]

129. Begley, M.; Gahan, C.G.M.; Hill, C. The interaction between bacteria and bile. *FEMS Microbiol. Rev.* **2005**, *29*, 625–651. [CrossRef]

130. Zhao, L.; Wei, J.; Zhao, H.; Zhu, B.; Zhang, B. Detoxification of cancerogenic compounds by lactic acid bacteria strains. *Crit. Rev. Food Sci. Nutr.* **2018**, *58*, 2727–2742.

131. Aziz, N.H.; Moussa, L.A.A. Influence of gamma-radiation on mycotoxin producing moulds and mycotoxins in fruits. *Food Control* **2002**, *13*, 281–288. [CrossRef]

132. Arzandeh, S.; Jinap, S. Effect of initial aflatoxin concentration, heating time and roasting temperature on aflatoxin reduction in contaminated peanuts and process optimisation using response surface modelling. *Int. J. Food Sci. Technol.* **2011**, *46*, 485–491. [CrossRef]

133. Lee, L.S.; Cucullu, A.F. Conversion of Aflatoxin B 1 to Aflatoxin D 1 in Ammoniated Peanut and Cottonseed Meals. *J. Agric. Food Chem.* **1978**, *26*, 881–884. [CrossRef] [PubMed]

134. Norred, W.P.; Morrissey, R.E. Effects of long-term feeding of ammoniated, aflatoxin-contaminated corn to Fischer 344 rats. *Toxicol. Appl. Pharmacol.* **1983**, *70*, 96–104. [CrossRef]

135. Rickman, J.C.; Barret, D.M.; Bruhn, C.M. Effect of different ozone treatments on aflatoxin degradation and physicochemical properties of pistachios. *J. Sci. Food Agric.* **2007**, *87*, 940–944.

136. Zhu, Y.; Hassan, Y.I.; Watts, C.; Zhou, T. Innovative technologies for the mitigation of mycotoxins in animal feed and ingredients-A review of recent patents. *Anim. Feed Sci. Technol.* **2016**, *216*, 19–29. [CrossRef]

137. Wang, L.; Wu, J.; Liu, Z.; Shi, Y.; Liu, J.; Xu, X.; Hao, S.; Mu, P.; Deng, F.; Deng, Y. Aflatoxin B1 Degradation and Detoxification by Escherichia coli CG1061 Isolated from Chicken Cecum. *Front. Pharmacol.* **2019**, *9*, 1–9. [CrossRef] [PubMed]

138. Sánchez, B.; Delgado, S.; Blanco-Míguez, A.; Lourenço, A.; Gueimonde, M.; Margolles, A. Probiotics, gut microbiota, and their influence on host health and disease. *Mol. Nutr. Food Res.* **2017**, *61*, 1–15. [CrossRef]

139. Santiago-López, L.; Hernández-Mendoza, A.; Mata-Haro, V.; Vallejo-Córdoba, B.; Wall-Medrano, A.; Astiazarán-García, H.; Estrada-Montoya, M.; González-Córdova, A. Effect of Milk Fermented with Lactobacillus fermentum on the Inflammatory Response in Mice. *Nutrients* **2018**, *10*, 1039. [CrossRef] [PubMed]

140. Giraffa, G.; Chanishvili, N.; Widyastuti, Y. Importance of lactobacilli in food and feed biotechnology. *Res. Microbiol.* **2010**, *161*, 480–487. [CrossRef]

141. Swain, M.R.; Anandharaj, M.; Ray, R.C.; Parveen Rani, R. Fermented Fruits and Vegetables of Asia: A Potential Source of Probiotics. *Biotechnol. Res. Int.* **2014**, *2014*, 1–19. [CrossRef]

142. Martin, N.H.; Boor, K.J.; Wiedmann, M. Symposium review: Effect of post-pasteurization contamination on fluid milk quality. *J. Dairy Sci.* **2017**, *101*, 861–870. [CrossRef]

143. Zeinab, A.M.; Elgadi, W.S.A.G.; Dirar, H.A. Isolation and Identification of Lactic Acid Bacteria and Yeast from Raw Milk in Khartoum State (Sudan). *Res. J. Microbiol.* **2008**, *3*, 163–168.

144. Muir, D.D. *The Stability and Shelf Life of Milk and Milk Products*; Woodhead Publishing Limited: Sawston, UK, 2011; ISBN 9781845697013.

145. Marrez, D.A.; Shahy, E.M.; El-Sayed, H.S.; Sultan, Y.Y. Detoxification of Aflatoxin B1 in milk using lactic acid bacteria. *J. Biol. Sci.* **2018**, *18*, 144–151. [CrossRef]

146. Foroughi, M.; Sarabi Jamab, M.; Keramat, J.; Foroughi, M. Immobilization of Saccharomyces cerevisiae on Perlite Beads for the Decontamination of Aflatoxin M1 in Milk. *J. Food Sci.* **2018**, *83*, 2008–2013. [CrossRef]

147. Gu, H.; Ren, D. Materials and surface engineering to control bacterial adhesion and biofilm formation: A review of recent advances. *Front. Chem. Sci. Eng.* **2014**, *8*, 20–33. [CrossRef]

148. Brackett, R.E.; Marth, E.H. Association of Aflatoxin M1 with Casein. *Z. Leb. Unters. Forsch.* **1982**, *174*, 439–441. [CrossRef]

149. Cattaneo, T.M.P.; Marinoni, L.; Iametti, S.; Monti, L. Behavior of Aflatoxin M1 in dairy wastes subjected to different technological treatments: Ricotta cheese production, ultrafiltration and spray-drying. *Food Control* **2013**, *32*, 77–82. [CrossRef]
150. Granados-Chinchilla, F. Insights into the Interaction of Milk and Dairy Proteins with Aflatoxin M1. In *Milk Proteins—From Structure to Biological Properties and Health Aspects*; InTechOpen: London, UK, 2016; pp. 265–286.

© 2019 by the authors. Licensee MDPI, Basel, Switzerland. This article is an open access article distributed under the terms and conditions of the Creative Commons Attribution (CC BY) license (http://creativecommons.org/licenses/by/4.0/).

 *toxins*

Article

Efficacy of the Combined Protective Cultures of *Penicillium chrysogenum* and *Debaryomyces hansenii* for the Control of Ochratoxin A Hazard in Dry-Cured Ham

Eva Cebrián, Mar Rodríguez *, Belén Peromingo, Elena Bermúdez and Félix Núñez

Food Hygiene and Safety, Meat and Meat Products Research Institute, Faculty of Veterinary Science, University of Extremadura, Avda. de las Ciencias, s/n, 10003 Cáceres, Spain; evcebrianc@unex.es (E.C.); belenperomingo@unex.es (B.P.); bermudez@unex.es (E.B.); fnunez@unex.es (F.N.)
* Correspondence: marrodri@unex.es

Received: 1 November 2019; Accepted: 3 December 2019; Published: 5 December 2019

Abstract: The ecological conditions during the ripening of dry-cured ham favour the development of moulds on its surface, being frequently the presence of *Penicillium nordicum*, a producer of ochratoxin A (OTA). Biocontrol using moulds and yeasts usually found in dry-cured ham is a promising strategy to minimize this hazard. The aim of this work is to evaluate the effect of previously selected *Debaryomyces hansenii* and *Penicillium chrysogenum* strains on growth, OTA production, and relative expression of genes involved in the OTA biosynthesis by *P. nordicum*. *P. nordicum* was inoculated against the protective cultures individually and combined on dry-cured ham for 21 days at 20 °C. None of the treatments reduced the growth of *P. nordicum*, but all of them decreased OTA concentration. The lower production of OTA could be related to significant repression of the relative expression of *otapks*PN and *otanps*PN genes of *P. nordicum*. The efficacy of the combined protective cultures was tested in 24 dry-cured hams in industrial ripening (an 8 month-long production). OTA was detected in nine of the 12 dry-cured hams in the batch inoculated only with *P. nordicum*. However, in the batch inoculated with both *P. nordicum* and the combined protective culture, a considerable reduction of OTA contamination was observed. In conclusion, although the efficacy of individual use *P. chrysogenum* is great, the combination with *D. hansenii* enhances its antifungal activity and could be proposed as a mixed protective culture to control the hazard of the presence of OTA in dry-cured ham.

Keywords: *Penicillium nordicum*; biocontrol agents; dry-cured ham; ochratoxin A (OTA)

Key Contribution: The use of a combined protective culture containing selected strains of *P. chrysogenum* CECT 20922 in combination with *D. hansenii* FHSCC 253H could be a promising strategy to reduce OTA production by *P. nordicum* in dry-cured hams.

---

1. Introduction

The ecological conditions reached during the ripening of dry-cured ham favour the growth of moulds on their surface, mainly belonging to the genera *Penicillium* and *Aspergillus* [1,2]. The presence of moulds is valuable due to their contribution to proteolytic and lipolytic changes observed during the ripening [3], and consequently, to the development of required sensory characteristics of dry-cured ham [4,5]. However, most of the moulds isolated in these products are potentially toxigenic [1,6,7].

Ochratoxin A (OTA) is the mycotoxin most commonly found in meat products, which contribute significantly to human exposure to this toxin [8,9]. OTA has nephrotoxic, immunosuppressive, genotoxic, carcinogenic, teratogenic and neurotoxic effects [10,11] and, it has been classified by the International Agency for Research on Cancer (IARC) as a possible human carcinogen in the 2B category [12].

The presence of OTA in dry-cured ham could be a consequence of the growth of *Aspergillus westerdijkiae* [13], *Penicillium verrucosum* and *Penicillium nordicum* [2,14–16]. Among them, *P. nordicum* is the most common ochratoxigenic mould isolated from dry-cured meats [2,15,17], due to its special adaptation to protein and NaCl-rich foods [2,16,18,19].

Due to the risk to human health, the European Union has set the maximum level permissible of OTA for different foodstuffs such as cereals, wine, fruits, coffee or baby foods, although not in meat products [20]. In addition, Italy has established the maximum guide value for OTA in pork and pork products at 1 µg/kg [21]. Therefore, it is important to implement strategies to control the growth of ochratoxigenic moulds and the production of this mycotoxin.

There are several effective antifungal treatments, including physical methods and chemical preservatives, but most of them are not adequate for mould-ripened foods, such as dry-cured ham, since they are non-selective and can interfere with the ripening, altering the sensory characteristics [22]. Thus, the use of microorganisms usually found in this type of foods as bioprotective agents has been proposed as a promising strategy to control toxigenic moulds in dry-cured meat products [23–32].

Yeasts are interesting as antagonistic microorganisms since they are part of the natural microbiota of dry-cured meat products, are phenotypically adapted to this environment and contribute to the development of sensory characteristics [4,33–35]. *Debaryomyces hansenii* is the predominant yeast species throughout the processing of dry-cured meat products [35,36], and some autochthonous strains have been proposed as biological control of OTA in these foods [24,27–32,37]. In this sense, *D. hansenii* FHSCC 253H isolated from dry-cured ham is able to reduce the production of OTA by *P. verrucosum* on this food [30]. Moreover, the species *D. hansenii* has been included in the European Qualified Presumption of Safety (QPS) list as a safe microorganism.

On the other hand, non-toxigenic moulds producers of antifungal proteins could be considered for their use as protective cultures in dry-cured meat products [38]. These antifungal proteins probably give producer moulds a selective advantage to compete with other moulds in its ecological niche [38]. In this sense, *Penicillium chrysogenum* CECT 20922, isolated from dry-cured ham and producer of the antifungal protein PgAFP [39], is able to reduce OTA contamination in dry-cured ham [40]. The protective effect of *D. hansenii* against OTA-producer moulds in meat substrates has been attributed to a combination of competition for space and nutrients, production of soluble metabolites [28], and interfering the secondary metabolism [41], mainly by the reduction relative expression of the OTA biosynthetic genes [30]. Meanwhile, the preventive activity of *P. chrysogenum* has been linked to the production of antifungal protein PgAFP [40], nutrient competition and hampering the secondary metabolism of *P. nordicum* [41]. Then, the combined use of these bicontrol agents on dry-cured ham could have an additive or synergistic effect against the growth of *P. nordicum* and OTA contamination, but this fact has not been demonstrated yet. Therefore, it is necessary to evaluate the implantation of these potential protective microorganisms in combined cultures in the presence of toxigenic moulds in dry-cured ham during the ripening process. In addition, the effect of these biocontrol agents over toxigenic mould growth and accumulation of mycotoxins should be also evaluated.

Although the biosynthetic pathway of OTA has not been fully clarified, two genes have been described as required by *P. nordicum* for the biosynthesis of OTA, the polyketide synthetase gene (*otapksPN*), and the non-ribosomal peptide synthetase gene (*otanpsPN*) [42]. Therefore, these two genes can be used as a target for the study of gene expression in *P. nordicum* [43]. Finally, among the genes of interest for endogenous control is the gene *β-tubulin* that encodes a globular protein that is the main component of the microtubules of fungal cells [7,44].

The main objective of the present work is to evaluate the effect of *D. hansenii* FHSCC 253H and *P. chrysogenum* CECT 20922 as a combined protective culture in the growth of *P. nordicum* FHSCC IB4, the OTA production and the expression of genes involved in the biosynthetic pathway of this mycotoxin in dry-cured ham.

2. Results

2.1. Effect of Bioprotective Agents on P. nordicum Growth in Dry-Cured Ham

The growth of toxigenic mould and bioprotective agents inoculated in dry-cured ham was evaluated by plate counting in MEA.

All microorganisms were able to grow properly on dry-cured ham, reaching counts higher than 10^6 cfu/cm^2 at the end of incubation in each batch. While the presence of each bioprotective agents separately did not affect *P. nordicum* counts, their use in combination produced a significant reduction ($p \leq 0.01$) in the growth of the toxigenic mould (Table 1).

Table 1. Growth of *P. nordicum*, *P. chrysogenum* and *D. hansenii* (log cfu/cm^2) in dry-cured ham after 21 days at 20 °C.

Batches	*P. nordicum*	*P. chrysogenum*	*D. hansenii*
Control	7.02 ± 0.14	-	-
D. hansenii	7.41 ± 0.38	-	8.28 ± 0.18
P. chrysogenum	7.25 ± 0.42	6.50 ± 0.18	-
D. hansenii + *P. chrysogenum*	6.71 ± 0.36*	7.19 ± 0.26	8.36 ± 0.03

-: growth < 1 log cfu/cm^2. * Significant differences with respect to control ($p \leq 0.01$).

2.2. Influence of Bioprotective Agents on OTA Production by P. nordicum

Table 2 shows the effect of the bioprotective agents on the production of OTA by *P. nordicum* in dry-cured ham after 21 days of incubation at 20 °C.

Table 2. Effect of *D. hansenii* and *P. chrysogenum* on OTA production by *P. nordicum* in dry-cured ham after 21 days at 20 °C.

Batches	OTA Concentration (μg/kg)*
Control	15.48 ± 8.87[a]
D. hansenii	2.78 ± 2.18[b]
P. chrysogenum	0.44 ± 0.63[bc]
D. hansenii + *P. chrysogenum*	0.23 ± 0.52[c]

* Values of OTA concentration followed by different letters are significantly different ($p < 0.05$).

The amount of OTA was significantly reduced ($p \leq 0.05$) when *P. nordicum* was inoculated with any of the bioprotective cultures individually as well as it was inoculated with both combined. The lower concentration of OTA was observed in the batches when *P. chrysogenum* was inoculated combined with *D. hansenii* (98.5%). Although no significant differences were detected between the individually inoculated batches, there was a greater reduction in the mean of OTA amount in the batch inoculated with *P. chrysogenum* (97.2%) than with *D. hanseni* (82%).

2.3. Effect of Bioprotective Agents in the Relative Expression of otapksPN and otanpsPN Genes in P. nordicum

The effect of bioprotective agents on the biosynthesis of OTA was assessed by studying the relative expression of the *otanpsPN* and *otapksPN* genes by *P. nordicum* over a 21-day incubation period in dry-cured ham (Figure 1). The threshold cycle (Ct) values of *β-tubulin* genes detected in qPCR analyses were always between 21 and 26 after 21 days of incubation, which is indicative that the moulds continued in active growth. The relative expression of the *otapksPN* genes was significantly reduced by both tested microorganisms. For *otanpsPN* a reduction of the expression was observed in the batches inoculated with *P. chrysogenum*. In the same way as in the production of OTA, the highest inhibition level of both gene expressions was observed when *P. chrysogenum* was inoculated individually or in combination with *D. hansenii*.

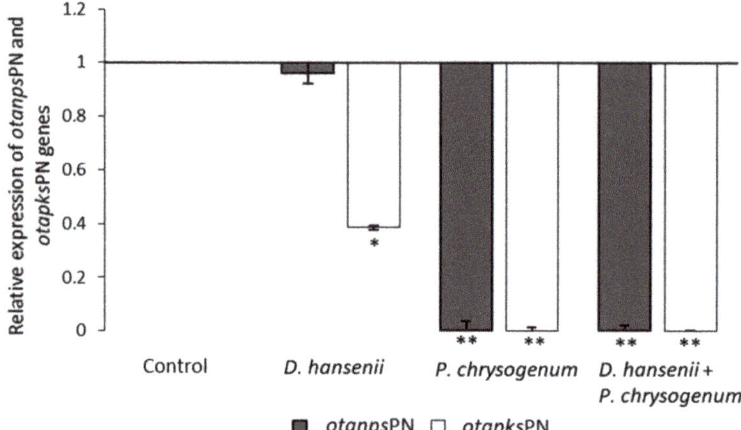

Figure 1. Relative expression of the *otanps*PN and *otapks*PN genes in *P. nordicum* inoculated in dry-cured ham without bioprotective cultures (control) with *D. hansenii* and *P. chrysogenum* individually, or both combined after incubation for 21 days at 20 °C. Significant differences of the gene expression of each batch compared to control are indicated by asterisk: * $p \leq 0.05$; ** $p \leq 0.01$.

2.4. Effect of Combined Protective Cultures on OTA Contamination in Dry-Cured Ham Inoculated with P. nordicum in Small Scale Manufacture

Table 3 shows the results of OTA concentration at the end of the processing of the 24 dry-cured hams. In the control batch, 9 of 12 hams were contaminated with OTA, and 6 of them showed OTA levels ranged from 7.71 to 1620 µg/kg, above the guide value permitted for pork meat and derived products in Italy [21]. However, the amount of OTA was substantially lower in the batch inoculated with the combined bioprotective culture. In this batch, 8 of 12 samples OTA was no detected (<0.58 ppb), and the 4 OTA-contaminated hams did not exceed levels of 11.73 µg/kg.

Table 3. Levels of ochratoxin A (OTA) detected in samples of dry-cured ham.

Batches	Sample Reference of Dry-Cured Ham	Average OTA Concentration (µg/kg)
	1	0.39
	2	7.71
	3	1226.92
	4	-
	5	1275.55
P. nordicum	6	176.86
	7	-
	8	-
	9	0.58
	10	284.13
	11	1620.00
	12	0.67

Table 3. Cont.

Batches	Sample Reference of Dry-Cured Ham	Average OTA Concentration (µg/kg)
	13	11.73
	14	3.67
	15	-
	16	-
	17	9.68
P. nordicum + *P. chrysogenum*	18	-
+ *D. hansenii*	19	-
	20	-
	21	-
	22	-
	23	-
	24	6.48

-: levels less than the limit of detection.

3. Discussion

In this work, the use of *D. hansenii* and *P. chrysogenum* strains isolated from dry-cured ham as biocontrol agents were evaluated. Although both strains used had proved their ability to reduce the concentration of OTA on dry-cured meat products individually, their potential synergistic effect had not yet been tested.

Both bioprotective agents, whether inoculated individually or combined, grew satisfactorily in the presence of *P. nordicum* on dry-cured ham in environmental conditions similar to those reached during the usual processing (Table 1). However, despite this strong growth, only when the bioprotective agents were used in combination, a statistically significant decrease in *P. nordicum* counts was observed. However, even in the batch inoculated with *D. hansenii* and *P. chrysogenum*, the count reached by *P. nordicum* was above the level estimated to be required for the accumulation of OTA in dry-cured ham [15]. This same strain of *D. hansenii* also failed to decrease the growth of the *P. verrucosum* producer of OTA in dry-cured ham incubated in similar conditions to those described in the present work [30]. However, in different culture media, this strain was able to delay the germination of *P. nordicum* spores [24] and decrease the growth rate of *P. verrucosum* [30] in the range of 0.90–0.97 a_w. This antifungal activity has been linked mainly to the competition for nutrients and the production of yeast metabolites [28]. The discrepancies in the antifungal activity of *D. hansenii* between that studies can likely be attributed to the different moulds tested [28] and the diverse environmental conditions, such as substrate or a_w [24,31].

Regarding *P. chrysogenum* CECT 20922, its inoculation did not have any noticeable influence on the total count of ochratoxigenic moulds observed throughout the ripening of dry-cured hams [40]. On the contrary, this *P. chrysogenum* strain reduced the levels of aflatoxigenic *Aspergillus flavus* in dry-cured ham [45] to non-detectable levels. Given that the antagonistic activity of *P. chrysogenum* CECT 20922 has been related with the production of the antifungal protein PgAFP [40], the differences in efficiency against different moulds in a similar substrate may be due to the greater sensitivity of *A. flavus* than OTA-producing moulds, such as *P. nordicum* and *P. verrucosum*, to PgAFP [46].

Concerning the production of OTA by *P. nordicum*, both bioprotective agents tested, used individually or combined, were effective in decreasing the concentration of this mycotoxin. However, in the batch inoculated with *D. hansenii*, the reduction of OTA amount is insufficient to meet the Italian guideline value of 1 µg/kg for pork derived products [21]. The maximum reduction was obtained with *P. chrysogenum* alone or in combination with *D. hansenii*. In these batches, the concentration of OTA is lower than 1 µg/kg, and they may be regarded as safe according to Italian regulation [21]. In addition, the study carried out in the dry-cured hams confirms the antifungal effect of the combined culture of *P. chrysogenum* and *D. hansenii*. In the batch inoculated with both protective microorganisms, OTA was detected just in 33% of dry-cured hams, whereas in the batch inoculated only with *P. nordicum*, 75% of hams were found to contain OTA (Table 3). In addition, in the four contaminated hams treated with protective cultures, the average level of OTA (7.89 µg/kg) was considerably lower than that of the nine contaminated hams from control batch inoculated only with *P. nordicum* (510.31 µg/kg). The differences found between the samples from the control batch can be attributed to the high heterogeneity in mould growth and mycotoxin distribution among different parts of solid foods. Since moulds often grow in certain spots on the food surface, it can be detected the presence of high contamination levels in a relatively confined or small part of a food [47].

Different strains of *D. hansenii* have demonstrated their ability to significantly decrease the production of OTA by *A. westerdijkiae* [37] and *P. nordicum* [32] in standard culture media and to completely inhibit the production of OTA by *P. nordicum* and *A. ochraceus* in dry-cured ham without altering its organoleptic properties [27]. In addition, the strain *D. hansenii* FHSCC 253H used in this work has previously shown its efficacy in reducing the production of OTA of *P. verrucosum* and *P. nordicum* [24,30] and aflatoxins B1 and G1 by *Aspergillus parasiticus* [29] in dry-fermented sausages and dry-cured ham.

In the same way, some non-toxigenic moulds have shown their capability to significantly decrease the presence of OTA in dry-cured meat products processed in meat industries, including *Penicillium nalgiovense* [48] and the strain of *P. chrysogenum* used in this work [40].

Although mould growth should not be considered the only parameter to predict OTA contamination in foods [16,49], a positive correlation has been described between the growth of ochratoxigenic moulds and the production of OTA on meat substrates [30]. However, considering the limited effect observed on the growth of *P. nordicum* (Table 1), it can be inferred that the recorded drastic decrease in OTA (Table 2) is not due to the influence of bioprotective agents on toxigenic mould development.

Among the mechanisms of action of yeasts to decrease the content of mycotoxins, the enzymatic degradation, the adsorption to the cell wall or the repression of the biosynthesis route has been described [22]. However, the *D. hansenii* strain tested does not degrade or adsorb OTA nor aflatoxins, but reduces the synthesis of these toxins by *P. verrucosum* [30] and *A. parasiticus* [29], respectively, in meat substrates. As in the present study, this reduction occurs without growth inhibition, and it was attributed to a blockage in the expression of genes related to the synthesis of mycotoxins [29,30]. The blockage of OTA-related genes was described in culture media against *P. verrucosum* [30], but this mechanism has not been proved with *P. nordicum* in dry-cured ham.

Then, to confirm this mechanism of action involved in the reduction of OTA in dry-cured ham, the effect of the protective agents on the expression of the *otapks*PN and *otanps*PN genes involved in the OTA biosynthetic pathway was evaluated. The expression of these two genes was repressed when *D. hansenii* and *P. chrysogenum* were inoculated both individually and combined against *P. nordicum*. In relation to *D. hansenii*, the *otapks*PN gene was more repressed than the *otanps*PN gene, obtaining 61% and 4% repression, respectively. This repression is consistent with studies that demonstrated that *D. hansenii* is able to reduce expression of the *pks* and *p450-B03* genes involved in the OTA production by *A. westerdijkiae* [37] and to inhibit the expression of the *otanps*PN gene of *P. verrucosum* in meat-based culture media [30].

With respect to *P. chrysogenum*, the repression was very similar in both *otapks* (99.99%) and *otanps* genes (99.62%). This repression could be complementary to the inhibition of the carbon catabolite repression (CCR) pathway as the consequence of the nutritional competition between *P. nordicum* and *P. chrysogenum* in meat substrates, evidenced by proteomic analysis [41]. This CCR pathway has been linked to secondary metabolism and with mycotoxins production [50]. These mechanisms may explain the effect of *P. chrysogenum* decreasing the OTA production by *P. nordicum* without affect the growth. However, *P. nordicum* showed the capacity to grow when it was co-inoculated with the bioprotective agents used in this work on dry-cured ham under the environmental conditions usually found through the ripening of this product. Nevertheless, the real hazard of *P. nordicum* is the production of OTA, and this was reduced in the presence of *D. hansenii* and *P. chrysogenum* used either individually or combined. This reduction is related to the repression of the *otapks*PN and *otanps*PN genes involved in the biosynthetic pathway of this mycotoxin. In addition, the efficacy of the bioprotective agents against the production of OTA by *P. nordicum*, on dry-cured hams processed in the usual environmental conditions set in the meat industries has been demonstrated. In conclusion, the use of *D. hansenii* FHSCC 253H and *P. chrysogenum* CECT 20922 could be proposed as an innovative combined protective culture for the prevention of OTA hazard in dry-cured ham.

4. Materials and Methods

4.1. Microorganisms

In this study the OTA-producer *P. nordicum* FHSCC IB4, the antifungal *D. hansenii* FHSCC 253H, both from the the Food Hygiene and Safety Culture Collection at the University of Extremadura (Cáceres, Spain), and the PgAFP-producer *P. chrysogenum* CECT 20922, from the Spanish Type Culture Collection (Valencia, Spain) were used. All three strains were isolated from dry-cured ham [16,28,38]. The ability of *P. nordicum* to produce OTA [16] and the antifungal activity of *D. hansenii* [24,28–30] and *P. chrysogenum* [40,41] have been previously assessed.

4.2. Preparation of Mould and Yeast Inocula

P. nordicum and *P. chrysogenum* were grown on potato dextrose agar (PDA) (Scharlab S.L., Sentmenat, Spain) at 25 °C for 7 days. Conidia were collected by adding 4 mL of saline phosphate buffer (PBS) and scraping the surface with a sterile glass rod. The spores were quantified by using a Thoma counting chamber Blaubrand® (Brand, Bremen, Germany), and adjusted to 10^5 spores/mL and used as inoculum.

The inocula of *D. hansenii* was obtained by inoculating 100 µL from a culture preserved at −80 °C in YES broth (Scharlab S.L., Sentmenat, Spain). It was incubated for 72 h at 150 rpm and 25 °C. The culture was centrifuged at 3500 rpm for 5 min, and the concentrated cells were resuspended in PBS. The yeast suspensions were counted by using the Thoma chamber and adjusted to 10^6 yeasts/mL to be used as inoculum.

4.3. Preparation of the Dry-Cured Ham Portions

Pieces of dry-cured ham were cut into cubes of approximately $3 \times 3 \times 3$ cm. To reduce microbial contamination of pieces, they were firstly submerged in ethanol for 30 s, and the surface was disinfected in a laminar flow hood with ultraviolet light for 2 h. After this time, each of the blocks was immersed in sterile distilled water for few seconds to be rehydrated. Then, the a_w of pieces was 0.88.

4.4. Experimental Setting

Four different batches were prepared: the control batch, inoculated only with *P. nordicum*, one batch inoculated with *P. nordicum* and *D. hansenii*, one batch inoculated with *P. nordicum* and *P. chrysogenum*, and one batch inoculated with *P. nordicum*, *D. hansenii* and *P. chrysogenum*.

Fifty µL of each microorganism was inoculated on the surface of dry-cured ham using a Drigalski spreader previously sterilized. When the piece of ham was inoculated with several microorganisms, we waited 10 min between each strain inoculation to allow the absorption.

The inoculated dry-cured ham cubes were incubated at 20 °C for 21 days in plastic containers previously sterilized. To simulate the dehydration that occurs during the ripening of dry-cured ham, a relative humidity of 94% was maintained inside the container by depositing 400 mL of oversaturated K_2SO_4 solution at the bottom of the containers. The experiment was carried out in quintupled.

4.5. Evaluation of Mould and Yeast Growth

At the end of incubation time, an area of approximately 5 cm^2 of the surface of each dry-cured ham piece was homogenized with 45 mL of 0.1% peptone water (Panreac Quimica S.L.U., Barcelona, Spain) in a filter bag BagPage (Interscience, Saint Nom, France) using a Stomacher (Stomacher® 400 Circulator, Seward, Worthing, UK). After performing decimal dilutions, the growth of the inoculated microorganisms was determined by plating in malt extract agar (MEA, Scharlab S.L., Sentmenat, Spain), and incubating at 25 °C for 7 days. The colonies of each strain were distinguished by their morphology, even the two species of mould presented clear differences in colour and size. *P. nordicum* showed a whitish colour with smaller colonies, while *P. chrysogenum* colonies were green and larger. The sampling was carried out in duplicate.

4.6. Extraction and Quantification of Ochratoxin A

4.6.1. OTA Extraction

OTA was extracted following the QuEChERS methodology [51] with modifications. Briefly, 3 g from the surface of each dry-cured ham cube was mixed with 2 mL of water acidified with 0.1% acetic acid and shaken for 30 s. Then, 2 mL acetonitrile acidified with 0.1% acetic acid was added and shaken for 1 min. Thereupon, 0.4 g NaCl and 1.6 g MgSO$_4$ were added and shaken manually for 30 s. The samples were centrifuged for 5 min at 5000 rpm, and an aliquot of 1 mL supernatant was filtered through a 0.45 μm pore size nylon membrane (MSI, Omaha, NE, USA), placed in a vial and analyzed by HPLC-FLD.

4.6.2. OTA Quantification

OTA quantification was performed in Agilent 1260 Infinity (Agilent Technologies, Santa Clara, CA, USA) equipment coupled to a fluorescence (FLD) detector (Agilent Technologies, Santa Clara, CA, USA). The column Zorbax SB C18 (2.1 mm × 50 cm × 1.8 μm; Agilent Technologies, Santa Clara, CA, USA) was used. The mobile phase was acetonitrile:water:acetic acid (50:100:1 *v/v/v*) with a flow rate of 0.1 mL/min. FLD detection was performed using 330 and 450 nm excitation and emission wavelengths, respectively. The calibration curve of OTA by HPLC-FLD revealed a linear relationship ($r^2 \geq 0.99$) between detector response and OTA standard (Sigma-Aldrich Co., San Luis, MO, USA) amounts between 0.5 and 50 ng/mL. The limit of detection (LOD) obtained in this study was 0.58 μg/kg, and the quantification limit (LOQ) was 1.94 μg/kg.

4.7. Relative Gene Expression

4.7.1. RNA Extraction

The mycelia were removed from half the surface of the dry-cured ham cubes and frozen quickly in liquid nitrogen, it was stored at −80 °C until the extraction of RNA. RNA was extracted using the commercial kit RNeasy® plant mini kit (QIAGEN, Madrid, Spain) [52]. For this, the mycelia were mixed with 500 μL of lysis solution buffer containing 10 μL of β-mercaptoethanol and then processed according to the manufacturer's instructions. Once the process was completed, the RNA obtained was diluted in 50 μL of RNAse-free water. The RNA samples obtained were treated with the commercial kit DNase I, RNase-free (Thermo Fisher Scientific, Waltham, MA, USA), to eliminate the possible residual genomic DNA. Finally, the concentration and purity (A_{260}/A_{280} ratio) of the RNA obtained was measured in a NanoDrop 2000c spectrophotometer (Thermo Fisher Scientific, Waltham, MA, USA).

4.7.2. Reverse Transcription Reaction

For the synthesis of the cDNA from the RNA obtained the commercial kit "PrimeScript™ RT Reagent" (Takara Bio Inc., Kusatsu, Japan) was used following the manufacturer's indications. The cDNA synthesis reaction was performed on a Mastercycler thermocycler (Eppendorf AG, Hamburg,

Germany) using the following conditions: 15 min at 37 °C; 5 s at 85 °C; and finally, a cooling to 4 °C. The resulting cDNA was stored at −20 °C until use.

4.7.3. RT-qPCR

Primers Used

Two primers (Table 4) were used to amplify the *β-tubulin* gene, which was used as an endogenous control [52]. For the relative expression of the genes *otapks*PN and *otanps*PN involved in the biosynthesis of OTA, two pairs of primers were used [19,52].

Table 4. Oligonucleotide sequences of primers used in this study.

Genes	Primers	Nucleotide Sequences (5′-3′)	Product Size (pb)	Positions
β-tubulin	β-tubF1	GCCAGCGGTGACAAGTACGT	93	279 [a]
	β-tubR1	TACCGGGCTCCAAATCGA		54 [a]
*otapks*PN	otapksF3	CGCCGCTGCGGTTACT	80	1816 [b]
	otapksR3	GGTAACAATCAACGCTCCCTCTT		1873 [b]
*otanps*PN	F-npstr	GCCGCCCTCTGTCATTCCAAG	113	5090 [b]
	R-npstr	GCCATCTCCAAACTCAAGCGTG		5181 [b]

[a] Positions are in accordance with the published sequence of *β-tubulin* gene of *P. nordicum* (GenBank accession no. AY674319.1). [b] Positions are in accordance with the published sequence of the *otapks* and *otanps* genes of *P. nordicum* (GenBank accession no. AY557343.2).

RT-qPCR Methods

The qPCR was carried out in the Applied Biosystems 7500 Fast real-time PCR system (Applied Biosystems, Waltham, MA, USA) and qPCR methods based on the SYBR Green methodology were used [45,53]. They were prepared in triplicates in MicroAmp optical 96-well plates which a total volume of 12.5 µL was added to each of them, consisting of 6.25 µl 2× SYBR® Premix Ex Taq™, 0.5 µL of 50 ROX™ reference dye, 2.5 µL of cDNA and 300 nM of each primer for the *otapks*PN, *otanps*PN and *β-tubulin* gene. In addition, three negative controls were incorporated without the target cDNA sequence. Finally, the plates were sealed with optical adhesive covers (Applied Biosystems, Waltham, MA, USA). The amplification conditions used for the qPCR reactions were: 1 cycle at 95 °C for 10 min, 40 cycles at 95 °C for 15 s and 60 °C for 1 min. After the final PCR cycle, the melting curve analysis of the PCR products was performed by heating to 60–95 °C and continuous measurement of the fluorescence to verify the PCR product.

The relative quantification of the expression of the *otapks*PN and *otanps*PN genes was performed using the housekeeping *β-tubulin* gene as an endogenous control to normalize the quantification of the mRNA target for differences in the amount of total cDNA added to each reaction. The expression ratio was calculated using the $2^{-\Delta\Delta CT}$ method [54]. The value of the relative expression of the control batch inoculated with *P. nordicum* without the presence of the other microorganisms was used as a calibrator.

4.8. Effect of Combined Protective Culture on OTA Contamination During Ripening of Dry-Cured Ham in Small Scale Manufacture.

The protective effect of the combined biocontrol agents *P. chrysogenum* and *D. hansenii* was tested in dry-cured ham ripened under the environmental conditions usually set in industrial ripening (an 8 month-long production).

A total of 24 dry-cured hams in the final post-salting stage (2 months of ripening approximately) were divided in two batches: 12 dry-cured hams were inoculated with only *P. nordicum* (control batch), and the other 12 dry-cured hams were inoculated with *P. nordicum* and the combined protective culture (*P. chrysogenum* and *D. hansenii*).

The inocula of both moulds and *D. hansenii* were prepared in the same way as in Section 4.2.

For the inoculation, each piece was immersed for 30 s in 25 liters solution with *P. nordicum* for the control batch, and with *P. nordicum, P. chrysogenum* and *D. hansenii* for the other batch. The dry-cured hams were hung up and ripened in two different maturation chambers for 6 months. During the first month, the temperature progressively increased from 4 to 15 °C and the relative humidity (RH) decreased from 80% to 75%. From the second to sixth month of ripening, the temperature was gently increased from 15 to 22 °C, and the RH was lowered below 70% at the end of the maturing. At the end of the processing, the dry-cured hams lost approximately 35% of their initial weight.

Sampling, Extraction and Quantification of OTA

Sampling was done at the end of ripening (an 8 month-long production). For that purpose, 25 cm^2 surface areas showing fungal growth and 1 cm of depth were obtained of each dry-cured ham. The procedure of extraction and quantification of OTA was carried out as described above in Section 4.6.

4.9. Statistical Analysis

All data were analyzed by IBM SPSS v.22 (Chicago, IL, USA, 2013). The data normality test was performed by Shapiro–Wilks. All data sets failed the normality test; a variable transformation was performed to improve normality or homogenize the variances. Without success, the analysis of non-parametric data was performed using the Kruskal–Wallis test to determine significant differences between the means. Subsequently, those samples that differed had the U Mann–Whitney test applied to compare the mean values obtained. Statistical significance was set at $p \leq 0.05$.

Author Contributions: The authors F.N and M.R. conceived, proposed the idea and designed the study. E.C. and B.P. performed the experiments and analyzed the data. E.C. and F.N. writing—original draft preparation. E.B., M.R. and F.N. writing—review and editing, F.N and M.R funding acquisition.

Funding: This work has been financed by the Spanish Ministry of Economy and Competitiveness, Government of Extremadura and FEDER (AGL2016-80209-P and GR18056). B. Peromingo are recipient of pre-doctoral fellowships (BES-2014-069484) from the Spanish Ministry of Economy and Competitiveness.

Acknowledgments: The authors acknowledge the technical support pro-vided by the Facility of Innovation and Analysis in Animal Source Foodstuffs (SiPA) of SAIUEx.

Conflicts of Interest: The authors declare no conflict of interest.

References

1. Núñez, F.; Rodríguez, M.M.; Bermúdez, M.E.; Córdoba, J.J.; Asensio, M.A. Composition and toxigenic potential of the mould population on dry-cured Iberian ham. *Int. J. Food Microbiol.* **1996**, *32*, 185–197. [CrossRef]

2. Battilani, P.; Pietri, A.; Giorni, P.; Formenti, S.; Bertuzzi, T.; Toscani, T.; Virgili, R.; Kozakiewicz, Z. *Penicillium* populations in dry-cured ham manufacturing plants. *J. Food Prot.* **2007**, *70*, 975–980. [CrossRef] [PubMed]

3. Martín, A.; Córdoba, J.J.; Núñez, F.; Benito, M.J.; Asensio, M.A. Contribution of a selected fungal population to proteolysis on dry-cured ham. *Int. J. Food Microbiol.* **2004**, *94*, 55–66. [CrossRef] [PubMed]

4. Martín, A.; Córdoba, J.J.; Aranda, E.; Córdoba, M.G.; Asensio, M.A. Contribution of a selected fungal population to the volatile compounds on dry-cured ham. *Int. J. Food Microbiol.* **2006**, *110*, 8–18. [CrossRef] [PubMed]

5. Perrone, G.; Rodriguez, A.; Magistà, D.; Magan, N. Insights into existing and future fungal and mycotoxin contamination of cured meats. *Curr. Opin. Food Sci.* **2019**, *29*, 20–27. [CrossRef]

6. Núñez, F.; Westphal, C.D.; Bermúdez, E.; Asensio, M.A. Production of secondary metabolites by some terverticillate penicillia on carbohydrate-rich and meat substrates. *J. Food Prot.* **2007**, *70*, 2829–2836. [CrossRef]

7. Rodríguez, A.; Rodríguez, M.; Martín, A.; Delgado, J.; Córdoba, J.J. Development of a multiplex real-time PCR to quantify aflatoxin, ochratoxin A and patulin producing molds in foods. *Int. J. Food Microbiol.* **2012**, *155*, 10–18. [CrossRef]

8. Mitchell, N.J.; Chen, C.; Palumbo, J.D.; Bianchini, A.; Cappozzo, J.; Stratton, J.; Ryu, D.; Wu, F. A risk assessment of dietary ochratoxin A in the United States. *Food Chem. Toxicol.* **2017**, *100*, 265–273. [CrossRef]

9. Sirot, V.; Fremy, J.-M.; Leblanc, J.-C. Dietary exposure to mycotoxins and health risk assessment in the second French total diet study. *Food Chem. Toxicol.* **2013**, *52*, 1–11. [CrossRef]

10. Bezerra da Rocha, M.E.; Oliveira Freire, F.C.; Feitosa Maia, E.F.; Florindo Guedes, M.I.; Rondina, D. Mycotoxins and their effects on human and animal health. *Food Control* **2014**, *36*, 159–165. [CrossRef]

11. Malir, F.; Ostry, V.; Pfohl-Leszkowicz, A.; Malir, J.; Toman, J. Ochratoxin A: 50 years of research. *Toxins* **2016**, *8*, E191. [CrossRef] [PubMed]

12. IARC (International Agency for Research on Cancer). OchratoxinA Some Naturally Occurring Substances: Food Items and Constituents, Heterocyclic Aromatic Amines and Mycotoxins. In *Monographs on the Evaluation of Carcinogenic Risk to Humans*; IARC: Lyon, France, 1993; Volume 56, pp. 489–521.

13. Vipotnik, Z.; Rodríguez, A.; Rodrigues, P. *Aspergillus westerdijkiae* as a major ochratoxin A risk in dry-cured ham based-media. *Int. J. Food Microbiol.* **2017**, *241*, 244–251. [CrossRef] [PubMed]

14. Iacumin, L.; Chiesa, L.; Boscolo, D.; Manzano, M.; Cantoni, C.; Orlic, S.; Comi, G. Moulds and ochratoxin A on surfaces of artisanal and industrial dry sausages. *Food Microbiol.* **2009**, *26*, 65–70. [CrossRef]

15. Rodríguez, A.; Rodríguez, M.; Martín, A.; Delgado, J.; Córdoba, J.J. Presence of ochratoxin A on the surface of dry-cured Iberian ham after initial fungal growth in the drying stage. *Meat Sci.* **2012**, *92*, 728–734. [CrossRef] [PubMed]

16. Sánchez-Montero, L.; Córdoba, J.J.; Peromingo, B.; Álvarez, M.; Núñez, F. Effects of environmental conditions and substrate on growth and ochratoxin A production by *Penicillium verrucosum* and *Penicillium nordicum*: Relative risk assessment of OTA in dry-cured meat products. *Food Res. Int.* **2019**, *121*, 604–611. [CrossRef]

17. Sonjak, S.; Ličen, M.; Frisvad, J.C.; Gunde-Cimerman, N. Salting of dry-cured meat. A potential cause of contamination with the ochratoxin A-producing species *Penicillium nordicum*. *Food Microbiol.* **2011**, *28*, 1111–1116. [CrossRef]

18. Delgado, J.; da Cruz Cabral, L.; Rodríguez, M.; Rodríguez, A. Influence of ochratoxin A on adaptation of *Penicillium nordicum* on a NaCl-rich dry-cured ham-based medium. *Int. J. Food Microbiol.* **2018**, *272*, 22–28. [CrossRef]

19. Rodríguez, A.; Medina, Á.; Córdoba, J.J.; Magan, N. The influence of salt (NaCl) on ochratoxin A biosynthetic genes, growth and ochratoxin A production by three strains of *Penicillium nordicum* on a dry-cured ham-based medium. *Int. J. Food Microbiol.* **2014**, *178*, 113–119. [CrossRef]

20. The Commission of the European Communities. Commission Regulation (EC) 1881/2006 of 19 December 2006 setting maximum levels for certain contaminants in foodstuffs. *Off. J. Eur. Union* **2006**, *L364*, 5–24.

21. Ministerio Della Sanità. Circolare 09/06/1999, n. 10. Direttive in materia di controllo ufficiale sui prodotti alimentari: Valori massimi ammissibili di micotossine nelle derrate alimentari di origine nazionale, comunitaria e Paesi terzi. *Gazz. Uff. della Repubb. Ital.* **1999**, *135*, 52–57.

22. Asensio, M.A.; Núñez, F.; Delgado, J.; Bermúdez, E. Control of toxigenic molds in food processing. In *Microbial Food Safety and Preservation Techniques*; Rai, V.R., Bai, A.J., Eds.; CRC Press (Taylor and Francis): Boca Raton, FL, USA, 2014; pp. 329–357.

23. Álvarez, M.; Rodríguez, A.; Peromingo, B.; Núñez, F.; Rodríguez, M. *Enterococcus faecium*: A promising protective culture to control growth of ochratoxigenic moulds and mycotoxin production in dry-fermented sausages. *Mycotoxin Res.* **2019**. [CrossRef]

24. Andrade, M.J.; Thorsen, L.; Rodríguez, A.; Córdoba, J.J.; Jespersen, L. Inhibition of ochratoxigenic moulds by *Debaryomyces hansenii* strains for biopreservation of dry-cured meat products. *Int. J. Food Microbiol.* **2014**, *170*, 70–77. [CrossRef] [PubMed]

25. Cebrián, E.; Núñez, F.; Alía, A.; Bermúdez, E.; Rodríguez, M. Effect of *Staphylococcus xylosus* on the growth of toxigenic moulds in meat substrates. Report of the IVth Workshop of the Spanish National Network on Mycotoxins and Toxigenic Fungi and Their Decontamination Processes (MICOFOOD), Pamplona, Spain, 29–31 May 2019. *Toxins* **2019**, *11*, 415. [CrossRef]

26. Delgado, J.; Peromingo, B.; Rodríguez, A.; Rodríguez, M. Biocontrol of *Penicillium griseofulvum* to reduce cyclopiazonic acid contamination in dry-fermented sausages. *Int. J. Food Microbiol.* **2019**, *293*, 1–6. [CrossRef]

27. Iacumin, L.; Manzano, M.; Andyanto, D.; Comi, G. Biocontrol of ochratoxigenic moulds (*Aspergillus ochraceus* and *Penicillium nordicum*) by *Debaryomyces hansenii* and *Saccharomycopsis fibuligera* during speck production. *Food Microbiol.* **2017**, *62*, 188–195. [CrossRef]

28. Núñez, F.; Lara, M.S.; Peromingo, B.; Delgado, J.; Sánchez-Montero, L.; Andrade, M.J. Selection and evaluation of *Debaryomyces hansenii* isolates as potential bioprotective agents against toxigenic penicillia in dry-fermented sausages. *Food Microbiol.* **2015**, *46*, 114–120. [CrossRef]

29. Peromingo, B.; Andrade, M.J.; Delgado, J.; Sánchez-Montero, L.; Núñez, F. Biocontrol of aflatoxigenic *Aspergillus parasiticus* by native *Debaryomyces hansenii* in dry-cured meat products. *Food Microbiol.* **2019**, *82*, 269–276. [CrossRef]

30. Peromingo, B.; Núñez, F.; Rodríguez, A.; Alía, A.; Andrade, M.J. Potential of yeasts isolated from dry-cured ham to control ochratoxin A production in meat models. *Int. J. Food Microbiol.* **2018**, *268*, 73–80. [CrossRef]

31. Simoncini, N.; Virgili, R.; Spadola, G.; Battilani, P. Autochthonous yeasts as potential biocontrol agents in dry-cured meat products. *Food Control* **2014**, *46*, 160–167. [CrossRef]

32. Virgili, R.; Simoncini, N.; Toscani, T.; Camardo Leggieri, M.; Formenti, S.; Battilani, P. Biocontrol of *Penicillium nordicum* growth and ochratoxin A production by native yeasts of dry cured ham. *Toxins* **2012**, *4*, 68–82. [CrossRef]

33. Andrade, M.J.; Córdoba, J.J.; Casado, E.M.; Córdoba, M.G.; Rodríguez, M. Effect of selected strains of *Debaryomyces hansenii* on the volatile compound production of dry fermented sausage "salchichón". *Meat Sci.* **2010**, *85*, 256–264. [CrossRef] [PubMed]

34. Cano-García, L.; Belloch, C.; Flores, M. Impact of *Debaryomyces hansenii* strains inoculation on the quality of slow dry-cured fermented sausages. *Meat Sci.* **2014**, *96*, 1469–1477. [CrossRef] [PubMed]

35. Núñez, F.; Rodríguez, M.M.; Córdoba, J.J.; Bermúdez, M.E.; Asensio, M.A. Yeast population during ripening of dry-cured Iberian ham. *Int. J. Food Microbiol.* **1996**, *29*, 271–280. [CrossRef]

36. Andrade, M.J.; Rodríguez, M.; Casado, E.M.; Bermúdez, E.; Córdoba, J.J. Differentiation of yeasts growing on dry-cured Iberian ham by mitochondrial DNA restriction analysis, RAPD-PCR and their volatile compounds production. *Food Microbiol.* **2009**, *26*, 578–586. [CrossRef]

37. Gil-Serna, J.; Patiño, B.; Cortés, L.; González-Jaén, M.T.; Vázquez, C. Mechanisms involved in reduction of ochratoxin A produced by *Aspergillus westerdijkiae* using *Debaryomyces hansenii* CYC 1244. *Int. J. Food Microbiol.* **2011**, *151*, 113–118. [CrossRef]

38. Acosta, R.; Rodríguez-Martín, A.; Martín, A.; Núñez, F.; Asensio, M.A. Selection of antifungal protein-producing molds from dry-cured meat products. *Int. J. Food Microbiol.* **2009**, *135*, 39–46. [CrossRef]

39. Rodríguez-Martín, A.; Acosta, R.; Liddell, S.; Núñez, F.; Benito, M.J.; Asensio, M.A. Characterization of the novel antifungal protein PgAFP and the encoding gene of *Penicillium chrysogenum*. *Peptides* **2010**, *31*, 541–547. [CrossRef]

40. Rodríguez, A.; Bernáldez, V.; Rodríguez, M.; Andrade, M.J.; Núñez, F.; Córdoba, J.J. Effect of selected protective cultures on ochratoxin A accumulation in dry-cured Iberian ham during its ripening process. *LWT—Food Sci. Technol.* **2015**, *60*, 923–928. [CrossRef]

41. Delgado, J.; Núñez, F.; Asensio, M.A.; Owens, R.A. Quantitative proteomics of *Penicillium nordicum* profiles and ochratoxin A repression by protective cultures. *Int. J. Food Microbiol.* **2019**, *305*, 108243. [CrossRef]

42. Karolewiez, A.; Geisen, R. Cloning a part of the ochratoxin A biosynthetic gene cluster of *Penicillium nordicum* and characterization of the ochratoxin polyketide synthase gene. *Syst. Appl. Microbiol.* **2005**, *28*, 588–595. [CrossRef]

43. Bernáldez, V.; Córdoba, J.J.; Andrade, M.J.; Alía, A.; Rodríguez, A. Selection of reference genes to quantify relative expression of ochratoxin A-related genes by *Penicillium nordicum* in dry-cured ham. *Food Microbiol.* **2017**, *68*, 104–111. [CrossRef]

44. Vela-Corcía, D.; Romero, D.; De Vicente, A.; Pérez-García, A. Analysis of β-tubulin-carbendazim interaction reveals that binding site for MBC fungicides does not include residues involved in fungicide resistance. *Sci. Rep.* **2018**, *8*, 1–12. [CrossRef]

45. Bernáldez, V.; Rodríguez, A.; Martín, A.; Lozano, D.; Córdoba, J.J. Development of a multiplex qPCR method for simultaneous quantification in dry-cured ham of an antifungal-peptide *Penicillium chrysogenum* strain used as protective culture and aflatoxin-producing moulds. *Food Control* **2014**, *36*, 257–265. [CrossRef]

46. Delgado, J.; Acosta, R.; Rodríguez-Martín, A.; Bermúdez, E.; Núñez, F.; Asensio, M.A. Growth inhibition and stability of PgAFP from *Penicillium chrysogenum* against fungi common on dry-ripened meat products. *Int. J. Food Microbiol.* **2015**, *205*, 23–29. [CrossRef]

47. Galaverna, G.; Dall'Asta, C. Sampling Techniques for the Determination of Mycotoxins in Food Matrices. In *Comprehensive Sampling and Sample Preparation*; Pawliszyn, J., Ed.; Academic Press: Oxford, UK, 2012; pp. 381–403.

48. Bernáldez, V.; Córdoba, J.J.; Rodríguez, M.; Cordero, M.; Polo, L.; Rodríguez, A. Effect of *Penicillium nalgiovense* as protective culture in processing of dry-fermented sausage "salchichón". *Food Control* **2013**, *32*, 69–76. [CrossRef]

49. Gil-Serna, J.; Patiño, B.; Cortés, L.; González-Jaén, M.T.; Vázquez, C. *Aspergillus steynii* and *Aspergillus westerdijkiae* as potential risk of OTA contamination in food products in warm climates. *Food Microbiol.* **2014**, *46*, 168–175. [CrossRef]

50. Fasoyin, O.E.; Wang, B.; Qiu, M.; Han, X.; Chung, K.R.; Wang, S. Carbon catabolite repression gene *creA* regulates morphology, aflatoxin biosynthesis and virulence in *Aspergillus flavus*. *Fungal Genet. Biol.* **2018**, *115*, 41–51. [CrossRef]

51. Kamala, A.; Ortiz, J.; Kimanya, M.; Haesaert, G.; Donoso, S.; Tiisekwa, B.; De Meulenaer, B. Multiple mycotoxin co-occurrence in maize grown in three agro-ecological zones of Tanzania. *Food Control* **2015**, *54*, 208–215. [CrossRef]

52. Bernáldez, V.; Córdoba, J.J.; Magan, N.; Peromingo, B.; Rodríguez, A. The influence of ecophysiological factors on growth, *aflR* gene expression and aflatoxin B1 production by a type strain of *Aspergillus flavus*. *LWT—Food Sci. Technol.* **2017**, *83*, 283–291. [CrossRef]

53. Rodríguez, A.; Rodríguez, M.; Luque, M.I.; Justesen, A.F.; Córdoba, J.J. Quantification of ochratoxin A-producing molds in food products by SYBR Green and TaqMan real-time PCR methods. *Int. J. Food Microbiol.* **2011**, *149*, 226–235. [CrossRef]

54. Livak, K.J.; Schmittgen, T.D. Analysis of relative gene expression data using real-time quantitative PCR and the $2^{(-\Delta\Delta CT)}$ method. *Methods* **2001**, *25*, 402–408. [CrossRef]

© 2019 by the authors. Licensee MDPI, Basel, Switzerland. This article is an open access article distributed under the terms and conditions of the Creative Commons Attribution (CC BY) license (http://creativecommons.org/licenses/by/4.0/).

Article

Development of an Antifungal and Antimycotoxigenic Device Containing Allyl Isothiocyanate for Silo Fumigation

Juan Manuel Quiles [1,†]**, Tiago de Melo Nazareth** [1,2,†]**, Carlos Luz** [1]**,
Fernando Bittencourt Luciano** [2]**, Jordi Mañes** [1] **and Giuseppe Meca** [1,*]

[1] Laboratory of Food Chemistry and Toxicology, Faculty of Pharmacy, University of Valencia, Av. Vicent
Andrés Estellés s/n, 46100 Burjassot, Spain; juan.quiles@uv.es (J.M.Q.); demena@alumni.uv.es (T.d.M.N.);
carlos.luz@uv.es (C.L.); jorge.manes@uv.es (J.M.)
[2] School of Life Sciences, Pontifícia Universidade Católica do Paraná, Rua Imaculada Conceição 1155,
Curitiba 80215-901, Brazil; fernando.luciano@pucpr.br
* Correspondence: giuseppe.meca@uv.es; Tel.: +34-963-544-959; Fax: +34-963-549-54
† Authors have contributed equally to the study and should be considered co-first authors.

Received: 29 January 2019; Accepted: 23 February 2019; Published: 1 March 2019

Abstract: The aims of this study were to evaluate the antifungal activity of the bioactive compound allyl isothiocyanate (AITC) against *Aspergillus flavus* (8111 ISPA) aflatoxins (AFs) producer and *Penicillium verrucosum* (D-01847 VTT) ochratoxin A (OTA) producer on corn, barley, and wheat. The experiments were carried out initially in a simulated silo system for laboratory scale composed of glass jars (1 L). Barley and wheat were contaminated with *P. verrucosum* and corn with *A. flavus*. The cereals were treated with a hydroxyethylcellulose gel disk to which 500 µL/L of AITC were added; the silo system was closed and incubated for 30 days at 21 °C. After that, simulated silos of 100 L capacity were used. Barley, wheat, and corn were contaminated under the same conditions as the previous trial and treated with disks with 5 mL of AITC, closed and incubated for 90 days at 21 °C. In both cases, the control test did not receive any antifungal treatment. The growth of the inoculated fungi and the reduction in the formation of AFs and OTA were determined. In the lab scale silo system, complete inhibition of fungal growth at 30 days has been observed. In corn, the reduction of aflatoxin B1 (AFB$_1$) was 98.5%. In the 100 L plastic drums, a significant reduction in the growth of *A. flavus* was observed, as well as the OTA formation in wheat (99.5%) and barley (92.0%).

Keywords: *Aspergillus flavus*; *Penicillium verrucosum*; AITC; fungal growth reduction; mycotoxin reduction

Key Contribution: The application of an allyl isothiocyanate device, as an antifungal system, to be applied in the storage of corn, wheat, and barley, to reduce the mycotoxigenic fungal growth and the mycotoxin biosynthesis.

---

1. Introduction

AFs are the foremost harmful category of mycotoxins naturally produced by *Aspergillus* species such as *Aspergillus nomius, Aspergillus flavus*, and *Aspergillus parasiticus* during pre- or postharvest of crops [1,2]. The most important and toxic aflatoxins are the AFB$_1$, AFB$_2$, AFG$_1$, and AFG$_2$ [3]. Among these compounds, the AFB$_1$ has been classified in Group 1 of the risk of the carcinogen molecules by the International Agency for Research on Cancer [4], and it has been implicated with the development of human hepatic and extra hepatic carcinogenesis [5]. Human exposure to AFs could be due to the intake of contaminated food or by the consumption of milk, meat, and eggs from animals that consumed contaminated feed [6]. The occurrence of AFs in foods from in the Spanish market has been previously reported in several products, such as cereals, pulses, dried fruits and nuts, snacks, breakfast cereals, bread, herbs, or spices [7,8].

OTA is one of the most dangerous mycotoxins and is produced by *Aspergillus* and *Penicillium* species, among which are *Aspergillus ochraceus* [9], *Aspergillus carbonarius* [10], and to a lesser extent *Aspergillus niger* [11] and *Penicillium verrucosum* species [12]. The contamination of food by the presence of OTA is common in Europe. In more than 50% of the 6476 foods analyzed, OTA amounts were detected above the detection limit of 0.01 mg/kg [13]. IARC considers OTA as a possibly carcinogenic compound in humans (Group 2B) [4]. OTA dietary average for citizens of the European Union has been experimentally established in a range of 0.9 (Germany) to 4.6 (Italy) ng/kg. Foods with the greatest OTA contamination are coffee, cereals, spices, and beer [14,15]. When these contaminated foods are ingested, OTA can cause a nephrotoxic, hepatotoxic, and teratogenic effects [16,17].

Methods for controlling mycotoxins are usually preventive, including good agricultural practice and drying of crops after harvest. Some researchers have reported that mycotoxins can be degraded by heat treatment, but the extent of mycotoxin degradation is dependent on temperature, time of exposure, and mainly the contamination level [18]. OTA is a stable molecule, which can resist roasting, brewing, baking, ammoniation, and heat treatment to some extent [19]. Likewise, AFB_1 seems to be stable up to 150 °C [20]. For this reason, other methods of detoxification have been developed to prevent these mycotoxins in food and feed.

Isothiocyanates (ITCs) are products originated from the enzymatic hydrolysis of glucosinolates, which are sulfur-containing glucosides present in plants of the *Brassicaceae* family. These compounds contribute to the characteristic pungent taste of these vegetables [20] and have been reported as potent antimicrobials [21]. Allyl isothiocyanate (AITC), which is the most studied ITCs, was found to inhibit the growth of yeast, mold, and bacteria at very low levels [22], including molds from the genera *Aspergillus*, *Penicillium*, and *Fusarium* [23,24]. ITCs are characterized by the presence of a –N.C.S group, whose central carbon atom is strongly electrophilic [25]. This electrophilic nature enables ITC to readily bind to thiol and amino groups of amino acids, peptides, and proteins, forming conjugates [21], dithiocarbamate, and thiourea structures [26]. OTA contains a free and readily available amino group and AFs contains a carboxylic group. Therefore, ITCs could be good candidates to react with these mycotoxins.

The objective of this research was to investigate the efficacy of an antifungal device based on the natural compound AITC to reduce the growth of *A. flavus* and *P. verrucosum* in cereals during storage and the mycotoxin production.

2. Results

2.1. AITC Concentration in Headspace and Cereals in Laboratory Scale Silo

Figure 1 shows the AITC present in the headspace of laboratory scale silos. The AITC concentration decreases gradually from 0.92 μL/L at day 1 to 0.25 μL/L at day 30. On the other hand, there was no significant difference in AITC concentration after day 7 up to day 30. These results suggest that either a fraction of AITC left the silo or it was absorbed to grains releasing a constant average concentration ranging from 0.37 to 0.25 μL/L for 30 days.

The residual concentration of AITC in barley, corn, and wheat was studied on day 1 and 30, and the results are shown in Figure 2. Corn was the most susceptible matrix to AITC penetration, showing 10.9 and 5.9 mg/kg of AITC at days 1 and 30, respectively. Barley grains showed a lower capacity to maintain AITC with 7.5 and 2.9 mg/kg absorbed at days 1 and 30, respectively. Wheat grains did not show a significant difference to barley and corn at day 1, with 9.6 mg/kg of AITC. However, at day 30, wheat grains showed 3 mg/kg more concentration of AITC than barley samples.

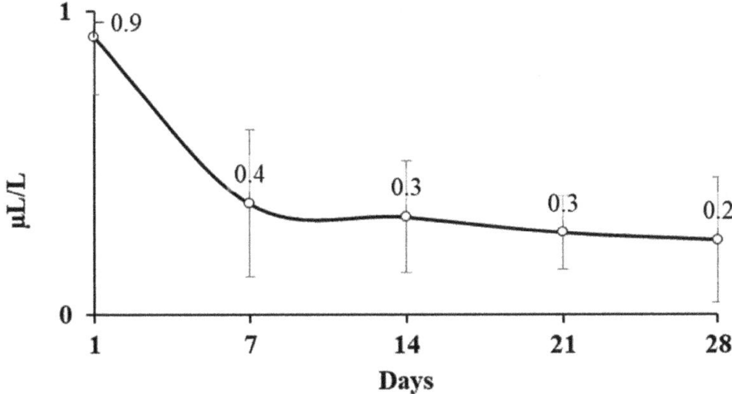

Figure 1. AITC detected in the headspace of the glass jar containing corn, wheat, and barley, used to simulate the storage of the cereals in a lab scale silo system.

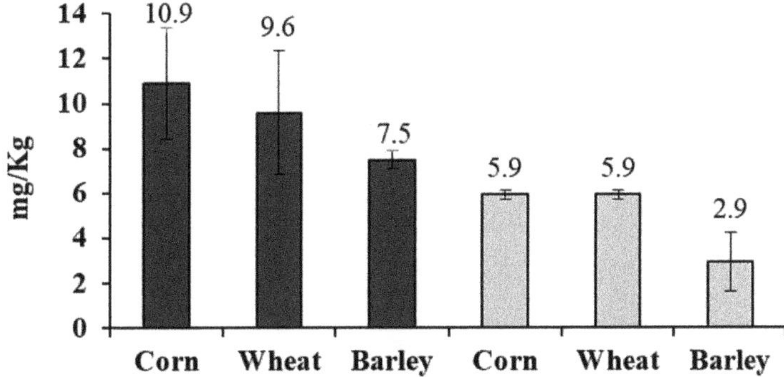

Figure 2. Concentrations of AITC detected in corn, wheat, and barley after 1 (gray) and 30 (white) days of incubation.

2.2. Validation Method for the Analysis of Mycotoxins in Cereals

To validate the analytical method, the following parameters such as linearity, recovery, repeatability, reproducibility, limits of detection (LOD) and quantification (LOQ), and the matrix effect for each mycotoxin analyzed were carried out. All the mycotoxins showed good linearity in the working range, with resolution determination coefficients (R2) greater than 0.9922.

Linearity was evaluated using paired matrix calibrations in triplicate at concentrations between 5 and 500 µg/kg. To calculate the matrix effect, the calibration slope from the matrix calibration curve was divided by the slope of the standard calibration curve and multiplied by 100. The value of the recovery was carried out in triplicate for three consecutive days using three addition levels: LOQ, 2 × LOQ, and 10 × LOQ.

The results were between 70.4% and 75.6% and the relative standard deviation (RSD) was less than 17%. The values for intraday repeatability ($n = 3$), expressed as the relative standard deviation of the repeatability (RSDr), varied from 7.5% to 11.6%; and the reproducibility between days ($n = 5$), expressed as the relative standard deviation of the reproducibility (RSDR), varied from 8.2% to 17.3% for the same linearity addition values. LODs and LOQs were calculated by analyzing blank samples enriched with the standard mycotoxins; these parameters have been assessed as the lowest concentration of the molecules studied that showed a chromatographic peak at a signal-to-noise ratio (S/N) of 3 and 10 for LOD and LOQ, respectively (Table 1).

Table 1. LODs, LOQs, recovery, and matrix effect (ME) (%) for AFB1, AFB2, AFG1, AFG2, and OTA in corn, wheat, and barley.

Mycotoxin	LOD (µg/Kg)	LOQ (µg/Kg)	Recovery (%)	ME (%)
AFB1	0.08	0.27	70.4	78.2
AFB2	0.08	0.27	64.2	76.5
AFG1	0.16	0.53	62.8	65.3
AFG2	0.30	1.00	66.1	60.9
OTA	0.05	0.17	75.6	89.7

2.3. Fungal Growth and Mycotoxin Production in Lab Scale Silo System

The results of *A. flavus* and *P. verrucosum* growth on barley, corn, and wheat at days 1 and 30 are shown in Figure 3. At day 1, AITC treatment demonstrated a significant reduction in the fungal population of corn, wheat, and barley, reducing in 1.5, 1, and 1.2 log CFU/g, respectively. After 30 days of storage, the fungus growth in the control groups remained stable. However, the treatment reduced the population of fungi in wheat and barley to levels below our limit of detection, while in the corn the reduction was of 4.4 log CFU/g.

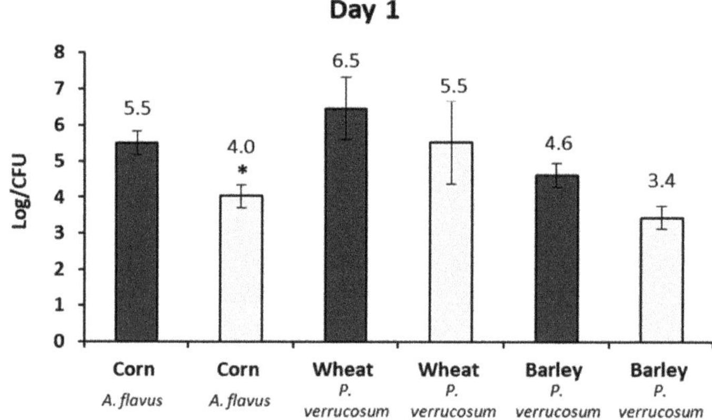

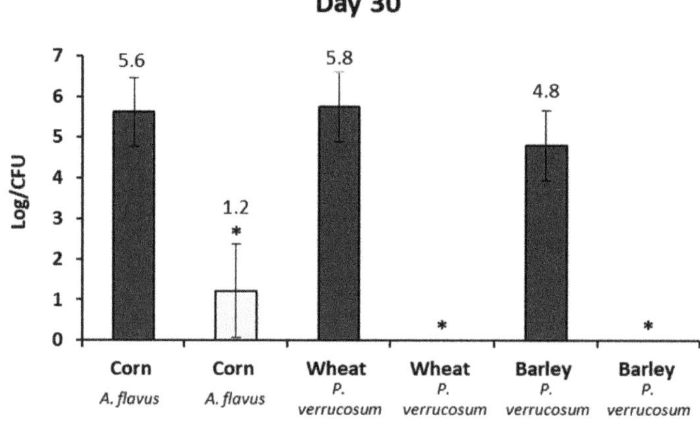

Figure 3. Growth, in lab scale silo system, of the *A. flavus* in corn and of *P. verrucosum* in wheat and barley exposed to the vapor of the AITC after 1 and 30 days of incubation. Samples control (dark gray) and treated samples (clear gray). Significantly different from untreated cereal, $p \leq 0.05$ (*).

In correlation to fungal growth, the mycotoxin production was determined after 30 days of storage and the results are shown in Figure 4. *A. flavus* produced in the control corn 8.07 µg/Kg of AFB_1 at day 30. This value is above the limit of AFB_1 in foodstuffs set by the European Commission (EC 165/2010). Therefore, this cereal is classified as inappropriate for human consumption. The AFB_1 present in the treated corn was 0.12 µg/Kg, representing a reduction of 98.51%. Regarding the values of OTA, the reduction was not significant in barley, whereas in the wheat samples the OTA was not produced even in the control group, reaching values below our limit of detection.

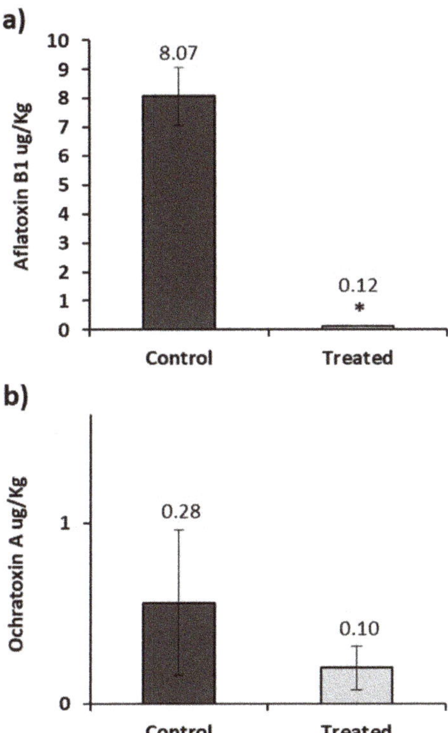

Figure 4. Aflatoxin B1 detected in corn **a**); and ochratoxin A detected in barley **b**) treated with the AITC device in the lab scale silo system at 30 days of incubation. Samples control (dark gray) and treated samples (clear gray). Significantly different from untreated cereal, $p \leq 0.05$ (*).

2.4. Fungal Growth and Mycotoxin Production in a Small-Scale Silo System

After laboratory scale analysis in silos of 100 L containing 50 Kg of cereal. Barley, corn, and wheat were contaminated and then treated with a gel dispositive developed with 5 mL of AITC. The sampling was realized monthly and the results for microbiological analysis are shown in Figure 5.

At day 1, AITC treatment did not demonstrate a significant reduction in fungal population in corn, wheat, and barley. Similarly, to the laboratory scale silo, after 30 days of storage, the treatment with AITC was able to reduce significantly the fungal growth of *A. flavus* in corn and *P. verrucosum* in barley and wheat. In addition, at the end of the experiment (after 60 days of storage), AITC treatment demonstrated a significant reduction in the fungal population of corn, wheat, and barley, reducing in 2, 0.9, and 1.1 log CFU/g in comparison to the control group, respectively (Figure 5).

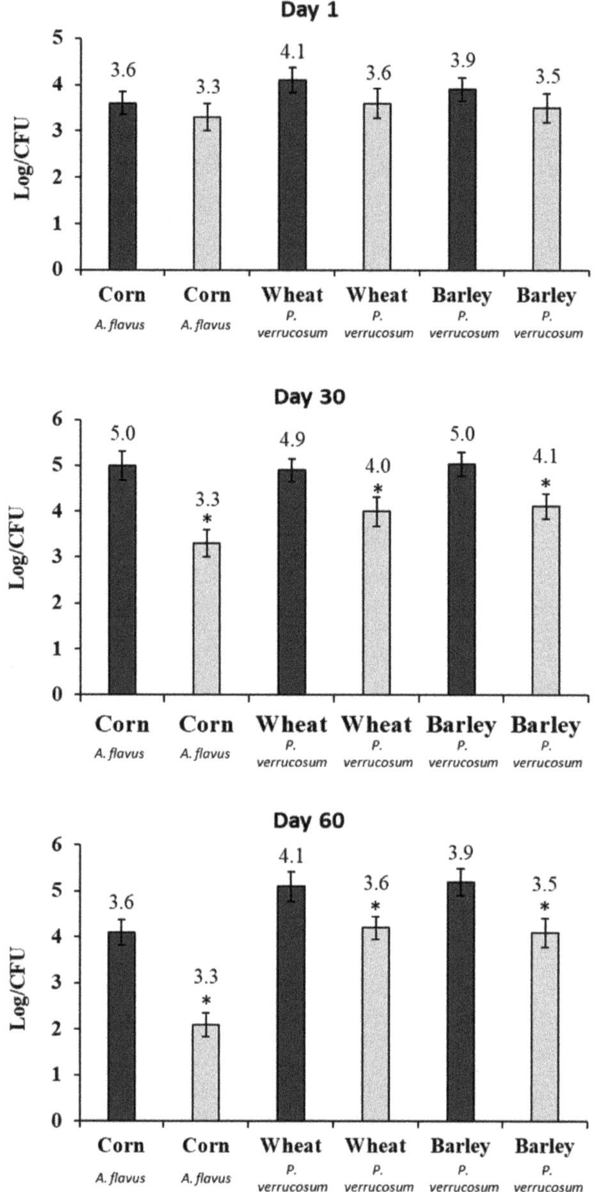

Figure 5. Growth, in small-scale silo system, of the *A. flavus* in corn and of *P. verrucosum* in wheat and barley exposed to the vapor of the AITC after 1, 30, and 60 days of incubation. Control (dark gray) and treated samples (light gray). Significantly different from untreated cereal, $p \leq 0.05$ (*).

Along with the fungal growth, the production of mycotoxins was determined at days 30 and 60 of storage (Figure 6). No AFs could be detected in corn contaminated with *A. flavus*. In the samples of wheat and barley contaminated with *P. verrucosum*, the OTA reduction was 90.0% and 99.5% for day 30 and 78.2% and 92.0% for day 60, respectively.

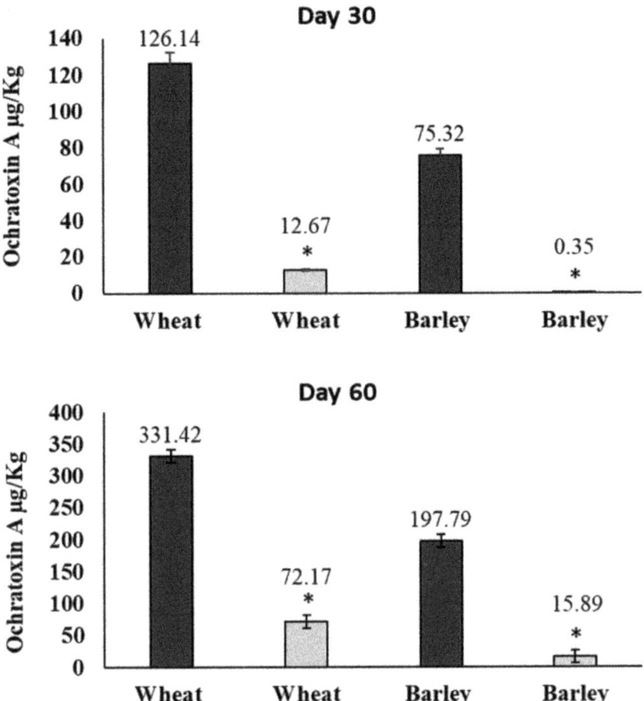

Figure 6. Ochratoxin A detected in wheat and barley treated with the AITC device in the small-scale silo system at 30 and 60 days of incubation. Samples control (dark gray) and treated samples (light gray). Significantly different from untreated cereal, $p \leq 0.05$ (*).

3. Discussion

Similarities and differences were identified comparing the results of the lab scale silo system with the results of small-scale silo system. Microbiologically, the exposure to AITC of cereals contaminated with *A. flavus* (corn) and *P. verrucosum* (wheat and barley) reduced, in all experiments, significantly the fungal population after 30 days. However, due to the lower concentration of AITC (50 µL/L) and the micro atmospheres generated by small-scale silo system, the AITC device could not completely inhibit the fungal growth when compared to the lab scale silo system results. In other words, a lower dose of AITC and higher headspace in the small silo system did not reduce the fungal population to the values below to our limit of detection (1.2 logs CFU/g). These results suggest that the effect of AITC is dose depending. In addition, the higher the grain volume, the higher should be the AITC concentration to achieve a total inhibitory effect.

Regarding the production of mycotoxins, differences between the two tests were observed, probably since a moderate inoculum was used to replicate actual contamination conditions of the field. Another difference among the experiments was the reduction of the potential maximum concentration of AITC in the headspace (500 µL/L in the lab silo and 50 µL/L in the small silo) due to issues of scaling and safety of the compound. Even so, in all analysis, a significant reduction in mycotoxin production could be observed among the control and the treated samples when both AFB_1 and OTA could be produced in matrices.

In the small-scale system, there was an increasing concentration of mycotoxins over time, even in the treated samples. These results could be explained by the presence of the fungal population in the cereals, which allows the mycotoxin production.

In particular, *A. flavus* and *P. verrucosum* depend on oxygen to grow. In our experiment, the headspace in the lab scale silo system and small-scale silo system was around 50% and 20%, respectively. The lower concentration of free oxygen could reduce the regular growth of *A. flavus* and consequently, avoid the AFB$_1$ production in the small-scale system. Moreover, the cereals in the small-scale system were not autoclaved, which increased the competitiveness among microorganisms by nutrients.

The application of the AITC to reduce the growth of the fungi mycotoxin producer has been studied previously by other authors. Manyes et al. studied the capacity of AITC produced by the volatilization of a standard solution of the oriental mustard essential oil to prevent the growth of the fungi *A. parasiticus* and *Penicillium expansum* [27]. In that study, Petri dishes were inoculated with the mycotoxigenic fungi *A. parasiticus* (producer of AFs) and *P. expansum* (producer of patulin), and the inhibition of micellar growth was observed when they were deposited in the center of the petri dish with 25 µL and 50 µL of AITC, respectively.

Okano et al. assessed the capacity of the AITC obtained by a commercial mustard seed extract to reduce the aflatoxins production by *A. flavus* during the corn storage [28], in a simulated silo condition. The AITC concentration in the headspace of the model system used by the authors reached the highest value of 54.6 µg/L on the day 14 of incubation and remaining stable until 21.8 µg/L until the end of the incubation period. Also, the AITC reduced completely the visible growth of *A. flavus* and the AFs production in both sterilized and unsterilized corn

Delaquis et al. evaluated the capacity of the AITC to reduce the growth of *A. flavus* and *P. expansum* at concentrations of 0.1 µL/L [29]. The experimental model used was the one used in this study. In that case, 2 L flasks were used and inocula of 10^5 conidia/mL were placed in the presence of different amounts of AITC. These antifungal properties of the AITC were also confirmed by Suhr and Nielsen who inoculated pieces of bread with 10^6 spores/mL of *Penicillium roqueforti*, *Penicillium corylophilum*, and *A. flavus* and arranged them in closed systems in the presence of mustard essential oil (99% of AITC) [30]. Fungal growth inhibition was observed at concentrations of 1 µL/L. Other studies on the fungicidal activity of AITC against food-disrupting fungi observed the ability of the compound to penetrate the matrix and extend its effect over time. Winther and Nielsen showed that cheeses treated with AITC could absorb this compound, increasing its useful life from 4 to 28 weeks [31].

Quiles et al. developed an active packaging dispositive based on the AITC to reduce the sporulation of *A. parasiticus* and AFs production in fresh pizza doughs during 30 days of inoculation [32]. The antifungal activity of the AITC was compared with untreated samples (fresh pizza doughs without any preservative treatment) and with samples treated with sodium propionate, the classical preservative used in bakery products. After 30 days, the growth of *A. parasiticus* was inhibited with the treatment of AITC at 5 µL/L and 10 µL/L. The reduction of AFs was total at the dose of 10 µL/L.

Nazaret et al. evaluated the capacity of AITC to reduce the production of AFs, beauvericin and enniatin, by *A. parasiticus* and *Fusarium poae* in wheat flour [24]. The analysis of the results showed that the AITC concentration of 0.1 µL/L reduced by 23% the production of the mycotoxins. Also, the application of the AITC at 10 µL/L completely reduced the biosynthesis of the mycotoxins studied during 30 days of incubation.

Tracz et al. evaluated the capacity of AITC at 50, 100, or 500 µL/L to avoid mycotoxin production in corn kernels [33]. Both treatments were able to avoid the production of 12 mycotoxins, including AFB$_1$ and Ochratoxin. Saladino et al. analyzed the fungal growth and AFB$_1$ reduction by AITC (0.5, 1, or 5 µL/L) in loaf bread [34]. As result, the treatments of 1 and 5 µL/L reduced the AFB$_1$ concentration by above 60%. Our results corroborate with these studies, since the AITC at 50 µL/L demonstrated a fungicide and antimicotoxigenic effect, inhibiting the AFB$_1$ and OTA synthesis and the fungal growth of *A. parasiticus* and *P.verrucosum* in our small-scale assays.

4. Conclusions

The results obtained in this study showed the capacity of the AITC to reduce the growth of the fungi *A. flavus* and *P. verrucosum* in corn, wheat, and barley. The volatilization of the AITC in the headspace of the lab scale silo system was enough to avoid the *A. flavus* and *P. verrucosum* growth in all cereals tested. Moreover, the treatment with AITC device was able to reduce the AFB1 and OTA production in corn and barley, respectively.

In the small-scale silo system, a significant reduction of the *A. flavus* and *P. verrucosum* growth was observed as well as an important reduction of the OTA produced by *P. verrucosum*. The application of the device based on the AITC could be an alternative method to reduce the growth of fungi mycotoxin producer in cereals during the storage phase.

For further studies, the tests carried out in this work will be staggered for 200-ton real silos with naturally contaminated barley and treated with AITC release devices.

5. Material and Methods

5.1. Chemicals and Microbial Strains

AFB_1, AFB_2, OTA (98% purity), AITC, and formic acid (HCOOH) were obtained from Sigma–Aldrich (St. Louis, MO, USA). Methanol and acetonitrile have been obtained by Fisher Scientific (Hudson, NH, USA). Deionized water (<18 MΩ cm resistivity) was produced by a water purification system (Millipore, Bedford, MA, USA). All the chromatographic solvents were filtered through a 0.22 µm membrane filter Scharlau (Barcelona, Spain). Barley, wheat, and corn were provided by Tot Agro (Barcelona, Spain). The peptone water and dextrose potato agar culture medium were obtained from Liofilchem (Teramo, Italy). The strains of *A. flavus* ITEM 8111 were provided by the Microbial Culture Collection of Institute of Sciences and of Food Production (ISPA, Bari, Italy) whereas the *P. verrucosum* VTT D-01847, was obtained from the VTT Culture Collection (Espoo, Finland).

5.2. Laboratory Scale Silo System and Antifungal Treatment with the AITC Device

The silo simulation was carried out as shown in Figure 7. Glass jars of 1 L containing 300 g of cereals were contaminated with 10^4 conidia/g of *P. verrucosum* (barley and wheat) and *A. flavus* (corn). The cereals were stored for three days to allow fungal adaptation, and treated with 500 µL/L (in the relation of volume of the jar) of AITC into a gel device.

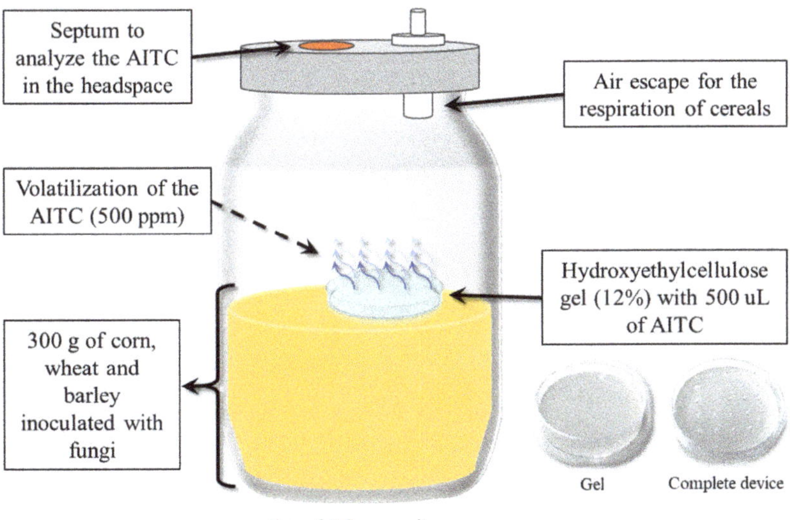

Figure 7. Lab scale silo system used for the treatment of corn, wheat and barley contaminated with *A. flavus* and *P. verrucosum*, and treated with the AITC device.

The gel device was manufactured mixing 1.2 g of hydroxyethyl cellulose (gelling agent), 10 mL of water and 500 µL/L of AITC into a Petri dish. The lid of the Petri dish was previously perforated to facilitate the AITC volatilization, as shown in Figure 7. Posteriorly, the antifungal device was placed inside the jars. The jars were closed with adapted lids that contained a septum and an air escape, which allowed AITC analysis in the headspace and cereal respiration, respectively. The samples were stored for 30 days at room temperature. After that, the fungal growth, the mycotoxins contained in the grains, the AITC in the headspace, and the AITC adsorbed by the grains were determined.

5.3. AITC Device Application in a Small-Scale Silo System

Fifty kilograms of barley, corn, and wheat were placed inside plastic drums (100 L) separately. Each cereal was contaminated with 10^4 conidia/g of *P. verrucosum* or *A. flavus*. The barley and wheat were contaminated with *P. verrucosum* and the corn was contaminated with *A. flavus*. Then, the device described in Section 5.2 was adapted to the small-scale silo system.

The petri dish was changed for a glass tapper wear, increasing the quantity of the pure AITC contained to 5 mL, in order to obtain 50 µL/L of this bioactive compound in the headspace of the silo. The device was located in the silo bottom and the grains were introduced in the upper part of the silo as shown in Figure 8. A metal grid separated the lower and the upper part of the silo in order to isolate the device and the AITC vapors from the stored cereals. The control group did not receive any antifungal treatment. The analysis carried out on the treated cereals was the same as described in Section 5.2.

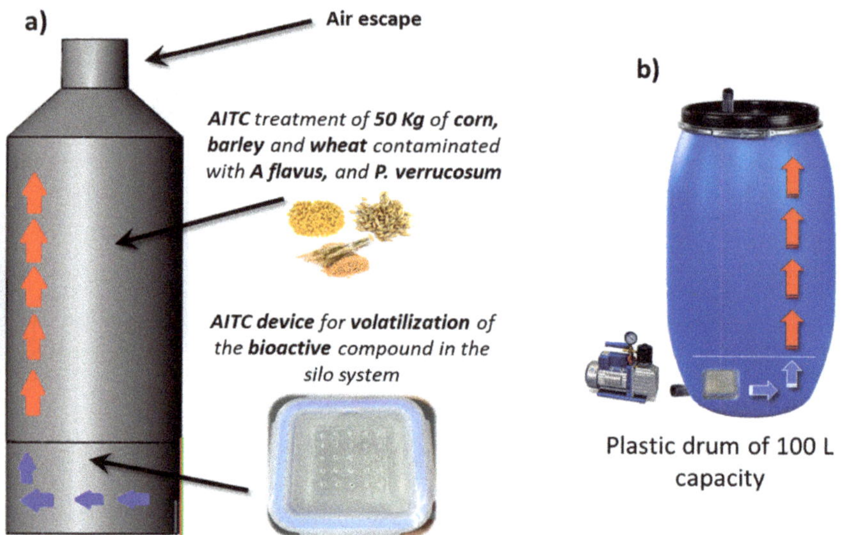

Figure 8. Small-scale silo system used for the treatment of corn, wheat, and barley contaminated with *A. flavus* and *P. verrucosum*, and treated with the AITC device. (**a**) Theoretical silo design; and (**b**) the plastic drum used in this study.

5.4. Determination of AITC Concentration in the Headspace of the Laboratory Scale Silo

The AITC content in headspace was determined through a septum localized in the lip of a laboratory silo system (Figure 7). The air was recovered using a syringe of 1 mL, and aliquots of 200 µL were injected in a gas chromatograph (GC) with flame ionization detector (FID) (GC 6890, Agilent Technologies Inc., Santa Clara, CA, USA.). The chromatograph was equipped with a 30 × 0.25 mm CP-SIL 88 fused capillary column (Varian, Middelburg, Netherlands). The temperature of the detector arrived at 200 °C with a gradient of temperature that starts at 60 °C. This temperature was maintained for 1 min and increased 8 °C per min up to 100 °C, then maintained for 5 min and finally increased in 15 °C per min up to 200 °C. The gas utilized as the carrier was H_2 at 5 mL/min. The ionization was realized with H_2 at 40 mL/min and purified air at 450 mL/min.

5.5. Determination of AITC Concentration in the Cereals of the Laboratory Scale Silo

Extraction of AITC from cereals samples was conducted as described by Tracz et al. with some modifications [33]. Five g samples were weighed into 15 mL polyethylene tubes to which 10 mL of methanol was added. The mixture was shaken for 30 min in a water bath (40 °C) and for 10 min in an ultrasonic bath. The samples were centrifuged at 4000 *g* for 5 min at 20 °C. The supernatant (8 mL) was collected and filtered through a 0.22 μm nylon membrane. 20 μL were injected in the LC system, 1220-Infinity (Agilent, Santa Clara, CA, USA) coupled with a diode array detector (LC-DAD) at 236 nm. A Gemini C18 column (Phenomenex, Torrance, CA, USA) 4.6 × 150 mm, 3 μm particle size at 30 °C was used as a stationary phase. The isocratic mobile phase consisted of water/acetonitrile (60:40, *v/v*) with a flow rate of 1 mL/min.

5.6. Mycotoxin Extraction and LC-MS/MS Analysis of Corn, Barley, and Wheat

The extraction of mycotoxins was carried out following the method described by Serrano et al. with some modifications [35]. Each cereal sample was crushed using a food grinder (Oster Classic Grinder 220e240 V, 50/60 Hz, 600 W, Oster, Valencia, Spain). The resulting particles were mixed, and three 5 g aliquots of each sample were taken in 50 mL plastic falcon tubes. 25 mL of methanol was added to each of these tubes and the samples were homogenized for 3 min by Ultra Ika T18 ultraturrax (Staufen, Germany) at 10,000 rpm. The extract was centrifuged at 4000 rpm during 5 min at 5 °C, and the supernatant was transferred to a plastic flask and evaporated to dryness with a Büchi Rotavapor R-200 (Postfach, Switzerland). The obtained residue was resuspended in 5 mL of methanol, transferred to a 15 mL plastic falcon tube and evaporated with nitrogen gas stream using a multi-sample Turbovap LV evaporator (Zymark, Hopkinton, MA, USA). Finally, the residue was reconstituted in 1 mL of methanol, filtered through a 13 mm/0.22 μm filter and transferred to a 1 mL glass chromatography vial. The liquid-chromatography system consisted of an LC-20AD pump coupled to a 3200QTRAP mass spectrometer (Applied Biosystems, Foster City, CA, USA) using an ESI interface in positive ion mode. The mycotoxins were separated on a Gemini NX C18 column (150 × 2.0 mm I.D, 3.0 mm, Phenomenex, Palo Alto, CA, USA). The mobile phases were the solvent A (5 mM ammonium formate and 0.1% formic acid in water) and solvent B (5 mM ammonium formate and 0.1% formic acid in methanol) at a flow rate of 0.25 mL/min. The elution was carried out using a linear gradient from 0 to 14 min. The injection volume set was of 20 mL, the nebulizer, the auxiliary and the auxiliary gas were set at 55, 50, and 15 psi respectively. The capillary temperature and the ion spray voltage were of 550 °C and 5500 V, respectively. The ions transitions used for the mycotoxin identification and quantification were: m/z 313.1/241.3 and 284.9 for AFB1, m/z 315.1/259.0 and 286.9 for AFB2, m/z 329.0/243.1 and 311.1 for AFG1, m/z 331.1/313.1 and 245.1 for AFG2, m/z 404.3/102.1 and 358.1 for OTA.

5.7. Determination of the Fungal Population

After the incubation time, 10 g of each sample was transferred to a sterile plastic bag containing 90 mL of sterile peptone water (Oxoid, Madrid, Spain) and homogenized with a stomacher (IUL, Barcelona, Spain) during 30 s. The suspensions formed were serially diluted in sterile plastic tubes containing 0.1% of peptone water. After that, aliquots of 0.1 mL were plated on Petri dishes containing acidified potato dextrose agar (pH 3.5) (Insulab, Valencia, Spain) and the plates were incubated at 25 °C for 7 d before microbial counting. The results were expressed in logs of colony-forming unit/g of cereal (log CFU/g). All analyses were conducted in triplicate.

5.8. Statistical Analysis

The Prism version 3.0 software (GraphPad corporation1, La Jolla, CA, USA, 1989) for Windows was used for the statistical analysis of data. The experiments were realized in triplicate and the differences among groups were analyzed by Student's *t*-test. The level of significance considered was $p \leq 0.05$.

Author Contributions: The authors (J.M.Q. and T.d.M.N.) consider that the first two authors should be regarded as joint first authors. F.B.L., J.M., and G.M. conceived and proposed the idea. J.M.Q. and G.M. designed the study. J.M.Q., C.L., and T.d.M.N. performed the experiments and analyzed the data. J.M.Q. wrote the paper. T.d.M.N. contributed to the writing of the manuscript.

Funding: This research was funded by [European Commission] grant number [GA 678781], by [Spanish Ministry of Economy and Competitiveness] grant number [AGL2016-77610R] by [Generalitat Valenciana] grant number [Prometeo (2018/126)] and by [Brazilian National Council for Scientific and Technological Development] grant number [CNPq Project 400896/2014-1].

Acknowledgments: The research was supported by the European Project (H2020-Research and Innovation Action) MycoKey "Integrated and innovative key actions for mycotoxin management in the food and feed chain" GA 678781, the Spanish Ministry of Economy and Competitiveness (AGL2016-77610R), the project Prometeo (2018/126) of Generalitat Valenciana, the Brazilian National Council for Scientific and Technological Development (CAPES/CNPq Project 400896/2014-1) and the pre-PhD program of University of Valencia (Atracció de Talent).

Conflicts of Interest: The authors declare that have no conflict of interest.

References

1. Majeed, S.; Iqbal, M.; Asi, M.R.; Iqbal, S.Z. Aflatoxins and ochratoxin A contamination in rice, corn and corn products from Punjab, Pakistan. *J. Cereal Sci.* **2013**, *58*, 446–450. [CrossRef]
2. Iqbal, S.Z.; Asi, M.R.; Ariño, A.; Akram, N.; Zuber, M. Aflatoxin contamination in different fractions of rice from Pakistan and estimation of dietary intakes. *Mycotoxin Res.* **2012**, *28*, 175–180. [CrossRef] [PubMed]
3. Pittet, A. Natural occurrence of mycotoxins in foods and feeds—An updated review. *Rev. Med. Vet.* **1998**, *149*, 479–492.
4. IARC-International Agency for Research on Cancer. Monographs on the evaluation of carcinogenic risks to humans. In *A Review of Biological Agents for Human Carcinogens*; IARC-International Agency for Research on Cancer: Lyon, France, 2012.
5. Iqbal, S.Z.; Asi, M.R.; Ariño, A. Aflatoxins. In *Brenner's Encyclopedia of Genetics*, 2nd ed.; Science Direct: London, UK, 2013; pp. 43–47.
6. Hammami, W.; Fiori, S.; Al Thani, R.; Ali Kali, N.; Balmas, V.; Migheli, Q.; Jaoua, S. Fungal and aflatoxin contamination of marketed spices. *Food Control* **2014**, *37*, 177–181. [CrossRef]
7. Blesa, J.; Soriano, J.M.; Moltó, J.C.; Mañes, J. Limited survey for the presence of aflatoxins in foods from local markets and supermarkets in Valencia, Spain. *Food Addit. Contam.* **2004**, *21*, 165–171. [CrossRef] [PubMed]
8. Van de Perre, E.; Jacxsens, L.; Lachat, C.; El Tahan, F.; De Meulenaer, B. Impact of maximum levels in European legislation on exposure of mycotoxins in dried products: Case of aflatoxin B1 and ochratoxin A in nuts and dried fruits. *Food Chem. Toxicol.* **2015**, *75*, 112–117. [CrossRef] [PubMed]
9. Van der Merwe, K.J.; Steyn, P.S.; Fourie, L. Mycotoxins. II. The constitution of ochratoxins A, B, and C, metabolites of *Aspergillus ochraceus* Wilh. *J. Chem. Soc. Perkin* **1965**, 7083–7088. [CrossRef]
10. Téren, J.; Varga, J.; Hamari, Z.; Rinyu, E.; Kevei, F. Immunochemical detection of ochratoxin A in black *Aspergillus* strains. *Mycopathologia* **1996**, *134*, 171–176. [CrossRef] [PubMed]
11. Abarca, M.L.; Bragulat, M.R.; Castellá, G.; Cabañes, F.J. Impact of some environmental factors on growth and ochratoxin A production by *Aspergillus niger* and *Aspergillus welwitschiae*. *Int. J. Food Microbiol.* **2018**, *291*, 10–16. [CrossRef] [PubMed]
12. Schmidt-Heydt, M.; Bode, H.; Raupp, F.; Geisen, R. Influence of light on ochratoxin biosynthesis by *Penicillium*. *Mycotoxin Res.* **2010**, *26*, 1–8. [CrossRef] [PubMed]
13. Wolff, J.; Bresch, H.; Cholmakov-Bodechtel, C.; Engel, G.; Garais, M.; Majerus, P.; Rosner, H.; Scheuer, R. Ochratoxin A: Contamination of foods and consumer exposure. *Arch. Lebensmittelhyg.* **2000**, *51*, 81–128.
14. Petzinger, E.; Weidenbach, A. Mycotoxins in the food chain: The role of ochratoxins. *Livestock Prod. Sci.* **2002**, *76*, 245–250. [CrossRef]

15. Rizzo, A.; Eskola, M.; Atroshi, F. Ochratoxin a in Cereals, Foodstuffs and Human Plasma. *Eur. J. Plant Pathol.* **2002**, *108*, 631–637. [CrossRef]

16. Denli, M.; Perez, J.F. Ochratoxins in feed, a risk for animal and human health: Control strategies. *Toxins* **2010**, *2*, 1065–1077. [CrossRef] [PubMed]

17. Reddy, L.; Bhoola, K. Ochratoxins-food contaminants: Impact on human health. *Toxins* **2010**, *2*, 771–779. [CrossRef] [PubMed]

18. Jackson, L.S.; Katta, S.K.; Fingerhut, D.D.; DeVries, J.W.; Bullerman, L.B. Effects of baking and frying on the fumonisin B1 content of corn-based foods. *J. Agric. Food Chem.* **1997**, *45*, 4800–4805. [CrossRef]

19. Meca, G.; Blaiotta, G.; Ritieni, A. Reduction of ochratoxin A during the fermentation of Italian red wine Moscato. *Food Control* **2010**, *21*, 579–583. [CrossRef]

20. Engel, E.; Baty, C.; Le Corre, D.; Souchon, I.; Martin, N. Flavor-active compounds potentially implicated in cooked cauliflower acceptance. *J. Agric. Food Chem.* **2002**, *50*, 6459–6467. [CrossRef] [PubMed]

21. Luciano, F.B.; Holley, R.A. Enzymatic inhibition by allyl isothiocyanate and factors affecting its antimicrobial action against Escherichia coli O157:H7. *Int. J. Food Microbiol.* **2009**, *131*, 240–245. [CrossRef] [PubMed]

22. Isshiki, K.; Tokuoka, K.; Mori, R.; Chiba, S. Preliminary examination of allyl-isothiocyanate vapor for food preservation. *Biosci. Biotechnol. Biochem.* **1992**, *56*, 1476–1477. [CrossRef]

23. Manyes, L.; Luciano, F.B.; Mañes, J.; Meca, G. In vitro antifungal activity of allyl isothiocyanate (AITC) against *Aspergillus parasiticus* and *Penicillium expansum* and evaluation of the AITC estimated daily intake. *Food Chem. Toxicol.* **2015**, *83*, 293–299. [CrossRef] [PubMed]

24. Nazareth, T.M.; Bordin, K.; Manyes, L.; Meca, G.; Mañes, J.; Luciano, F.B. Gaseous allyl isothiocyanate to inhibit the production of aflatoxins, beauvericin and enniatins by *Aspergillus parasiticus* and Fusarium poae in wheat flour. *Food Control* **2016**, *62*, 317–321. [CrossRef]

25. Zhang, Y. Cancer-preventive isothiocyanates: Measurement of human exposure and mechanism of action. *Mutat. Res.* **2004**, *555*, 173–190. [CrossRef] [PubMed]

26. Cejpek, K.; Valusek, J.; Velísek, J. Reactions of allyl isothiocyanate with alanine, glycine, and several peptides in model systems. *J. Agric. Food Chem.* **2000**, *48*, 3560–3565. [CrossRef] [PubMed]

27. Okano, K.; Ose, A.; Takai, M.; Kaneko, M.; Nishioka, C.; Ohzu, Y.; Odano, M.; Sekiyama, Y.; Mizukami, Y.; Nakamura, N.; et al. Inhibition of aflatoxin production and fungal growth on stored corn by allyl isothiocyanate vapor. *J. Food Hyg. Soc. Jpn.* **2015**, *56*, 1–7. [CrossRef] [PubMed]

28. Delaquis, P.J.; Sholberg, P.L. Antimicrobial activity of gaseous allyl isothiocyanate. *J. Food Prot.* **1997**, *60*, 943. [CrossRef]

29. Suhr, K.I.; Nielsen, P.V. Antifungal activity of essential oils evaluated by two different application techniques against rye bread spoilage fungi. *J. Appl. Microbiol.* **2003**, *94*, 665–674. [CrossRef] [PubMed]

30. Winther, M.; Nielsen, P.V. Active packaging of cheese with allyl isothiocyanate, an alternative to modified atmosphere packaging. *J. Food Prot.* **2006**, *69*, 2430–2435. [CrossRef] [PubMed]

31. Quiles, J.M.; Manyes, L.; Luciano, F.; Manes, J.; Meca, G. Antimicrobial compound allylisothiocyanate against the *Aspergillus parasiticus* growth and its aflatoxins production in pizza crust. *Food Chem. Toxicol.* **2015**, *83*, 222–228. [CrossRef] [PubMed]

32. Tracz, B.L.; Bordina, K.; Nazareth, T.M.; Costa, L.B.; Macedo, R.E.F.; Meca, G.; Luciano, F.B. Assessment of allyl isothiocyanate as a fumigant to avoid mycotoxin production during corn storage. *LWT* **2017**, *75*, 692–696. [CrossRef]

33. Saladino, F.; Quiles, J.M.; Luciano, F.B.; Mañes, J.; Fernandez-Franzón, M.; Meca, G. Shelf life improvement of the loaf bread using allyl, phenyl and benzyl isothiocyanates against *Aspergillus parasiticus*. *LWT* **2017**, *78*, 208–214. [CrossRef]

34. Serrano, A.B.; Font, G.; Mañes, J.; Ferrer, E. Emerging Fusarium mycotoxins in organic and conventional pasta collected in Spain. *Food Chem. Toxicol.* **2013**, *51*, 259–266. [CrossRef] [PubMed]

35. Pitt, J.I.; Hocking, A.D. *Fungi and FoodSpoilage*; Blackie Academic and Professional: London, UK, 1997; Volume II.

© 2019 by the authors. Licensee MDPI, Basel, Switzerland. This article is an open access article distributed under the terms and conditions of the Creative Commons Attribution (CC BY) license (http://creativecommons.org/licenses/by/4.0/).

Article

Influence of Two Garlic-Derived Compounds, Propyl Propane Thiosulfonate (PTS) and Propyl Propane Thiosulfinate (PTSO), on Growth and Mycotoxin Production by *Fusarium* Species *In Vitro* and in Stored Cereals

Kalliopi Mylona [1,†], **Esther Garcia-Cela** [1], **Michael Sulyok** [2], **Angel Medina** [1] and **Naresh Magan** [1,*]

[1] Applied Mycology Group, Environment and AgriFood Theme, Cranfield University, Cranfield MK43 0AL, UK
[2] Institute of Bioanalytics and Agro-Metabolomics, Department of Agrobiotechnology (IFA-Tulln), University of Natural Resources and Life Sciences, Vienna, Konrad Lorenzstr. 20, A-3430 Tulln, Austria
* Correspondence: n.magan@cranfield.ac.uk
† Present address: Centre for Agriculture, Food and Environmental Management (CAFEM) School of Life and Medical Sciences, University of Hertfordshire College Lane, Hatfield, Hertfordshire AL10 9AB, UK.

Received: 29 July 2019; Accepted: 24 August 2019; Published: 27 August 2019

Abstract: Two garlic-derived compounds, Propyl Propane Thiosulfonate (PTS) and Propyl Propane Thiosulfinate (PTSO), were examined for their efficacy against mycotoxigenic *Fusarium* species (*F. graminearum*, *F. langsethiae*, *F. verticillioides*). The objectives were to assess the inhibitory effect of these compounds on growth and mycotoxin production *in vitro*, and *in situ* in artificially inoculated wheat, oats and maize with one isolate of each respectively, at different water activity (a_w) conditions when stored for up to 20 days at 25 °C. *In vitro*, 200 ppm of either PTS or PTSO reduced fungal growth by 50–100% and mycotoxin production by >90% depending on species, mycotoxin and a_w conditions on milled wheat, oats and maize respectively. PTS was generally more effective than PTSO. Deoxynivalenol (DON) and zearalenone (ZEN) were decreased by 50% with 80 ppm PTSO. One-hundred ppm of PTS reduced DON and ZEN production in wheat stored at 0.93 a_w for 20 days, although contamination was still above the legislative limits. Contrasting effects on T-2/HT-2 toxin contamination of oats was found depending on a_w, with PTS stimulating production under marginal conditions (0.93 a_w), but at 0.95 a_w effective control was achieved with 100 ppm. Treatment of stored maize inoculated with *F. verticilliodies* resulted in a stimulation of total fumonsins in most treatments. The potential use of such compounds for mycotoxin control in stored commodities is discussed.

Keywords: *Fusarium*; mycotoxins; garlic-derived extracts; green chemistry; fungi; EU limits; abiotic factors; storage; wheat; maize; oats

Key Contribution: PTSO was more effective than PTS in inhibiting *Fusaria in vitro*. *In vitro* efficacy for control of growth/mycotoxin production was influenced by water activity. *In situ* efficacy by these garlic-derived compounds was less effective as in *in vitro* studies. At times, lower doses triggered mycotoxin production *in situ*.

---

1. Introduction

There has been interest in the use of essential oils (EOs) and extracts derived from plants to control food spoilage microorganisms, especially mycotoxigenic moulds, as an alternative to traditional preservatives based on aliphatic acids [1]. However, an important aspect to consider when utilizing natural plant extracts for control of mycotoxigenic spoilage fungi is whether they are classed as food grade. In addition, many studies have studied effects on germination and growth of spoilage mycotoxigenic fungi, while neglecting impacts on mycotoxin production, especially *in situ*.

Onion and garlic, both members of the *Allium* family, have received attention as extracts from these two plant species have been found to have significant antimicrobial properties [1]. Their antifungal efficacy has been studied despite the relative instability of some of their compounds or their strong odour. Yin and Tsao [2] found garlic, out of seven *Allium* plants, to be the most effective against three *Aspergillus* species. Some inter-species differences in efficacy were previously noted, with higher concentrations of plant extracts required for control of *Aspergillus flavus* and *Aspergillus fumigatus* than *Aspergillus niger*. Yoshida et al. [3] suggested that the antifungal activity of garlic was due to the compounds allicin and ajoene. Benkeblia [4] observed growth inhibition of *A. niger* and *Penicillium cyclopium* by red onion and garlic EOs, while higher concentrations (200–500 mL/L) of green and yellow onion EOs were required for control the fungal pathogen *Fusarium oxysporum*. Singh and Singh [5] studied the effect of *Allium sativum* extracts and other plant extracts on the growth of *A. flavus* and aflatoxin production, but only in liquid cultures. Addition of the extract at the beginning of the incubation period showed 85% inhibition in mycelial biomass and complete inhibition of aflatoxin production. However, a later addition only gave marginal control of toxin biosynthesis. Liquid culture systems are relatively artificial as the aim should be to try and develop intervention strategies to control mycotoxigenic fungi under similar conditions found in different agrifood commodities or under simulated nutritional conditions relevant to where the target control measures are going to be instituted.

In the last decade, some compounds have been extracted from *Allium sativus* and been successfully utilised as antimicrobial preservatives in a number of products, especially in cheese production and as a fruit coating. The two extracted compounds being used are propyl propane thiosulfonate (PTS) an organosulphate, and propyl propane thiosulfinate (PTSO). Formulations of these compounds are now commercially available as Proallium, especially as a coating for extension of fruit shelf-life. While they are considered effective anti-microbials, very little detailed information is available on the efficacy of these compounds against mycotoxigenic fungi, either *in vitro* or *in situ*.

Previous studies to compare some EOs and antioxidants to control *P. verrucosum* and ochratoxin A (OTA) production showed that of those tested only resveratrol was effective at controlling populations of the mycotoxigenic species and inhibiting OTA production in stored wheat grain under different storage conditions [6]. Environmental factors were also shown to have a significant influence on the relative control achieved. In addition, this study showed that sometimes under sub-optimal concentrations of these compounds, some stimulation of toxin production occurred despite growth being significantly reduced. This has been suggested to be due to a combination of water and physiological stress caused by the antifungal agent itself which may stimulate secondary metabolite production as a defence response [7–10].

Some very early studies suggested thiosulfinates had potential applications as anti-microbials, although very focused on their use in post-harvest grain preservation [11]. Some extracts of garlic and onion have been shown to have promising efficacy against *Aspergillus* and *Penicillium* species and in some cases mycotoxin production. However, very limited information is available on the efficacy of such EOs on *Fusarium* species. In addition, such compounds have rarely been assessed on naturally contaminated grains where a range of different species may be encountered.

The objectives of this study were to study the effect of two garlic-derived compounds (PTS, PTSO) for the control of (a) fungal growth and (b) mycotoxin production by *Fusarium graminearum* (Deoxynivalenol, Zearalenone; wheat), *F. langsethiae* (T-2/HT-2 toxins; oats) and *F. verticilioides* (Fumonisins; maize) for the first time. Experiments were initially conducted *in vitro* for a preliminary assessment of these two compounds to determine the effective concentrations for control of growth and mycotoxin production. Subsequently, *in situ* experiments were carried out with inoculation of wheat, oats and maize with the relevant mycotoxigenic species and treatment with each of these two compounds based on *in vitro* results and stored at 25 °C under different moisture content conditions for up to 20 days to examine effects on mycotoxin production.

2. Results

2.1. In Vitro Efficacy of PTS and PTSO Garlic-Derived Compounds against Fungal Growth

Figure 1 shows the effect of 0–200 ppm PTS and PTSO on the *in vitro* radial growth rates of *F. graminearum*, *F. verticillioides* and *F. langsethiae*. Both PTS and PTSO had very good inhibitory effects on the growth of the isolate of each of these species studied and this increased with concentration. Complete inhibition of the growth of *F. langsethiae* was observed with 100 ppm PTS and of *F. verticillioides* with 200 ppm. Growth of *F. graminearum* was significantly inhibited, although complete inhibition was not achieved in the range of concentrations examined. Two hundred ppm PTSO completely inhibited the growth of *F. langsethiae* and was very effective against *F. graminearum* and *F. verticillioides*. PTS was always more effective than PTSO when applied at the same concentration against the same fungal species. Table 1 shows the concentration necessary for effective dose (ED) 50 and 90% control of growth of the three *Fusarium* species.

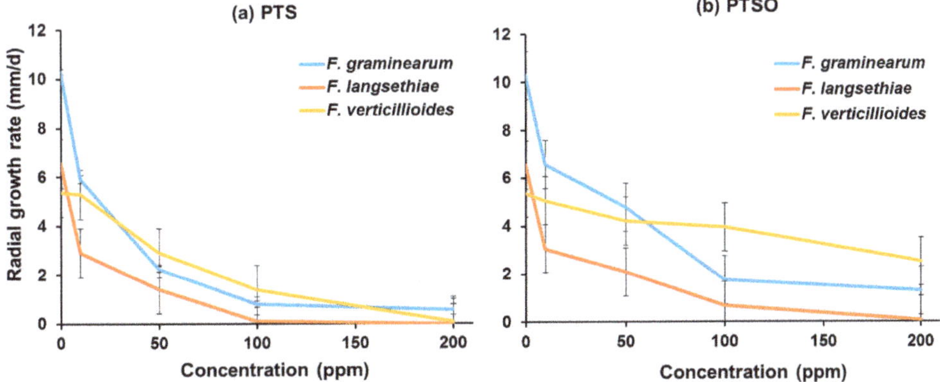

Figure 1. Effect of 0–200 ppm (**a**) Propyl Propane Thiosulfonate (PTS) and (**b**) Propyl Propane Thiosulphonate (PTSO) on the *in vitro* (2% wheat agar medium) on the radial growth rate (n = 6) of an isolate of *F. graminearum*, *F. langsethiae* and *F. verticillioides*. Vertical bars indicate the standard error of the means. For PTS effects on growth: *F. graminearum*: H (4, N = 30) = 26.612, $p < 0.001$; *F. langsethiae*: H (4, N = 30) = 27.647, $p < 0.001$; *F. verticillioides*: H (4, N = 30) = 26.522, $p < 0.001$; for fungal species: H (2, N = 90) = 3.176, $p = 0.2004$. For PTSO effects on growth: *F. graminearum*: H (4, N = 30) = 27.451, $p < 0.001$; *F. langsethiae*: H (4, N = 30) = 27.877, $p < 0.001$; fungal species: H (2, N = 90) = 12.926, $p = 0.002$.

Table 1. Effective dose of PTS and PTSO for 50% and 90% control (ED_{50}; ED_{90} values) of growth, when compared to the untreated control, for inhibition of *F. graminearum*, *F. langsethaie* and *F. verticillioides* on 2% milled wheat agar medium at 25 °C.

Treatment (ppm)	PTS		PTSO	
Species	ED_{50}	ED_{90}	ED_{50}	ED_{90}
F. graminearum	33	144	52	186
F. langsethiae	12	80	21	113
F. verticillioides$_0$	64	172	>200	>200

The statistical analyses (ANOVA) showed that the effect of PTS and PTSO concentration was highly significant on the growth of the isolate of each *Fusarium* species, with significant intra-isolate differences for growth rate in the presence of the either of these two compounds.

Additional studies were carried out to assess the efficacy of the two compounds at different water activity (a_w) levels on growth of the three *Fusarium* species. Figure 2 shows an example of the effect of 0–100 ppm PTSO on the growth of *F. graminearum* and *F. langsethiae* on wheat agar media modified to three different a_w levels. The radial growth of both isolates of the two species was significantly inhibited as the PTSO concentration was increased and water stress was imposed (0.94, 0.92 a_w), when compared to the unmodified control medium. The maximum percentage (%) inhibition was observed at the highest PTSO concentration and 0.995 a_w (~83% for *F. graminearum* and ~90% for *F. langsethiae*). ANOVA showed that the effects of either PTS or PTSO concentration, a_w and fungal species were highly significant on the log-transformed radial growth rate data, while the effects of the interactions between these factors were not significant.

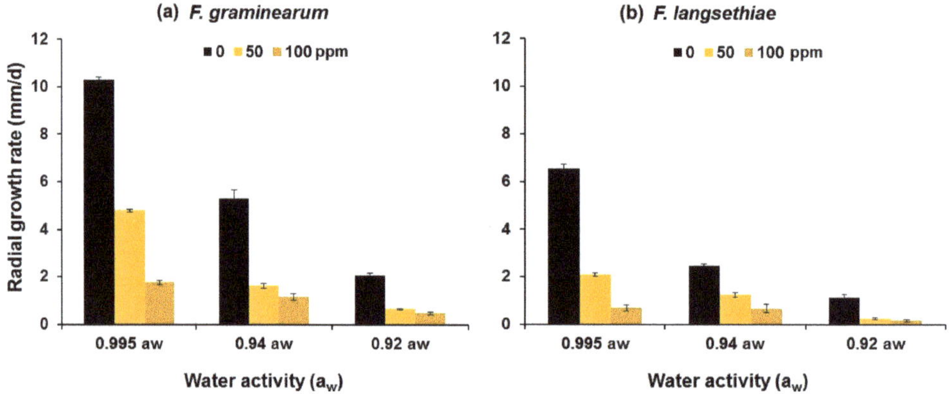

Figure 2. Effect of 0–100 ppm Propyl Propane Thiosulfinate (PTSO) on the radial growth rates (n = 6) of an isolate of (**a**) *F. graminearum* and (**b**) *F. langsethiae in vitro* (2% wheat agar medium) in relation to water activity (a_w). Vertical bars indicate the standard error of the means. Statistical analyses for (**a**) *F. graminearum*: H (4, N = 30) = 27.451, $p < 0.001$; and for (**b**) *F. langsethiae*: H (4, N = 30) = 27.877, $p < 0.001$.

Table 2 summarises the effective dose (ED_{50}) concentrations of PTSO only for the isolates of *F. graminearum*, *F. langsethiae* and *F. verticillioides* used in this study grown on wheat agar media modified to different a_w levels and 25 °C. Overall, lower concentrations of each compound were required for 50% inhibition of growth of the isolate of *F. langsethiae* than for the isolates of the other two *Fusarium* species.

Table 2. Effective dose ED_{50} values (ppm) of PTSO for 50% inhibition of the growth of *F. graminearum*, *F. verticillioides* and *F. langsethiae* on wheat agar media of different water activities (a_w) at 25 °C.

Treatments	*F. graminearum*	*F. verticillioides*	*F. langsethiae*
PTSO (0.98 a_w)	42	189	9.5
PTSO (0.94 a_w)	36	>100	50
PTSO (0.92 a_w)	37	>100	32

2.2. Effects of the Compounds on In Vitro Mycotoxin Inhibition

Figure 3 shows the effect of different concentrations of PTS and PTSO on fumonisins B_1 and B_2 (FB$_1$ and FB$_2$) production by the isolate of *F. verticillioides* on wheat-based medium at 25 °C. Two-hundred ppm of PTS inhibited FB$_1$ and FB$_2$ toxins by up to 90%. For PTSO, the production of both toxins was slightly stimulated with up to 100 ppm. However, the production of the two fumonisins was inhibited with 200 ppm. ANOVA showed that the effect of PTS concentration was highly significant on the log-transformed data of both FB_1 and FB_2. For PTSO concentration, analyses showed that there was a significant effect on FB_1 production by the isolate of *F. verticillioides in vitro* but it was not significant for the production of FB_2 (see Supplementary Table S1).

The production of deoxynivalenol (DON) by *F. graminearum* in response to exposure to either of these compounds was reduced when compared to the untreated control (Supplementary Figure S1). DON production was below the limit of detection in the 200 ppm PTS treatments. However, DON production in the 250 ppm PTSO was ~4 times more than in the control. ANOVA showed that the concentration of PTS significantly affected DON production. PTSO had no significant effect on *in vitro* DON production by *F. graminearum*.

There was a decrease in T-2 production by *F. langsethiae* as the PTS concentration was increased, and this was below the limit of detection with 200 ppm concentration (data not shown). HT-2 toxin in the same samples was not detected at ≥50 ppm PTS (Supplementary Figure S2; Supplementary Table S2). With PTSO, T-2 toxin production by *F. langsethiae* was inhibited compared to the control, although no specific pattern was observed with concentration. HT-2 toxin production was below the limit of detection in all PTSO concentrations. Statistically, the effect of PTS concentration was significant on T-2 toxin production but not for HT-2 toxin. The effect of PTSO was not significant for either of these two related type A trichothecenes.

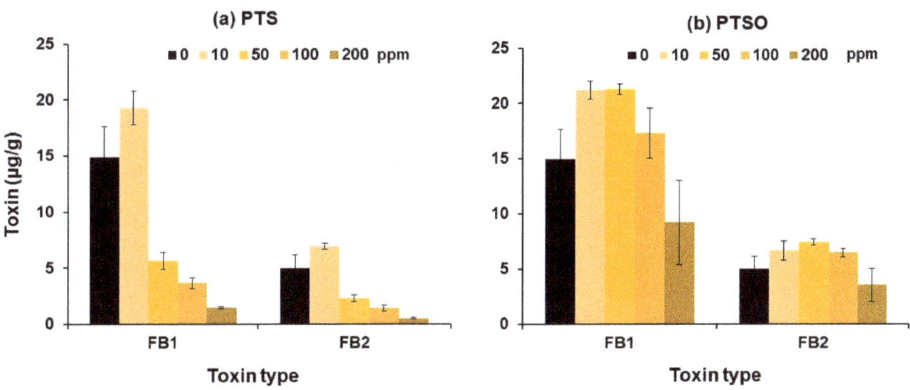

Figure 3. Effect of 0–200 ppm of (**a**) PTS and (**b**) PTSO on the production of fumonisins B_1 and B_2 by *F. verticillioides in vitro* at 25 °C. Vertical bars indicate the standard error of the means. For statistical analyses see Supplementary Table S2.

2.3. In Situ Mycotoxin Control in Stored Wheat, Oats and Maize Using PTS and PTSO

Figures 4 and 5 show the effects of either PTS or PTSO on DON and ZEN production in wheat modified to two different a_w levels, inoculated with *F. graminearum* and stored for 10 and 20 days at 25 °C. The red lines show the EU regulatory limits for each of the toxins in wheat [12] (EC 1881/2006). DON production was reduced in all the PTS-treated samples stored at 0.93 a_w for up to 20 days at 25 °C compared to the control. Maximum inhibition was obtained with 100 and 200 ppm PTS after 10 and 20 days storage (76% and 95%) respectively when compared with the control. In the wetter wheat samples at 0.95 a_w stored for 10 days DON production increased with increasing PTS concentration, while after 20 days irregular results were obtained ranging from 90% inhibition of DON production with 200 ppm PTS to >100% stimulation at 300 ppm. In the PTSO treatments, a small reduction in DON contamination (~33%) was observed with 80 ppm PTSO in wheat stored for 10 days at both a_w levels. With lower concentration of PTSO (40 ppm) there was a stimulation of toxin production when compared to the control. After 20 days storage stored wheat treated with PTSO had toxin levels higher than in the controls at both a_w levels. ANOVA showed that the effects of PTS concentration, a_w and storage time significantly affected log(DON) production by *F. graminearum* in the stored wheat treatments (see Supplementary Table S3). However, interactions between these factors was not significant. For PTSO, ANOVA showed that concentration and aw significantly affected log (DON) production, while the effects of storage time and interactions between factors were not significant.

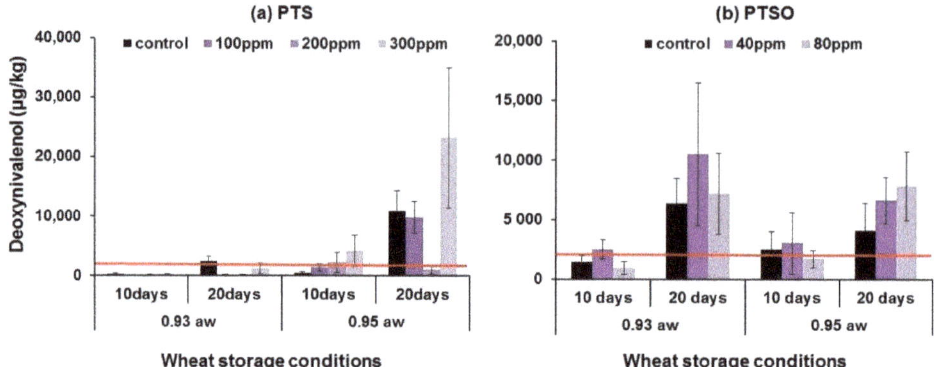

Figure 4. Effect of 0–300 ppm of PTS (**a**) and 0–80 ppm PTSO (**b**) on deoxynivalenol production by *F. graminearum* in artificially inoculated wheat of 0.93 and 0.95 a_w stored for 10 and 20 days at 25 °C (n = 2 × 3). Vertical bars indicate the standard error of the means. The red lines show the EU legislative limits (EC 1881/2006) for deoxynivalenol in unprocessed wheat for feed use (1750 µg/kg).

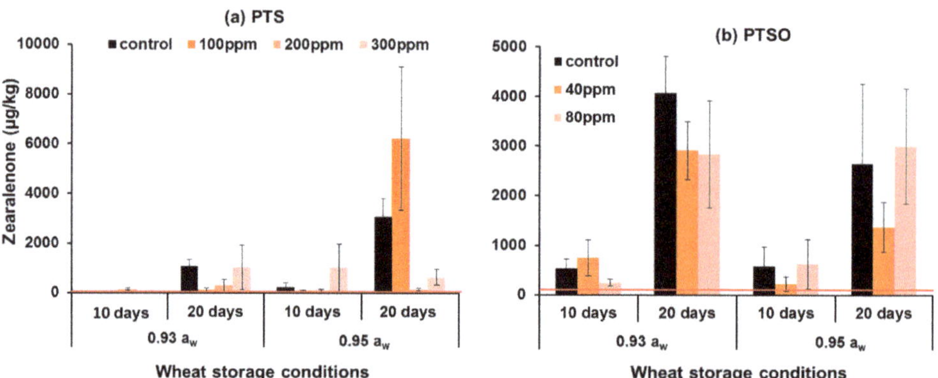

Figure 5. Effect of (**a**) 0–300 ppm PTS and (**b**) 0–80 ppm PTSO on zearalenone production by *F. graminearum* in artificially inoculated stored wheat at 0.93 and 0.95 a_w for 10 and 20 days at 25 °C (n = 2 × 3). Vertical bars indicate the standard error of the means. The red lines show the EU legislative limits (EC. 1881/2006) for zearalenone in unprocessed wheat for human use (100 µg/kg). Kruskal–Wallis ANOVA for Statistical effects for PTS concentration: H (3, N = 48) = 1.39, p = 0.707, grain a_w: H (1, N = 48) = 3.44, p = 0.064, storage time: H (1, N = 48) = 12, p < 0.001; For PTSO conc.: H (2, N = 36) = 0.47, p = 0.789, a_w: H (1, N = 36) = 1.48, p = 0.223, storage time: H (1, N = 36) = 21.34, p < 0.001.

Figure 5 shows the effect of PTS and PTSO on ZEN production by *F. graminearum* in stored wheat. The most effective control of ZEN production by PTS was 100ppm at 0.93 a_w with 89% control after 20 days storage. In the wetter 0.95 a_w grain 100 ppm of PTS inhibited ZEN production by about 64% after 10 days storage. However, after 20 days storage, 200 ppm PTS was required for 96% ZEN control. Overall, the most efficient PTS concentrations for the control of both toxins (DON, ZEN) was 100 ppm PTS in the 0.93 a_w stored wheat, and 200 ppm for the wetter stored wheat (0.95 a_w). PTSO at 40 and 80 ppm reduced ZEN contamination of stored wheat inoculated with *F. graminearum* at both 0.93 and 0.95 a_w by 30–60% after 20 days storage. Statistically, the Kruskal–Wallis analyses (non-parametric analyses) showed that there was no significant effect of PTS concentration or grain a_w on ZEN production, while storage time was highly significant. For PTSO, concentration or storage a_w were not significant for ZEN production, while storage time was highly significant.

Figure 6 shows the T-2 + HT-2 toxins in stored oats inoculated with *F. langsethiae* and treated with 0–300 ppm PTS and 0–80 ppm PTSO for up to 20 days. The red line shows the indicative directive with regard to limits more commonly established in Europe for the sum of T-2 + HT-2 toxins [13] (EC 165/2013). In all cases, the toxin levels produced were below the indicative levels suggested by the EU. Statistical analyses showed that the effects of PTS concentration, and storage time had no significant effect on log(T-2 + HT-2) toxin contamination of stored oats by *F. langsethiae* (see Supplementary Table S3) The effect of a_w, and interaction between a_w × storage time were significant. For PTSO, ANOVA showed that concentration and a_w had significant effects on the production of these two toxins while the effect of storage time and interactions between the factors were not significant.

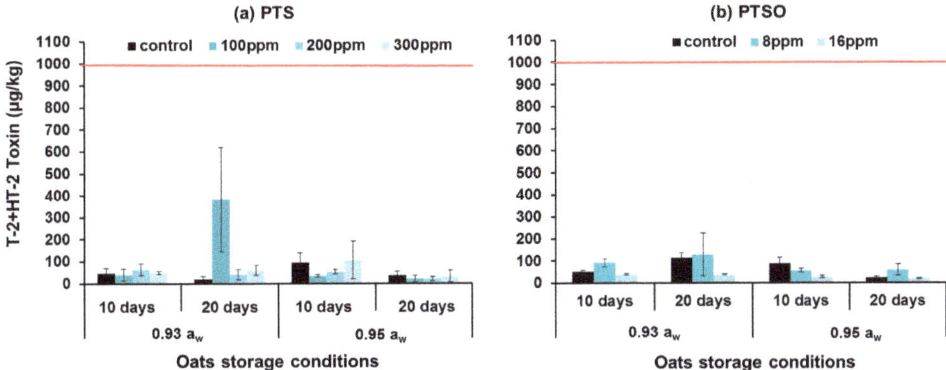

Figure 6. Effect of (**a**) 0–300 ppm aqueous PTS and (**b**) 0–16 ppm PTSO on T-2+HT-2 toxin production by *F. langsethiae* in artificially inoculated oats of 0.93 and 0.95 a_w and stored for 10 and 20 days at 25 °C. Vertical bars indicate the standard error of the means. The red lines show the EU recommended for maximum limits (EC. 165/2013) for sum T-2 and HT-2 toxin in unprocessed oat for human use (1000 µg/kg).

Figure 7 shows the effect of treatments on the production of total fumonisins B_1 + B_2 (FUMs) in maize rewetted to 0.91 and 0.94 a_w and inoculated with *F. verticillioides* spores and treated with 0–300 ppm PTS, Only PTS was studied in these assays because of the limited efficacy of PTSO in controlling FUMs production *in vitro* (see Figure 3). The red line shows the EU legislative limits established for the sum of FUMs in maize. At 0.91 a_w PTS was effective with up to 80% control achieved after 20 days with 200–300 ppm treatment. This was also below the relevant EU limit. However, in wetter maize stored at 0.94 a_w most treatments were ineffective in controlling FUMs contamination. Indeed there was a stimulation in FUMs, especially with 100–200 ppm PTS, regardless of a_w or storage time. There was a statistically significant effect of PTS concentration, maize a_w and storage time on the logarithm of total FUMs production (see Supplementary Table S3). The interaction of a_w×storage time and a_w × PTS concentration were also significant. However, the interactions of PTS × storage time and interaction between all three factors was not significant. For PTSO, none of the single factors or interacting factors were significant except for a_w × storage time was significant.

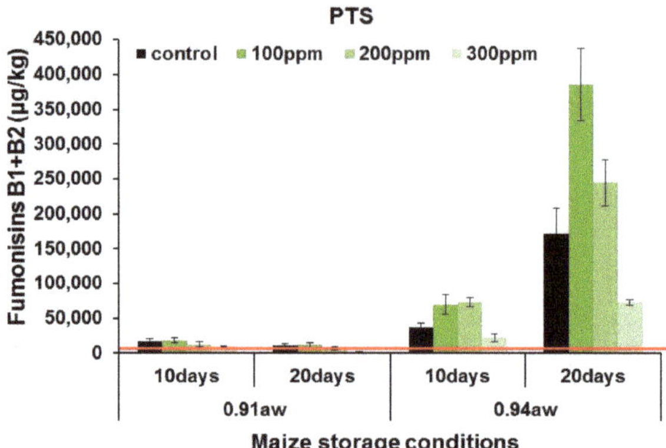

Figure 7. Effect of 0–300 ppm aqueous PTS on total fumonisins production by *F. verticillioides* in artificially inoculated maize stored at 0.91 and 0.94 a_w for 10 and 20 days at 25 °C. Vertical bars indicate the standard error of the means. The red lines show the EU legislative limits (EC. 1881/2006) for fumonisins (FB_1 + FB_2) in unprocessed maize for human use (2000 µg/kg).

3. Discussion

3.1. In Vitro Efficacy of PTS and PTSO

This is the first study to evaluate the efficacy of the garlic-derived compounds PTS and PTSO for control of fungal growth and mycotoxin production by isolates of the three important mycotoxigenic *Fusarium* species *in vitro*. The present study has shown that solutions of either PTS or PTSO were effective at inhibiting the *in vitro* growth of each isolate of the three *Fusarium* species studied. The inhibitory effect increased with concentration and generally, the level of inhibition was species-dependent. *F. langsethiae* was the most sensitive *Fusarium* species to both compounds with complete inhibition of growth with 100 and 200 ppm of PTS and PTSO respectively.

Our results suggest that there is differential sensitivity of *Fusarium* species to PTS and PTSO. Thus, higher ED_{50} concentrations of the two garlic-derived compounds were required for the isolate of *F. verticilloides* when compared to *F. gaminearum* and *F. langshetiae*. Previously, 200 ppm of star anise extract was needed to inhibit growth of *F. verticillioides* completely when compared to effects on *F. solani*, *F. oxysporum* and *F. graminearum* where only 100 ppm was required [14]. Similarly, Morcia et al. [15] observed lower ED_{50} values of seven EOs were needed in the case of *F. langsethiae* when compared to effects on *F. graminearum* and *F. sporotrichioides*. However, most of these studies did not assess interactions with water availability which is critical for determining efficacy, especially in relation to toxin control [14–22]. It is also difficult to compare the dose-effect relationships obtained between species across the studies because different times and doses, and different matrices were used, often nutritionally unrelated to the commodity in which *in situ* control was required.

In addition, different EO extracts have different active ingredients and thus do not have the same efficacy against the same species. For example, the effect of EOs belonging to five botanical families (Umbeliferae, Labiateae, Compositeae, Rosaceae and Lauraceae) against *F. moniliforme* (=*F. verticillioides*) was evaluated [16]. Growth was completely inhibited with 500 ppm of anise, thyme and cinnamon, 2000 ppm of marigold, and 3000 ppm of the other extracts examined. Overall, no significant differences between these botanical families were found. Also, complete inhibition of this *Fusarium* species was achieved with 600 ppm of thyme, 800 ppm of basil and lemongrass and 1000–2500 ppm of ginger [17,18]. These differences could be due to the different tolerances between strains, but also because of the different compositions and purity of EOs obtained from the different plants. Elhouiti et al. [23] studied the chemical composition of leaves and flowers of *Rhanterium adpressum*, harvested at different times over three years. They observed that the percentage of the leading chemical groups changed according to the month of extraction. Also, the EOs produced by the flower had better inhibitory activity than the leaf extracts (MICs, 6–10 ppm vs 11–14 ppm) against *F. culmorun* and *F. graminearum* respectively. Kurita and Koike et al. [24] pointed out that the magnitude of the antifungal effect was related to the functional groups and they proposed a scale of antifungal potency of chemical groups from the best to worst being phenols > alcohols > aldehydes > ketones > ethers > hydrocarbons. However, these previous studies did not evaluate the impact that changing water availability might have on the relative control achieved.

In terms of control of mycotoxin production, both PTS and PTSO reduced DON, ZEN, T-2, HT-2 and FUMs production by the relevant *Fusarium* species when compared to the controls. There were some differences in efficacy between PTS and PTSO. The latter compound only controlled FB_1 and FB_2 production at 200 ppm, with >4 times the amount of PTS required, compared to PTSO to obtain the same inhibitory effects. Generally, PTS was more effective in controlling mycotoxin production than PTSO, except in the case of T-2 toxin where the latter compound was more effective at <200 ppm. Similarly, HT-2 toxin was completely inhibited at all PTSO concentrations, while for PTS ≥50 ppm was required. It was also generally observed that higher PTS concentrations (200 ppm) completely inhibited the production of all toxins, except in the case of the FUMs.

Previous studies have suggested that complete inhibition of FUMs could be achieved with 6 ppm of 3-carene, D-limonene and B-ocimene [21]. Also, ginger EOs inhibited FB_1 (2500 ppm), FB_2 (2000 ppm) and DON (2000 ppm) production [17,22]. Lower doses of extracts of *R. adpressum* (0.25 ppm) significantly inhibited production of type B tricothecenes (3-acetyl deoxynivalenol, 15-acetyl deoxynivalenol and fusarenon X) [23]. Morcia et al. [15] reported that the seven EOs (cuminaldehyde, cinnamaldehyde, lemon oil, citral, limonene, bergamot and citroneall) had a variable effect on the biosynthesis of T-2 and HT-2 by *F. langsethiae* and *F. sporotrichioides*. 0.1 mL of bergamoil/mL reduced T-2 and HT-2 toxin produced by the former species but stimulated production by the latter species. However, again, few of these studies included the impact of a_w stress on the efficacy of the EOs.

3.2. In Situ Efficacy of PTS and PTSO against Mycotoxin Production in Stored Grain

The a_w levels chosen for the wheat, oats and maize storage studies was based on the marginal and optimum a_w levels for growth and mycotoxin production by the isolates of these three species from previous studies [25–27]. Drier conditions of <0.90 a_w are considered very marginal for colonisation by *Fusarium* species of wheat, oats or maize. Treatment of wheat with 100 ppm of PTS resulted in 76–94% control of DON, depending on the storage a_w. Better results were obtained in the 0.93 a_w treatment where colonisation was slower than in the 0.95 a_w treatment. PTS was also effective in controlling ZEN production, by reducing the production to below or near the applicable EU limits after both 10 and 20 days storage respectively. However, it should be noted that under some storage conditions intermediate PTS concentrations stimulated mycotoxin production.

Overall, PTS was not efficient in controlling the production of T-2 and HT-2 toxins in artificially inoculated oats at 0.93 a_w, particularly in samples stored for 20 days. In contrast, at 0.95 a_w reasonably good control with 100 ppm PTS for T-2 and HT-2 production was achieved after 10 days storage, and with 200 ppm after 20 day storage. For control of FUMs, 300 ppm PTS was necessary in stored maize at 0.91 a_w and 0.94 a_w resulting in 39–80% inhibition when compared to the control samples. The control achieved was below the EU legislative limits but only in the 0.91 a_w treatments. In moist maize, much higher concentrations would be required to try and control FUMs to below the EU legislative limits, especially if destined for feed use.

Treatment of wet grain with PTSO was generally capable of reducing mycotoxin contamination after storage at 0.93–0.95 a_w for 10–20 days compared to the untreated control samples. Thus, where water ingress might occur in a silo this compound would be effective at controlling toxin contamination in stored cereals produced by *Fusaria*, in the short to medium term. DON contamination in wheat treated with 80 ppm PTSO and stored for 10 days had 1/3rd less toxin than the control, which was also below the relevant EU limits. However, this treatment was not efficient for extended storage beyond

20 days with the exception of ZEN, which was more effectively reduced, even after 20 days storage. Better control was observed with 40 ppm PTSO (48–60%) in relatively moist wheat stored at 0.95 a_w. However, all the treatments contained ZEN levels above the applicable EU limits.

Overall, PTSO was more efficient in controlling the production of T-2/HT-2 toxins by *F. langsethiae* in oats, with as little as 16 ppm required. This treatment was generally more effective against HT-2 toxin (8–78%) than T-2 toxin (18–42%) and this may be important, as often T-2 toxin is converted rapidly to HT-2 toxin and thus control of this toxin is important.

Previously, Soliman et al. [16] examined the efficacy of low concentrations (0.1–2.0 ppb) of the more efficient EOs tested *in vitro* (anise, cinnamon, spearmint and thyme) on FUMs contamination of stored wheat over 8 week storage periods. They claimed that low doses (0.1 ppb) completely inhibited the biosynthesis of FUMs after 2 weeks. Thyme EO was shown to have the highest anti-mycotoxigenic activity and the best control of growth. Although water availability was not considered, the concentrations appear to be very low for achieving control. In addition, *F. verticillioides* colonisation is more important in maize where it is primarily responsible for FUM contamination. Venkatesh et al. [21] suggested that use of guggul EO at 10 ppm treatment of maize at 28 °C for 10 days reduced FUMs from 42.5 to 2.6 ppm; with complete inhibition of FB_1 achieved with 200 ppm of star anise or 50 ppb of allyl isothiocyanate [14,28]. Allyl isothiocyanate was found to inhibit the production of FB_1 by *F. verticilloides* [2,28].

These cereals are naturally contaminated with a mixture of toxigenic fungi as part of the mycobiota. Thus, the differential effect on different Fusarium species would also apply to other toxigenic species such as *Penicillium verrucosum* (ochratoxin A producer) or *Aspergillus* species. Thus, consideration should be given to changes in the ratio of mycotoxins which might occur when treated with different preservatives. Recently, Giorni et al. [29] showed that in ripening maize co-inoculated with mixtures of *F. graminearum*, *F. verticillioides* and *Aspergillus flavus* influenced the relative contamination of the maize cobs with deoxynivalenol, FUMs and aflatoxins. Interactions between non-toxigenic mycobiota and has also been shown to influence the relative contamination with different mycotoxins in both wheat grain and in grape-based matrices [30].

The *in situ* storage studies have been done for a maximum of 20 days. For longer term storage periods of 6–9 months it may be necessary to use a slightly increased treatment concentration of such compounds for ensuring that control can be maintained. This would have an impact on the relative economic costs of treatment which would have to be considered in the context of the overall inputs into management of grain for food and feed use post-harvest.

4. Conclusions

This was the first detailed examination of these two compounds for control of growth and mycotoxin contamination by *Fusarium* species *in vitro* and in artificially inoculated stored wheat, oats and maize under different temporal and water availability conditions. Overall, *in vitro* efficacy should include important parameters such as water availability, temperature and perhaps pH stress to identify the most effective candidates for control of colonisation, and more importantly, mycotoxin production. Potential efficacy was demonstrated and identified against isolates of three different *Fusarium* species. However, efficacy *in situ* was not as effective as that observed *in vitro*. In addition, the efficacy of the treatments depended on the specific "*Fusarium* species-toxin" pathosystem. Differences were observed with regard to the production of different mycotoxins by an isolate of a single fungal species when inoculated into naturally contaminated cereals. However, the right concentrations need to be used for effective control to be achieved. This depended on the water availability and the mycotoxigenic species involved.

Both garlic-derived compounds tested in this study (PTS and PTO) are liquids and water soluble and can thus be applied to grain prior to storage for post-harvest control of spoilage mycotoxigenic fungi. Garlic extracts have been approved for use as a pesticide (although currently under re-evaluation) and commercial products are available based on such extracts. Certainly, the use of odourless versions of these compounds could be effectively used for food applications. In addition, the existing compounds could also be used for animal feed applications where often moist grain needs to be preserved for the short to medium term prior to use.

5. Materials and Methods

5.1. Preparation of Stock Solutions of Chemical Compounds

Stock solutions of the following compounds were prepared in sterile distilled water.

(a) Propyl propane thiosulfonate (PTS): This organosulfonate compound is obtained from the decomposition of initial compounds present in garlic bulbs (*Allium sativum*) and was kindly provided by DOMCA SA, Granada, Spain. A stock solution of 5000 ppm was prepared by dissolving 1.1 g PTS (90% PTS, Domca, S.A., Granada, Spain) (Mousala SL., 2006) into a 200 mL container containing 200 mL of sterile distilled water and vigorously shaking. Due to the oily nature of PTS the stock solution had the appearance of a stable water emulsion. A second stock solution of 20000 ppm was prepared by dissolving 2.2 g PTS into 100 mL of a mixture of ethanol:H_2O (80:20). This solution was clear.

(b) Propyl propane thiosulfinate (PTSO): A stock solution of 10000 ppm was prepared by dissolving 1.1 g PTSO (90% PTSO, Domca, S.A., Granada, Spain) (Mousala SL., 2006) into a 100 mL container containing 100 mL sterile distilled water and vigorously shaking. Due to the oily nature of PTSO the stock solution had the appearance of a water emulsion. A second stock solution of 20,000 ppm was prepared as for PTS.

5.2. In Vitro Studies: Fungal Species, Media, Inoculation and Measurements of Growth and Mycotoxin Production

Fusarium graminearum isolate L1-2/2D (wheat; DON, ZEN), *F. langsethiae* strain 2004/59 (oats; T-2 + HT-2) and *F. verticillioides* isolate MPVP 294 (maize; FUMS) were used in this study. The strains were all maintained on Malt Extract Agar (MEA) media (OXOID, malt extract, 30; mycological peptone, 5; agar, 15 g/L). These isolates were kindly supplied by Prof. S. Edwards, Harper Adams University and Prof. P. Battilani, Catholic University of Italy, Piacenza, Italy). They have all been examined previously for mycotoxin production and shown to be high producers of the respective mycotoxins in previous studies [31,32].

For the *in vitro* trials a 2% milled wheat medium was prepared by adding 2% milled wheat and 2% agar (OXOID Ltd, Basingstoke, England) to water to obtain the basal medium. For the initial screening, concentrations of PTS and PTSO in the range 10–200 ppm were used by adding the necessary stock solutions to the molten cooled medium, shaking vigorously and then pouring the media into 9 cm Petri plates (15 mL per plate). The basal 2% media had a water activity (a_w) value of 0.995. This basal medium was modified by replacing water with different amounts of mixtures of glycerol/water solution to the milled cereal + agar to obtain the target a_w values of 0.92, 0.94 and 0.98 (20.7, 15.4 and 4.9 g glycerol/50 mL of water, respectively) without diluting the nutritional status of the media. These a_w levels represent the range over which these *Fusaria* can effectively grow [33]. The media were all checked with an Aqualab TE4 to confirm the actual a_w levels were achieved.

Agar plugs (4 mm diameter) cut from the margin of 10-day-old cultures with a sterile cork borer, were used as an inoculum for the *in vitro* trials. Three replicate per treatment and replicate plates were centrally inoculated with the inoculum agar plugs. All experiments were repeated once. Each a_w treatment and replicates were stored in separate polyethylene bags to maintain the environmental conditions over the experimental period.

The treatments and replicates were all incubated at 25 °C for 10 days, or until the Petri dishes were completely colonised by the fungi. Each concentration of the PTS and PTSO treatments were kept separately in polyethylene bags to avoid cross-contamination. Two diameters of the colonies formed (at right angles of each other) were measured daily and compared against the diameters of the controls. From these data the relative growth rates were calculated and the effect of different concentrations was calculated. The percentage inhibition of mycelial growth of the *Fusarium* species was determined at each different chemical compound concentration and the different water activities.

For mycotoxin analyses on the tenth day of incubation, agar plugs (5, 5-mm diameter) were cut out from each of the replicate plates diagonally across the colony. The agar plugs were placed in 2-mL safe-lock Eppendorf®tubes (Eppendorf AG, Hamburg, Germany), their weight was recorded and frozen at −40 °C for subsequent toxin analysis. The extraction and analysis of the relevant toxins for each fungal species were performed according to the methods described later.

5.3. In Situ Studies with Stored Cereal Grain

Fungal inoculum: Cultures of the above fungal species were prepared on MEA and incubated at 25 °C for 10 days. A Tween 80 solution was prepared by addition of one drop of Tween 80 (ACROS organics) in 100 mL sterile water. Spore suspensions were prepared by gently scraping the culture surface with a sterile spatula and transferring the spores into sterile 25 mL Universal glass vials containing the water + Tween 80 solution. The spore suspensions were filtered through glass wool in order to remove any mycelial fragments. The spore concentration was determined using a haemocytometer (Olympus BX40 microscope, Microoptical Co.; slide Marienfeld superior, Germany; microscope glass cover slips, No 3, 18 × 18mm, Chance Proper LTD, Smethwick, UK) and adjusted by dilution to 10^7 spores/mL. Naturally contaminated wheat, oats and maize were artificially inoculated with these suspensions for the *in situ* storage experiments.

Grain equilibration: Water adsorption curves were prepared for each grain type. The amount of water required to accurately modify these cereals to 0.91, 0.94 (maize) and 0.93 and 0.95 a_w (wheat, oats) was determined from these curves. Initially, each grain type (approx. 1 kg) was taken from a 25-kg bag and placed in a Duran bottle (2.5 L). The initial moisture content was known from the moisture adsorption curves. The grain was randomly divided into batches of 100 g in Duran flasks (1 L). These were labeled for each treatment condition and the required amounts of sterile water added to each one including the treatment PTS or PTSO stock solution, shaken vigorously and sealed. They were placed at 4°C to equilibrate overnight. The treatments were then inoculated with 1-mL spore suspension containing ~10^5 spores/mL of the individual *Fusarium* species and thoroughly mixed using a roller mixer in order for the spores to become dispersed throughout the grain mass. For each grain type, 15 g was weighed into surface sterilized 40 mL vials (Chromacol Ltd., London, UK) with microporous lids to obtain six replicates per a_w treatment and stored in sandwich boxes for up to 20 days. The equilibrium relative humidity (ERH) was maintained by including 2 × 500 mL of a glycerol-water solution in beakers to maintain the treatment a_w levels. The experiments were carried out twice with six replicates per treatment.

For each experiment, after 10 and 20 days storage, three replicates were destructively removed from storage chambers and frozen at −20 °C for subsequent toxin analysis. Grain samples were oven-dried for 24–48 h at 60 °C, milled and then extracted and analysed as described later.

5.4. Mycotoxin Analyses

5.4.1. Equipment Description

High performance liquid chromatography (HPLC) were used to quantify DON, T-2 and HT-2 from media. In addition, liquid chromatography tandem mass spectrometry (LC-MS/MS) were used to quantify FUMS from media and *Fusarium* toxins from grains. HPLC used consisted of an Agilent 1200 Series system equipped with a UV diode array detector (DAD) set at 220.4 nm (Agilent Technologies, Palo Alto, CA, USA). The column used for the chromatographic separation was a Phenomenex® Gemini C_{18}, 150 mm × 4.6 mm, 3 μm (Phenomenex, Macclesfield, UK) preceded by a Phenomenex® Gemini 3 mm guard cartridge and the column temperature was set at 25 °C. LC-MS/MS used consisted of an QTrap 5500 LC-MS/MS System (Applied Biosystems, Foster City, CA, USA) equipped with a TurboIonSpray electrospray ionization (ESI) source and an 1290 Series HPLC System

(Agilent, Waldbronn, Germany). Chromatographic separation was performed at 25 °C on a Gemini®
C_{18}-column, 150 × 4.6 mm i.d., 5 μm particle size, equipped with a C_{18} 4 × 3 mm i.d. security guard
cartridge (all from Phenomenex, Torrance, CA, USA).

5.4.2. *In Vitro* Extraction and Analysis

Agar plugs were removed from media and placed in the 2-mL safe-lock Eppendorf®tubes. After
that, the weight of the agar plugs were register.

Deoxynivalenol

The extraction was performed using 1-mL acetonitrile:water (AcN:H_2O) (84:16), the mixture more
commonly used for trichothecenes extraction [34] and the tubes were shaken in an orbital shaker at
200 rpm in the dark for 60 min at 25 °C. The extract was transferred to a new tube and oven-dried
overnight at 60 °C. Subsequently, it was redissolved in 1 mL 90:10 (H_2O: AcN) and vortexed for a few
seconds. The cleaning step involved the addition of 150 mg/mL Alumina directly into the redissolved
extract followed by vortexing the mixture for 15 s. The treated extract was then filtered through a
0.22 μm Millipore filter (Minisart, Sartorius, Germany) into an amber silanised LC vial and inserted
into the LC-DAD for analysis. The chromatographic analysis was performed in the gradient mode,
using water (solvent A) and acetonitrile (solvent B). The starting composition of the mobile phase was
5% B, at a flow rate of 0.5 mL/min held for 2 min. The composition was then gradually changed to
25% B over 15 min and maintained for further 3 min. Then it increased gradually to 30% B over 3 min
at the same flow rate. The composition was then changed to 99% B during 1 min with a flow rate of
1 mL/min in this case, in order to achieve a fast cleaning step and maintained at 99% B for 4 more
minutes. Afterwards the composition of the mobile phase was changed linearly to 5% B in 1 min at a
flow rate of 1 mL/min and held for 4 min for further cleaning. In the last step, the composition was
maintained at 5% B, but the flow rate changed to 0.5 mL/min for 1 min, in order to be the same as the
starting composition of the mobile phase for the following chromatographic run. The injection volume
was 50 μL. The total time for the analysis of each sample was 35 min. DON was eluted from the column
at 16.2 min. The LOD was 4 μg/kg. The mean recovery for DON using this method was 63.2 ± 2.8%.

T-2 and HT-2

Extraction method used was Medina et al. [32] with modifications. The extraction was performed
using 1 mL acetonitrile:water (AcN:H_2O) (84:16) and the tubes were shaken for 1 h at 150 rpm at 25 °C
in the dark in an orbital shaker. The samples were then centrifuged at 1150× *g* for 15 min. The extract
was filtered through a 0.2 μm Millipore filter (Minisart, Sartorius) directly into an HPLC silanised
amber vial and injected in the chromatograph. The analysis was performed in the gradient mode
with a mobile phase of AcN:H_2O at a flow rate of 1 mL/min and the conditions were 3 min 30% AcN,
changed linearly to 55% AcN over 18 min, changed to 99% AcN in 1 min and held to 99% AcN for
5 min. The LOD was 4 and 5 μg/kg for T-2 and HT-2. The mean recoveries for this method were
99 ± 1.53% for T-2 toxin and 101.28 ± 3.11% for HT-2 toxin.

Fumonisins

The extraction was performed using 1 mL of extraction solvent AcN:H_2O:Acetic acid (79:20:1)
and the tubes were shaking 250 rpm in the dark for 1 h. The extracts were filtered through a 0.22 μm
Millipore filter (Minisart, Sartorius) into new tubes, and dried in an oven at 60 °C for 24 h. The dried
extracts were redissolved in a mixture of AcN:H_2O (1:1) containing 1% acetic acid. The individual
fumonisins were quantified using LC-MS/MS according to the method of Vishwanath et al. [35]. LOD
was 25 μg/kg with a recovery rate of 57% (FB_1) and 70% (FB_2).

5.4.3. *In Situ* Extraction and Analysis

The initial mycotoxin contamination levels were quantified and this was taken into account when calculating the results of the treatments and appropriately corrected. The mean contamination levels were: DON, 0.233 µg/kg; T-2 toxin, 9.07 µg/kg (no HT-2 toxin present); ZEN, 8.42 µg/kg, and FB_1 0.14 µg/kg. The grain samples were oven-dried at 60 °C for 24–48 h and then milled in a small laboratory blender (Waring Commercial, Christison, UK). Samples were analysed for *Fusarium* toxins by LC-MS/MS at the Centre for Analytical Chemistry, Department of Agrobiotechnology (IFA-Tulln, Tulln, Austria), University of Natural Resources and Life Sciences, Vienna, Austria. The analysis was performed according to the methods described by Sulyok et al. and Vishwanath et al. [35,36]. The accuracy of the method was externally checked by participation in proficiency testing organised by the Bureau Inter Professionel d'Etudes Analytiques (BIPEA; Gennevilliers, France). Z-scores were 0.4 and 0.62 for DON in two wheat samples, −0.8 and −1.09 for ZON in two wheat samples and 1.36 and 1.55 for FB_1 and FB_2, respectively in a sample of maize.

5.5. Statistical Analysis

All experiments have been performed in triplicate and repeated once. Data were analysed with Microsoft Office Excel 2007 and with the package STATISTICA 9 (StatSoft®, Inc. 2010. STATISTICA (data analysis software system), version 9.1. www.statsoft.com, (Tulsa, OK, USA). The standard error of the mean was calculated in all trials and it is denoted with vertical bars in the figures.

Datasets were tested for normality and homoscedasticity using the Shapiro–Wilk and Levene test, respectively. When data failed the normality test, variable transformation was performed to try to improve normality or homogenise the variances. If still not normally distributed, it was analysed using the Kruskal–Wallis test by ranks.

Supplementary Materials: The following are available online at http://www.mdpi.com/2072-6651/11/9/495/s1, Figure S1: Effect of 0–250 ppm PTSO and PTS on *in vitro* DON production by *F. graminearum* in wheat agar media at 25 °C, Figure S2: Effect of 0–250 ppm (**a**) PTSO and (**b**) PTS on the production of T-2 and HT-2 toxins by *F. langsethiae in vitro* at 25 °C, Table S1: One-way ANOVA for the effect of different PTS and PTSO concentrations on *in vitro* fumonisins (B_1 and B_2) production by *F. verticillioides*, Table S2: Kruskal-Wallis ANOVA by ranks for the effect of PTSO concentration and substrate water activity on the production of T-2 and HT-2 toxins by *F. langsethiae*, Table S3: Summary of statistical *P*-values of effects of (**a**) PTS and (**b**) PTSO on mycotoxin contamination of stored cereals.

Author Contributions: K.M. and E.G.-C. carried out the practical work and statistical analyses, M.S. carried out some of the mycotoxin analyses in naturally contaminated cereals and with preservatives. A.M. and N.M. supervised the research work and wrote the draft and final manuscript.

Funding: Parts of this work were funded by the European Union via the FP7 MYCORED Project (Grant Agreement No. 222690).

Acknowledgments: The authors would like to thank S. Edwards, Harper Adams University and P. Battilani, Catholic University of Italy, Piacenza, Italy) for supplied the strains.

Conflicts of Interest: The authors declare no conflict of interest.

References

1. Prakash, B.; Mishra, P.K.; Kedia, A.; Dwivedy, A.K.; Dubey, N.K. Efficacy of some essential oil components as food preservatives against food contaminating molds, aflatoxin b_1 production and free radical generation. *J. Food Qual.* **2015**, *38*, 231–239. [CrossRef]
2. Yin, M.; Tsao, S. Inhibitory effect of seven *Allium* plants upon three *Aspergillus species*. *Int. J. Food Microbiol.* **1999**, *49*, 49–56. [CrossRef]
3. Yoshida, S.; Kasuga, S.; Hayashi, N.; Ushiroguchi, T.; Matsuura, H.; Nakagawa, S. Antifungal Activity of Ajoene Derived from Garlic. *Appl. Environ. Microbiol.* **1987**, *53*, 615–617.
4. Benkeblia, N. Antimicrobial activity of essential oil extracts of various onions (*Allium cepa*) and garlic (*Allium sativum*). *LWT - Food Sci. Technol.* **2004**, *37*, 263–268. [CrossRef]

5. Singh, I.; Singh, V.P. Effect of plant extracts on mycelial growth and aflatoxin production by *Aspergillus flavus*. *Indian J. Microbiol.* **2005**, *45*, 139–142.

6. Aldred, D.; Cairns-Fuller, V.; Magan, N. Environmental factors affect efficacy of some essential oils and resveratrol to control growth and ochratoxin A production by *Penicillium verrucosum* and *Aspergillus westerdijkiae* on wheat grain. *J. Stored Prod. Res.* **2008**, *44*, 341–346. [CrossRef]

7. Magan, N.; Aldred, D. Post-harvest control strategies: Minimizing mycotoxins in the food chain. *Int. J. Food Microbiol.* **2007**, *119*. [CrossRef]

8. Garcia, D.; Garcia-Cela, E.; Ramos, A.J.; Sanchis, V.; Marín, S. Mould growth and mycotoxin production as affected by *Equisetum arvense* and *Stevia rebaudiana* extracts. *Food Control* **2011**, *22*, 1378–1384. [CrossRef]

9. Garcia-Cela, E.; Gil-Serna, J.; Marin, S.; Acevedo, H.; Patino, B.; Ramos, A.J. Effect of preharvest anti-fungal compounds on *Aspergillus steynii* and *A. carbonarius* under fluctuating and extreme environmental conditions. *Int. J. Food Microbiol.* **2012**, *159*, 167–176. [CrossRef]

10. Magan, N.; Hope, R.; Colleate, A.; Baxter, E. Relationship Between Growth and Mycotoxin Production by *Fusarium* species, Biocides and Environment. *Eur. J. Plant Pathol.* **2002**, *108*, 685–690. [CrossRef]

11. Small, L.D.; Bailey, J.H.; Cavallito, C.T. Alkyl thiosulfinates. *J. Am. Chem. Soc.* **1947**, *69*, 1710–1713. [CrossRef]

12. European Commission. Commission Regulation (EC) No. 1881/2006 of 19 December 2006, Setting maximum levels for certain contaminants in foodstuffs. *Off. J. Eur. Union* **2006**, *364*, 5–24.

13. European Commission. Commission Recommendation No 2013/165/EU of 27 March 2013 on the presence of T-2 and HT-2 toxin in cereals and cereal products. *Off. J. Eur. Union* **2013**, *91*, 12–15. [CrossRef]

14. Aly, S.E.; Sabry, B.A.; Shaheen, M.S.; Hathout, A.S. Assessment of antimycotoxigenic and antioxidant activity of star anise (*Illicium verum*) in vitro. *J. Saudi Soc. Agric. Sci.* **2016**, *15*, 20–27. [CrossRef]

15. Morcia, C.; Tumino, G.; Ghizzoni, R.; Bara, A.; Salhi, N.; Terzi, V.; Morcia, C.; Tumino, G.; Ghizzoni, R.; Bara, A.; et al. *In vitro* evaluation of sub-lethal concentrations of plant-derived antifungal compounds on FUSARIA growth and mycotoxin production. *Molecules* **2017**, *22*, 1271. [CrossRef]

16. Soliman, K.; Badeaa, R. Effect of oil extracted from some medicinal plants on different mycotoxigenic fungi. *Food Chem. Toxicol.* **2002**, *40*, 1669–1675. [CrossRef]

17. Yamamoto-Ribeiro, M.M.G.; Grespan, R.; Kohiyama, C.Y.; Ferreira, F.D.; Mossini, S.A.G.; Silva, E.L.; de Abreu Filho, B.A.; Mikcha, J.M.G.; Machinski Junior, M. Effect of *Zingiber officinale* essential oil on *Fusarium verticillioides* and fumonisin production. *Food Chem.* **2013**, *141*, 3147–3152. [CrossRef]

18. Nguefack, J.; Leth, V.; Amvam Zollo, P.H.; Mathur, S.B. Evaluation of five essential oils from aromatic plants of Cameroon for controlling food spoilage and mycotoxin producing fungi. *Int. J. Food Microbiol.* **2004**, *94*, 329–334. [CrossRef]

19. Xing, F.; Hua, H.; Selvaraj, J.N.; Zhao, Y.; Zhou, L.; Liu, X.; Liu, Y. Growth inhibition and morphological alterations of *Fusarium verticillioides* by cinnamon oil and cinnamaldehyde. *Food Control* **2014**, *46*, 343–350. [CrossRef]

20. Sumalan, R.-M.; Alexa, E.; Poiana, M.-A. Assessment of inhibitory potential of essential oils on natural mycoflora and *Fusarium* mycotoxins production in wheat. *Chem. Cent. J.* **2013**, *7*, 32. [CrossRef]

21. Venkatesh, H.N.; Sudharshana, T.N.; Abhishek, R.U.; Thippeswamy, S.; Manjunath, K.; Mohana, D.C. Antifungal and antimycotoxigenic properties of chemically characterised essential oil of *Boswellia serrata* Roxb. ex Colebr. *Int. J. Food Prop.* **2017**, 1–13. [CrossRef]

22. Ferreira, F.M.D.; Hirooka, E.Y.; Ferreira, F.D.; Silva, M.V.; Mossini, S.A.G.; Machinski, M., Jr. Effect of *Zingiber officinale* Roscoe essential oil in fungus control and deoxynivalenol production of *Fusarium graminearum* Schwabe *in vitro*. *Food Addit. Contam. Part A* **2018**, *35*, 2168–2174. [CrossRef]

23. Elhouiti, F.; Tahri, D.; Takhi, D.; Ouinten, M.; Barreau, C.; Verdal-Bonnin, M.-N.; Bombarda, I.; Yousfi, M. Variability of composition and effects of essential oils from *Rhanterium adpressum* Coss. & Durieu against mycotoxinogenic *Fusarium* strains. *Arch. Microbiol.* **2017**, *199*, 1345–1356. [CrossRef]

24. Kurita, N.; Koike, S. Synergistic antimicrobial effect of ethanol, sodium chloride, acetic acid and essential oil components. *Agric. Biol. Chem.* **1983**, *47*, 67–75. [CrossRef]

25. Marin, S.; Magan, N.; Ramos, A.J.; Sanchis, V. Fumonisin-producing strains of *Fusarium*: A review of their ecophysiology. *J. Food Prot.* **2004**, *67*, 1792–1805. [CrossRef]

26. Hope, R.; Aldred, D.; Magan, N. Comparison of the effect of environmental factors on deoxynivalenol production by *F. culmorum* and *F. graminearum* on wheat grain. *Lett. Appl. Microbiol.* **2005**, *40*, 295–300. [CrossRef]

27. Medina, A.; Magan, N. Water availability and temperature affects production of T-2 and HT-2 by *Fusarium langsethiae* strains from north European countries. *Food Microbiol.* **2011**, *28*, 392–398. [CrossRef]
28. Tracz, B.L.; Bordin, K.; de Melo Nazareth, T.; Costa, L.B.; de Macedo, R.E.F.; Meca, G.; Luciano, F.B. dAssessment of allyl isothiocyanate as a fumigant to avoid mycotoxin production during corn storage. *LWT* **2017**, *75*, 692–696. [CrossRef]
29. Giorni, P.; Bertuzzi, T.; Battilani, P. Impact of fungi co-occurrence on mycotoxin contamination in maize during the growing season. *Front. Microbiol.* **2019**, *10*, 1265. [CrossRef]
30. Magan, N.; Aldred, D.; Hope, R.; Mitchell, D. Environmental factors and interactions with mycoflora of grain and grapes: effects on growth and deoxynivalenol and ochratoxin production by *Fusarium culmorum* and *Aspergillus carbonarius*. *Toxins* **2010**, *2*, 353–366. [CrossRef]
31. Mylona, K.; Sulyok, M.; Magan, N. Fusarium graminearum and *Fusarium verticillioides* colonisation of wheat and maize, environmental factors, dry matter losses and mycotoxin production relevant to the EU legislative limits. *Food Addit. Contam.* **2012**, *29*, 1118–1128. [CrossRef]
32. Medina, A.; Valle-Algarra, F.M.; Jiménez, M.; Magan, N. Different sample treatment approaches for the analysis of T-2 and HT-2 toxins from oats-based media. *J. Chromatogr. B* **2010**, *878*, 2145–2149. [CrossRef]
33. Sanchis, V.; Magan, N. Mycotoxins in food: Detection and control. In *Environmental Profiles for Growth and Mycotoxin Production*; Magan, N., Olsen, M., Eds.; Woodhead Publishing: Cambridge, UK, 2004; pp. 174–189.
34. Krska, R.; Welzig, E.; Boudra, H. Analysis of *Fusarium* toxins in feed. *Anim. Feed Sci. Technol.* **2007**, *137*, 241–264. [CrossRef]
35. Vishwanath, V.; Sulyok, M.; Labuda, R.; Bicker, W.; Krska, R. Simultaneous determination of 186 fungal and bacterial metabolites in indoor matrices by liquid chromatography/tandem mass spectrometry. *Anal. Bioanal. Chem.* **2009**, *395*, 1355–1372. [CrossRef]
36. Sulyok, M.; Berthiller, F.; Krska, R.; Schuhmacher, R. Development and validation of a liquid chromatography/tandem mass spectrometric method for the determination of 39 mycotoxins in wheat and maize. *Rapid Commun. Mass Spectrom.* **2006**, *20*, 2649–2659. [CrossRef]

© 2019 by the authors. Licensee MDPI, Basel, Switzerland. This article is an open access article distributed under the terms and conditions of the Creative Commons Attribution (CC BY) license (http://creativecommons.org/licenses/by/4.0/).

Article

Phytochemicals of Apple Pomace as Prospect Bio-Fungicide Agents against Mycotoxigenic Fungal Species—In Vitro Experiments

Marta Oleszek [1,*], **Łukasz Pecio** [2], **Solomiia Kozachok** [2,3], **Żaneta Lachowska-Filipiuk** [1], **Karolina Oszust** [1] and **Magdalena Frąc** [1]

[1] Institute of Agrophysics, Polish Academy of Sciences, Doświadczalna 4, 20-290 Lublin, Poland; zaneta_lachowska@wp.pl (Ż.L.-F.); k.oszust@ipan.lublin.pl (K.O.); m.frac@ipan.lublin.pl (M.F.)
[2] Institute of Soil Science and Plant Cultivation, State Research Institute, Czartoryskich 8, 24-100 Puławy, Poland; lpecio@iung.pulawy.pl (Ł.P.); skozachok@iung.pulawy.pl (S.K.)
[3] I. Horbachevsky Ternopil National Medical University, Maidan Voli 1, 46001 Ternopil, Ukraine
* Correspondence: m.oleszek@ipan.lublin.pl

Received: 10 May 2019; Accepted: 18 June 2019; Published: 20 June 2019

Abstract: The phytochemical constituents of apple waste were established as potential antifungal agents against four crops pathogens, specifically, *Botrytis* sp., *Fusarium oxysporum*, *Petriella setifera*, and *Neosartorya fischeri*. Crude, purified extracts and fractions of apple pomace were tested in vitro to evaluate their antifungal and antioxidant properties. The phytochemical constituents of the tested materials were mainly represented by phloridzin and quercetin derivatives, as well as previously undescribed in apples, monoterpene–pinnatifidanoside D. Its structure was confirmed by 1D- and 2D-nuclear magnetic resonance (NMR) spectroscopic analyses. The fraction containing quercetin pentosides possessed the highest antioxidant activity, while the strongest antifungal activity was exerted by a fraction containing phloridzin. Sugar moieties differentiated the antifungal activity of quercetin glycosides. Quercetin hexosides possessed stronger antifungal activity than quercetin pentosides.

Keywords: mycotoxins; *Fusarium* sp., *Botrytis* sp., apple pomace; phloridzin; quercetin glycosides; pinnatifidanoside D

Key Contribution: The results of the study showed that apple pomace could be a good source of natural bio-fungicide agents against mycotoxigenic fungal species, such as *Botrytis* sp., *Fusarium oxysporum*, *Petriella setifera* and *Neosartorya fischeri*.

---

1. Introduction

Billions of tons of agricultural waste are generated every year. A substantial part of them causes pollution problems, when they are not managed properly [1]. Apples are one of the crops with the largest annual production worldwide. Poland is one of the major producers of apples, with ca. 3.6 million metric tons of apples produced every year [2]. On the other hand, crop residues are a rich source of biologically active compounds and may become the important raw materials for obtaining various valuable by-products [3]. Apple pomace consists of apple skin, seeds, and flash, and represents about 25% of a fruit's fresh weight [4]. The main bioactive compounds of apple processing by-products are, in particular, flavonoids (phloretin and quercetin glycosides, flavone derivatives and catechins) as well as organic acids [5,6]. Their applications have been addressed to exploit antioxidant and pharmacological properties. Kołodziejczyk et al. [7] stated that polyphenols from industrial pomace were good antimicrobial agents against human pathogens such as *Salmonella* spp., *Escherichia coli* and

Listeria spp. Two main flavonoids of apple, phloretin and quercetin, have previously been isolated from apple fruits and tested against various fungi [8,9]. However, to the best of our knowledge, there is a lack of evidence on the usage of phytochemicals of industrial pomaces, particularly from apples, as natural bio-pesticides or bio-fungicides for organic farming [1]. There is a wide variety of chemically synthetized pesticides [10], but their application leads to a resistance and causes the selection of less-sensitive isolates [11]. Resistance to antimicrobial agents is consistently increasing and becoming a global problem. Moreover, many industrial fungicides are harmful to humans and detrimental to animal health [9]. For this reason, there is an urgent need to find new or more efficient, safe and ecologically friendly antifungal agents, especially against toxigenic fungi, that could be applied in organic farming [12].

A set of the following fungi: *Botrytis* sp., *Fusarium oxysporum*, *Petriella setifera* and *Neosartorya fischeri* is posing a worldwide threat in farming, gardening and food processing. *Fusarium* species may cause plant diseases of both underground and aboveground parts and can produce mycotoxins [10,13]. *Botrytis* sp. is an important pathogen in many economically important crops [14]. Additionally, *P. setifera* was classified as a potential plant pathogen [15]. The economic importance of this pathogen is connected with forest and especially with oak trees [16]. The other relevant magnitude of *Petriella* sp. is participation in wood decay as soft rot fungi or sometimes as brown rot fungi [17,18]. On the other hand, contamination by heat-resistant fungi such as *N. fischeri* is a major problem for the fruit-processing industry in many countries, due to mycotoxins such as verruculogen and fumitremorgins [19].

Data from the literature proved that flavonoids participate in the reaction against pathogen, both as components of plants tissues, but also when they are applied externally [12,20–22]. Sanzani et al. [20] showed that quercetin is very effective in reduction of *Penicilium expansum* growth and patulin accumulation in stored apples. The inhibited effect of low concentration of quercetin and rutin was observed also in vitro on *F. oxysporum* [21]. Parvez et al. [23] proved inhibitory effect of quercetin-3-O-glucoside (isoquercitrin) and quercetin-3-methyl ether, as well as its glycosides, on the conidial germination of *Neurospora crassa*.

Therefore, the aim of the presented study was to determine the antifungal and antioxidant activity of apple pomace's crude and purified extracts, chromatographic fractions and to evaluate their suitability as a source of natural bio-fungicides against *Botrytis* sp., *Fusarium oxysporum*, *Petriella setifera* and *Neosartorya fischeri*. For antifungal activity determination, a new, fast and simple instrumental method utilising BIOLOG MT2 Plates® was applied and optimised in the place of conventional hole-plate method [10]. Phytochemical constituents of the studied object were established by means of ultra high performance liquid chromatography-photodiode array detection-mass spectrometry (UHPLC-PDA-MS) analysis. Furthermore, for the first time, the undescribed constituent of apple-monoterpene pinnatifidanoside D was isolated and structurally elucidated.

2. Results and Discussion

Among all identified compounds, hyperoside, quercitrin and phloridzin were the most abundant in crude extract (CE) and purified extract (PE); (Table 1, Figure S1–S6. These results were in accordance with previous study [24–26]. Additionally, other flavonoids such as isoquercetin, rutin, reynoutrin, quercetin-3-O-pentosyls, avicularin, quercitrin and quercetin were determined. Furthermore, one monoterpene, not detected previously in apples-pinnatifidanoside D was isolated and structurally elucidated by extensive 1D and 2D nuclear magnetic resonance (NMR) spectroscopic analyses (Figure S7–S15). Characteristic data of this compound are as follows: pinnatifidanoside D: white amorphous solid; ultraviolet (UV) Λ_{max} (UPLC-PDA) 240 nm; electrospray ionization-in-source collision-induced dissociation mass spectrometry (ESI-isCID MS) (% of base peak) m/z 541 [M + Na]$^+$ (22), 519 [M + H]$^+$ (16), 387 [M − 132 + H]$^+$ (27), 225 [M − 132 − 162 + H]$^+$ (13), 207 [M − 132 − 162 − 18 + H]$^+$ (100), 189 [M − 132 − 162 − 2 × 18 + H]$^+$ (13), 161 (11), 149 (17), 123 (37); ^{1}H-NMR (500 MHz, MeOH-d$_4$), δ_H 5.89 (1H, t-like, J = 1.3 Hz, H-4), 5.85 (2H, m, H-7, 8), 4.43 (1H, qd, J = 6.4, 1.9 Hz, H-9), 4.35 (1H, d, J = 7.8 Hz, H-1′), 4.28 (1H, d, J = 7.5 Hz, H-1″), 4.06 (1H, dd, J = 11.3, 1.8 Hz, H-6a′), 3.86 (1H, dd, J = 11.5, 5.3 Hz, H-5a″), 3.69 (1H, dd, J = 11.3, 4.8 Hz, H-6b′), 3.49 (1H, ddd, J = 10.1, 8.7, 5.3 Hz,

H-4''), 3.35 (2H, m, H-4', 5'), 3.34 (1H, m, H-3'), 3.31 (1H, t, J = 8.8 Hz, H-3''), 3.22 (1H, dd, J = 9.0, 7.5 Hz, H-2''), 3.18 (1H, *t-like*, J = 8.4 Hz, H-2'), 3.18 (1H, t, J = 10.8 Hz, H-5b''), 2.51 (1H, d, J = 16.9 Hz, H-2a), 2.16 (1H, d, J = 16.9 Hz, H-2b), 1.92 (3H, d, J = 1.4 Hz, H-13), 1.29 (3H, d, J = 6.4 Hz, H-10), 1.04 (3H, s, H-11), 1.03 (3H, s, H-12); ^{13}C-NMR (125 MHz, MeOH-d$_4$), δ_C 201.2 (C-3), 167.2 (C-5), 134.9 (C-8), 131.7 (C-7), 127.2 (C-4), 105.6 (C-1''), 102.6 (C-1'), 80.0 (C-6), 77.9 (C-3'), 77.7 (C-3''), 76.9 (C-9), 76.8 (C-5'), 75.2 (C-2'), 74.8 (C-2''), 71.3 (C-4'), 71.2 (C-4''), 69.8 (C-6'), 66.9 (C-5''), 50.8 (C-2), 42.5 (C-1), 24.7 (C-12), 23.5 (C-11), 19.7 (C-13).

In turn, pinnatifidanoside D was the main component of fraction 1 (F1). This compound has been isolated for the first time from *Crataegus pinnatifida* [27]. Li et al. [27] stated also that pinnatifidanoside D exhibited small antiplatelet aggregation activity.

The LH20 fractions F2 and F3 contained many unknown compounds of various structures. Their identification was left for separate investigation. Analysing the UV-spectres, tandem mass spectrometry (MS/MS) fragmentation pattern and literature data allows us to identify the main components of the F4, F5 and F6. The major compound of F4 was phloridzin, flavonoid belonging to chalcones group. Fractions F5 and F6 consisted of quercetin derivatives, while F5 contained mostly quercetin with hexoside moieties, and F6 included mainly quercetin with pentoside moieties (Table 1).

Table 1. Quantification of compounds in crude and purified extracts, as well as selected fractions of apple pomace.

R_t (min)	Compound	MW (g mol^{-1})	% *w/w* (Relative)					
			[1] CE	PE	F1	F4	F5	F6
3.77	Pinnatifidanoside D (vomifoliol-9-O-[β-D-Xyl(1→6)-β-D-Glc])	518	0.11 (14)	1.23 (16)	5.9 (100)	-	-	-
6.29	Hyperoside (Q-3-O-β-D-Gal)	464	0.16 (21)	1.55 (20)	-	-	33.4 (43)	1.91 (3)
6.33	Rutin (Q-3-O-α-L-Rha(1→6)-β-D-Glc)	610	*	*	-	2.3 (5)	-	-
6.64	Isoquercetin (Q-3-O-β-D-Glc)	464	0.02 (3)	0.25 (3)	-	-	5.2 (7)	-
7.20	Reynoutrin (Q-3-O-β-D-Xyl)	434	0.05 (6)	0.52 (7)	-	-	-	20.17 (32)
7.53	Q-3-O-pentosyl	434	-	0.05 (1)	-	-	-	1.82 (3)
7.91	Avicularin (Q-3-O-α-L-Ara)	434	0.1 (13)	1.02 (13)	-	-	3.5 (5)	34.10 (53)
8.19	Q-3-O-pentosyl	434	0.02 (2)	0.16 (2)	-	-	-	5.86 (9)
8.47	Quercitrin (Q-3-O-α-L-Rha)	448	0.15 (19)	1.58 (20)	-	-	34.9 (45)	-
10.31	Phloridzin (phloretin-2'-O-β-D-Glc)	436	0.17 (22)	1.29 (17)	-	44.7 (95)	-	-
12.53	Quercetin	302	*	0.09 (1)	-	-	-	-
	Total, % *w/w*		0.77	7.75	5.9	47.00	77.04	63.86

[1] CE—crude extract; PE—purified extract; F—LH20 fractions; Q—quercetin; *—traces; R_t—retention time; MW—molecular weight.

The presence of phloretin and quercetin derivatives (glucoside, galactoside, xyloside, arabinoside, rhamnoside) in apples and their residues, particularly skins, has already been well recognised and confirms the results of the present study [24,25,28]. Moreover, many previous reports have also shown the presence of procyanidin B, epicatechins and chlorogenic acid as the major phenolic compounds in apple [29–32]. Tested crude extract did not contain catechins, probably because they are sensitive to oxidation by heat and light [33]. For the further investigation of antioxidants and antifungal activity, the LH20 fractions with established composition (F1, F4–F6) were selected.

The results of reducing power and radical-scavenging activity showed that apple pomace contained strong antioxidants. The antioxidant activity of the CE was low, due to the high content of the polar fraction (PF) containing mainly simple sugars, which do not exhibit antioxidant properties (Table 2). Nevertheless, the PE presented much higher values of the tested parameters. When it comes to LH20 fractions, values of EC_{50} and IC_{50} decreased along with subsequent fraction number. At the same time, F5 and F6 were not significantly different in terms of IC_{50} value of radical scavenging activity. The antioxidant activity depends on the structure of compounds, primarily the presence of hydroxyl, 4-oxo and catechol group as well as 2–3 double bond [34]. For this reason, the F4, containing mainly phloridzin, exhibited lower antioxidant properties (higher EC_{50} and IC_{50}) than F5 and F6, which included quercetin derivatives. Quercetin is known as a strong antioxidant, mainly due to the presence of catechol group in ring B [35].

Table 2. Antioxidant activity of tested samples.

Sample	Reducing Power EC_{50} (μg mL^{-1})	Radical-Scavenging Activity IC_{50} (μg mL^{-1})
CE	>1500	>1500
PE	298.33 ± 5.84	444.65 ± 10.57
PF	>1500	>1500
F1	460.60 ± 28.84	1117.21 ± 59.10
F4	137.38 ± 1.61	188.54 ± 7.95
F5	100.83 ± 1.62	105.92 ± 1.23
F6	93.94 ± 2.68	107.22 ± 1.77
Ascorbic acid	27.82 ± 0.07	73.61 ± 6.35

[1] CE—crude extract; PE—purified extract; PF—polar fraction; F—LH-20 subfraction.

Microbiological assays showed that apple pomace contained compounds with antifungal activity (Figure 1). In the case of *P. setifera* all tested formulations caused inhibition of the mycelium growth even at quite low doses. The exception was F6, which exhibited antifungal properties only at the highest concentration of 500 μg mL^{-1}. The CE of apple pomace caused also inhibition of the growth of *Botrytis* sp. at concentration in the range of 5–100 μg mL^{-1}, it stimulated the growth of *F. Oxysporum* and did not influence significantly the growth of *N. fischeri*.

Generally, the purification of crude extract increased its antifungal activity or weakened its stimulating effect, though the differences between CE and PE was not significant. Among all fractions, F1 showed no significant influence on *N. fischeri*, *Botrytis* sp., *F. oxysporum*, even regardless of the dose. Its effect on *P. setifera* was negative, but it was also independent from the concentration. On the contrary, F4 exhibited the strongest activity against all fungal strains. Moreover, it can be noticed, that A_1/A_0 absorbance ratios in the case of higher concentrations of F4 were below 100% (Figure 1). It means that absorbance for the control (the solution of tested substance without fungi) was higher than absorbance for tested sample (solution of tested substance with fungi). It can be supposed that such a low absorbance ratio was due to the fact that fungi intensively utilized the tested sample. Consequently, the concentration of the tested sample was decreased and its influence on the value of absorbance of the tested sample was reduced. To confirm this supposition, the ratio of absorbance at 490 nm and 750 nm was measured (A_{490}/A_{750}); (Table 3).

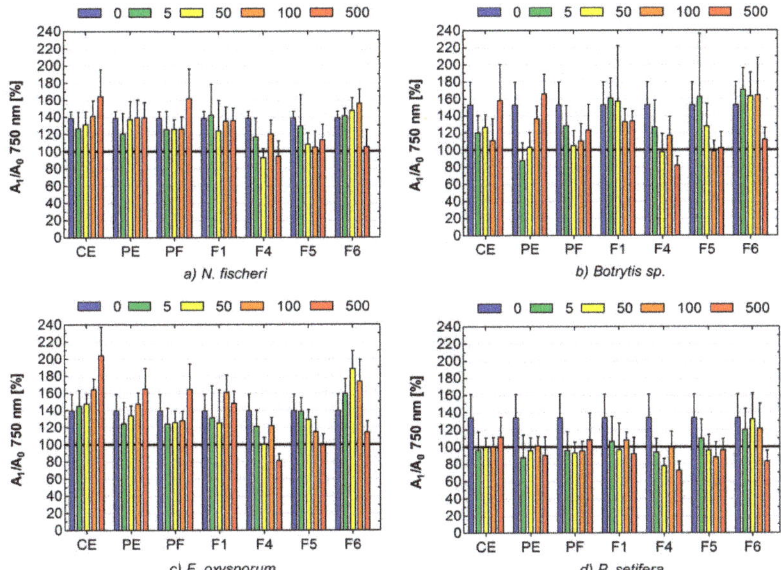

Figure 1. The impact of the apple pomace crude extract, purified extract and its fractions on the growth of fungi: (**a**) *Neosartoria fischeri*, (**b**) *Botrytis* sp., (**c**) *Fusarium oxysporum*, (**d**) *Petriella setifera*. The concentrations of solutions was 0, 5, 50, 100, 500 μL ml^{-1}. A1/A0—the ratio of absorbance for tested samples with fungi and absorbance for adequate control (sample alone, without fungi), CE—crude extract, PE—purified extract, PF—polar fraction and F—LH20 subfraction. Bars represent the mean of 24 replicates ± standard deviation.

The absorbance at 490 nm reflects the respiration rate, so also substrate use, while the value of absorbance at 750 nm informs us about biomass/turbidity production (growth pattern) [36]. According to the above, A_{490}/A_{750} ratio much higher than 1 indicates stressful metabolic situation, when a small biomass (low absorbance at 750 nm) yielding high respiration rates (high absorbance at 490 nm). The highest values of A_{490}/A_{750} were noted for the highest concentration of F4: 1.44, 1.20 and 1.26 for *F. oxysporum*, *Botrytis* sp., *P. setifera*, respectively (Table 3). In the case of *N. fischeri*, this ratio was 1.03, and no significant drop of A_1/A_0 below 100% was observed (Figure 1a).

The main compound of F4—phloridzin plays a major role in apple in the resistance to fungal infection. It is metabolized to phloretin and then, to the next oxidation products such as *o*-quinone, which are fungitoxic [37–39]. Antifungal activity of phloridzin and its aglycone, phloretin, was previously described [8,40]. The first report on the antifungal activity of phloretin against plant pathogenic fungi was done by Shim et al. [8], who investigated the influence of phloretin isolated from apple against *B. cinerea*, *F. oxysporum* and five other fungi. The results showed that phloretin could be used as biopesticide for control of rice blast as well as tomato late blight.

The F5 and F6 consisted of quercetin derivatives (Table 1). Quercetin, similarly to phloretin, affects the resistance of plants to fungal diseases. Lee et al. [41] observed the increase in concentration of quercetin glycosides in onion infected by *F. oxysporum*. Sanzani et al. [20] stated that quercetin in apple is responsible for the resistance on *P. expansum* and inhibition of patulin synthesis. For this reason, it can be considered as a natural compound to be used as alternative strategy to chemical fungicides in post-harvest control of *P. expansum* infections [9].

Table 3. The ratio of absorbance at 490 nm to absorbance at 750 nm (A490/A750).

Tested Sample	Concentration (µL mL^{-1})	A490/A750			
		N. fischeri	*F. oxysporum*	*Botrytis sp.*	*P. setifera*
Crude extract	0	0.93	0.96	0.89	0.89
	5	0.97	1.05	1.09	0.91
	50	1.00	1.07	1.10	0.80
	100	0.99	1.07	1.08	0.91
	500	1.04	1.15	1.14	0.74
Purified extract	0	0.93	0.96	0.89	0.89
	5	1.02	1.07	1.06	1.07
	50	0.99	1.03	1.06	0.92
	100	0.98	1.01	1.11	0.71
	500	0.95	1.01	1.03	0.52
Polar fraction of the extract	0	0.93	0.96	0.89	0.89
	5	0.96	1.02	1.02	1.01
	50	1.00	1.06	1.03	0.94
	100	0.99	1.05	1.03	0.90
	500	1.04	1.10	1.04	0.94
Fraction 1	0	0.93	0.96	0.89	0.89
	5	0.96	1.00	0.98	0.67
	50	0.98	1.02	1.01	0.59
	100	0.97	1.01	1.02	0.75
	500	0.94	0.97	0.95	0.59
Fraction 4	0	0.93	0.96	0.89	0.89
	5	0.95	0.97	1.04	0.67
	50	0.93	1.04	0.97	0.78
	100	0.98	1.13	1.01	0.96
	500	1.03	[illegible]	1.20	[illegible]
Fraction 5	0	0.93	0.96	0.89	0.89
	5	0.93	0.95	0.95	0.58
	50	0.93	0.94	0.93	0.62
	100	0.93	0.95	0.94	0.75
	500	1.01	1.02	1.02	0.89
Fraction 6	0	0.93	0.96	0.89	0.89
	5	1.01	1.07	1.09	0.69
	50	1.04	1.13	1.11	0.85
	100	1.02	1.12	1.05	0.79
	500	1.00	1.06	1.10	0.82

□—A490/A750 < 0.75, ▨—0.75 ≤ A490/A750 ≤ 0.95, ▨—0.95 ≤ A490/A750 < 1.05, ▩—1.05 ≤ A490/A750 ≤ 1.10, ▦—1.10 < A490/A750 ≤ 1.20, ■—1.20 < A490/A750.

Despite the fact that both F5 and F6 contained quercetin glycosides, they differed meaningfully in their antifungal properties. F5 containing mostly quercetin hexosides almost completely inhibited the growth of *N. fischeri*, *Botrytis* sp., *P. setifera* at the concentration of 100 µg mL^{-1}. No growth of *F. oxysporum* was observed only at the highest concentration of 500 µg mL^{-1}. F6 including mostly quercetin pentosides, rather, stimulated the fungal growth until the dose of 100 µg mL^{-1}, although in the case of *N. fischeri*, *Botrytis* sp. and *P. setifera* the effect was not significant. Antifungal activity of F6 was detected only at the highest dose in the case of all tested isolates. The difference in antifungal properties between quercetin hexosides and pentosides is not entirely clear. Dissimilarity in the antifungal action of various quercetin glycosides was stated previously [23]. The authors reported that quercetin-3-*O*-glucoside (isoquercitin) was the only non-methylated flavonoid to inhibit conidial germination of *Arabidopsis thaliana* and *Neurospora crassa*. Among the tested quercetin derivatives were: quercetin-3-O-galactoside (hyperoside), quercetin-3-O-arabinoside (avicularin), quercetin-3-O-rhamnoside and quercetin. The antifungal effect was not noted for those quercetin derivatives, despite seemingly being very similar chemical structure. Various actions in spite of the same aglycone may result from the fact that glycosides rarely were metabolized to aglycone, but very often to higher molecules by glycosylation, sulfonation or methylation [42,43]. Simultaneously, a wide range of metabolic activity towards flavonoids exists in different fungal strains [43]. Generally, it is widely known that bioactive action of compounds depends on their structures and the bioactivity of flavonoids is ascribed to their aglycone moiety [43,44]. Nonetheless, there is a plethora of studies reporting a stronger antifungal effect of substituted flavonoids than unsubstituted ones [23,44]. The result of the present study showed also that antifungal activity do not always go hand in hand with antioxidant activity, because both properties depend on different structural conditions of the compounds.

Gauthier et al. [45] reported that antifungal activity of flavonoids directly resulted from their ability to combine irreversibly with nucleophilic aminoacids in fungal proteins. Moreover, they inhibit proteins to form several hydrogen and ionic bonds and disturb three-dimensional structure transporters [12]. As was mentioned by Lourenço et al. [12], flavonoids may be of interest in the agriculture as they can enhance the activity of pesticides, as well as reverse resistance to synthetic preparations.

In conclusion, the results of the study showed that apple pomace could be a good source of natural bio-fungicides, due to inhibition of mycotoxigenic fungal growth. The crude extract, contained mainly polar compounds like sugars, as well as phloridzin and quercetin glycosides, but also monoterpene—pinnatifidanoside D, which was for the first time isolated from apple waste. The effect of crude extract and its fractions was similar towards all tested fungi species. The strongest antifungal activity was exhibited by fraction F4 containing phloridzin, while the highest antioxidant activity was showed by fraction F6 containing mainly quercetin pentosides. Sugar moiety significantly determines the antifungal activity of quercetin glycosides. Despite the same aglycone of constituents of F5 and F6, they differed in their antifungal properties. Both antioxidant and antifungal activities of fraction F1, containing pinnatifidanoside D, were rather low. That means that the screening of proper bioactivity for this poorly studied compound is required. The antifungal and antioxidant effects did not go hand in hand, probably because of the differences in structural conditions of the compounds determining these properties.

3. Materials and Methods

3.1. Materials

3.1.1. Apple Pomace

The tested material—apple pomace—was supplied from local apple juice-processing factory. Raw material was lyophilised, powdered and subjected to extraction. The CE of apple pomace, PE and F1-F6 obtained by gel-filtration of the PE using Sephadex LH20 were tested as potential bio-fungicides against four selected fungi: *Botrytis* sp., *F. oxysporum*, *P. setifera* and *N. fischeri*.

3.1.2. Fungal Strains and Culture Conditions

Four fungal isolates were taken for the experiments. Three isolates (*Botrytis* sp., *F. oxysporum*, *P. setifera*) were selected from the Laboratory of Molecular and Environmental Microbiology (LMEM), Institute of Agrophysics, Polish Academy of Sciences (Lublin, Poland) and one strain (*N. fischeri* G90/14) was obtained from the National Institute of Technology and Evaluation, Biological Research Centre, NITE (NBRC). *Botrytis sp.* G669/16 and *F. oxysporum* G648/16 were isolated from strawberry, while *P. setifera* G11/16 was obtained from compost for agricultural usage. All strains were cultured on 90 mm Petri dishes with potato dextrose agar (PDA) at 27 °C for 5 days in the dark prior to DNA isolation. The isolates from (LMEM) collection were identified on the basis of the D2 domain of large-subunit ribosomal DNA (D2 LSU rDNA) or internal transcribed spacer 1 rRNA (ITS1) sequencing (Thermo Fisher Scientific, United States) according to methodology [18,46]. The following universal primers D2LSU2_F (5′-AGA CCG ATA GCG AAC AAG-3′) and D2LSU2_R (5′-CTT GGT CCG TGT TTC AAG-3′) [18] and ITS1 (5′-TCC GTA GGT GAA CCT GCG G-3′) and ITS2 (5′-GCT GCG TTC TTC

ATC GAT GC-3′) [47], were used for D2 LSU and ITS1, respectively. The run was performed in final volume of 20 μL using a Veriti 96-Well Fast Thermal Cycler (Applied Biosystems, Foster City, CA, USA) in the following conditions: 95 °C for 10 min, then 35 cycles at 95 °C for 15 s, 53 °C for 20 s and 72 °C for 20 s and followed by a final step at 72 °C for 5 min. Nucleotide sequences of the strains were deposited in the National Centre for Biotechnology Information (NCBI) under the following accession numbers: KX639294.1, KX639319.1, MG594608, respectively. The fourth isolate *N. fischeri* obtained from the NBRC collection was designed as isolate number NBRC 31895. Prior to antifungal analysis, strains were cultured for 14 days on 90 mm Petri dishes with potato dextrose agar (PDA) in the dark at 27 °C to obtain conidial spores. Next, the cultures were harvested into sterile BagPage® membrane filters containing IF-FF liquid and processed using an Ultra Turax IKA® homogenizer for 30 s and then filtered to extricate spores. Spores of each strain were used to set up 75% transmittance inoculum measured with a turbidimeter (Biolog®) to serve as inoculum for 96-well MT2 microplates (Biolog®) to analyze antifungal activity.

3.2. Extract Preparation and Fractionation

The CE was obtained according to the method described by Oleszek and Krzemińska [48]. Briefly, 30 g of powdered, dried material was defatted with chloroform in a Soxhlet apparatus, and then extracted (3 × 300 mL, 20 min. each) by sonication with 70% aq. MeOH at room temperature in the dark place. The CE was concentrated using the rotary evaporator under reduced pressure (at 40 °C) and freeze-dried to yield 9.95 g (33.17% of the dry plant material).

In the next step, the CE was dissolved in Milli-Q water and purified on a short self-packed RP-C$_{18}$ column (60 mm × 100 mm, 75 μm, Cosmosil 75C18-PREP). The polar fraction (PF) of CE, included sugars and simple organic acids and was eluted by acidified water (0.1% formic acid, *v/v*), while purified extract (PE) containing plant specific metabolites was eluted with methanol-water (95:5, *v/v*) solution. Obtained solutions were evaporated, suspended in *t*-butanol-water solution and freeze-dried to obtain 9.11 g of PF, and 0.84 g of PE. Afterwards, the PE was fractionated on a Sephadex LH-20 (40–120 μm) glass column (95 cm × 3.2 cm) and connected to a Gilson prep-HPLC (high-performance liquid chromatography) system with ELS™ II detector. The separation was achieved by the flow of acidified 95% MeOH (0.1% formic acid) at a flow rate of 2.4 mL min^{-1} [49]. Six LH-20 fractions were collected according to the ELS chromatogram, evaporated and freeze-dried to obtain: F1 (0.18 g), F2 (0.17 g), F3 (0.01 g), F4 (0.03 g), F5 (0.06 g), F6 (0.03 g); (Figure S16). The fractions were kept at freezer for further analysis.

3.3. Phytochemical Analysis

3.3.1. Identification and Quantification of Individual Compounds in Crude, Purified Extracts and Its Fractions

The CE and PE, as well as LH20 fractions (F1-F6) were analysed by Waters ACQUITY UPLC system (Waters Corp., Milford, MA, USA) equipped with a binary pump system, sample manager, column manager, and MS and PDA detectors (Waters Corp). For acquisition and data processing, Waters MassLynx software v.4.1 was used. The separation was carried out on the ACQUITY UPLC BEH C$_{18}$ column (100 mm × 2.1 mm, 1.7 μm, Waters Corp., Milford, MA, USA) at temperature of 40 °C and flow rate adjusted to 400 μL min^{-1}. The injection volume of the sample was 2.5 μL. The mobile phase was composed of 0.1% (v/v) formic acid in Milli-Q water (solvent A) and acetonitrile with 0.1% (*v/v*) formic acid (solvent B). Gradient program was as follows: 0–1.5 min, 10% B; 1.5–15.0 min, 10–25% B; 15.0–15.10 min, 25–100% B; 15.1–16.6 min, 100% B; 16.6–16.7 min, 100–10% B; 16.7–20.0 min, 10% B. The MS analyses were carried out on a Waters ACQUITY TQD (tandem quadrupole detector) (Waters Corp) equipped with a Z-spray electrospray interface. The parameters for ESI source were: capillary voltage 2.8 kV, cone voltage 45 V, desolvation gas N$_2$ 800 L h^{-1}, cone gas N$_2$ 100 L h^{-1}, source temp. 140 °C, desolvation temp. 350 °C.

Peaks were assigned based on their retention times, mass to charge ratio (m/z), and ESI-MS/MS fragmentation pattern, as well as their comparison to the previously isolated standards, Department of Biochemistry and Crop Quality, IUNG. The individual compounds were quantified by the external standard method using the calibration curves of pinnatifidanoside D (240 nm, 0.010–0.482 µmol/mL), rutin (355 nm, 0.008–0.410 µmol/mL) for quercetin glycosides calculation, with five different concentration levels (R^2 ranged between 0.9923 and 0.9997). The molar concentration was plotted against peak area. Due to the lack of a phloridzin standard, structurally similar (αS)-4′-O-β-D-glucopyranosyl-α,2′,4-trihydroxydihydrochalcone with 436 MW, previously isolated from lentil root [50], was used for constructing the calibration curve at 284 nm (0.002–0.401 µmol/mL).

3.3.2. Isolation Process of Pinnatifidanoside D

LH20 F1 was subjected to semi-preparative HPLC, equipped with a Gilson 321 pump, a Gilson GX-271 liquid handler with a 2 mL sample loop and a Gilson Prep ELS™ II detector. Pinnatifidanoside D (9.1 mg) was isolated in an isocratic mode using $CH_3CN:H_2O:FA$ (13:87:0.1, v/v), at 4 mL min^{-1}, on Atlantis Prep T3 at 40 °C.

3.3.3. Nuclear Magnetic Resonance (NMR) Analysis

The pure isolates were analysed at 25 °C in methanol-d_4 using Bruker Ascend III HD 500 MHz NMR spectrometer (Bruker BioSpin GmbH-Rheinstetten, Germany). Standard 1D (^{1}H, ^{13}C) and 2D (gCOSY, TROESY, gHSQC, gHMBC) pulse programs were used for data acquisition. NMR data was processed using Topspin 3.2 pl7.

3.3.4. Antioxidant Activity

Antioxidant activity of CE, PE, F1 and F4–F6, such as reducing power and DPPH radical-scavenging activity, was determined according to the methods described by Oleszek and Kozachok [51]. Briefly, tested samples were dissolved in methanol in the range of concentrations from 0 to 1500 µg mL^{-1}. For reducing power analysis, phosphate buffer (2.5 mL, 0.2 M and pH 6.6) and potassium ferricyanide [$K_3Fe(CN)_6$] (2.5 mL, 1%, w/v) were adjusted to 1 mL of the solution of tested samples. Next, the samples were incubated at 50°C for 30 min., after which trichloroacetic acid (TCA); (2.5 mL, 10%, w/v) was added. The obtained solutions (2.5 mL) were mixed with deionised water (2.5 mL) and ferric chloride ($FeCl_3$); (0.5 mL, 0.1%, w/v). The absorbance was measured at 700 nm. The results were expressed as EC_{50}, which was the concentration that gave absorbance equal to 0.5. Ascorbic acid was used as the reference sample.

Radical-scavenging activity was determined by the reaction of the solutions of the samples (3′mL) with 1,1-diphenyl-2-picrylhydrazyl (DPPH) radical (1 mL, 0.1 mM). Purple radical solution was discoloured and the colour change was stated by measurement of the absorbance at 517 nm. DPPH radical-scavenging activity was calculated according to the following formula:

$$\%\text{Inhibition} = [(A_0 - A_1)/A_0] \times 100 \tag{1}$$

where: A_0 was the absorbance for the reference sample (DPPH solution) and A_1 was the absorbance for the tested sample. The results were presented as IC_{50}, which was the concentration, which corresponded to 50% of inhibition.

The values of EC_{50} and IC_{50} were expressed as means ± standard deviations from three replicates. The significance of differences between tested samples were evaluated by the Tukey post-hoc test at $p < 0.05$.

3.3.5. Antifungal Activity

Antifungal activity analysis was performed using 96-well MT2 microplates (Biolog®, Hayward, CA, USA) according to the method of Frąc et al. [10] with modifications. The aqueous solutions of tested samples were prepared in the concentrations of 0, 5, 50, 100 and 500 µg mL^{-1}. One hundred microliters of each solution was added to each well inoculated previously with 50 µL (containing ca. 5–17.5 × 10^4 spores) of the fungal mycelium suspended in filamentous fungi inoculating fluid (IF-FF) (Biolog®, Hayward, CA, USA). Before inoculation, the suspension was standardized for each isolates into 75% transmittance (1–3.5 × 10^6 spores/mL, depending on the fungal strain). Wells filled with each tested solution or water with the IF-FF fluid without fungus were used as the controls. Three experimental replicates for each test were used. The MT2 plates were inoculated with 100 µL of inoculum per well. The plates were incubated at 26 °C for 8 days. The absorbance was measured every day at the wavelength of 490 nm as mitochondrial activity (substrate utilization) and 750 nm as mycelial growth (growth pattern) using microstation (Biolog®). The results were expressed as the ratio of absorbance for tested samples with fungi and absorbance for adequate control (sample alone, without fungi); (A$_1$/A$_0$). Moreover, the ratio of absorbance at 490 nm and at 750 nm (A490/A750), indicating the metabolic intensity compared to biomass production, were analysed to better explain the metabolisms of tested fungi [36].

For data analysis, the mean value of all days was taken and expressed as means ± standard deviations from 24 replicates (3 replicates for each of 8 days). The significance of differences between tested formulations and control were evaluated by Tukey's post-hoc test at $p < 0.05$.

Supplementary Materials: The following are available online at http://www.mdpi.com/2072-6651/11/6/361/s1, Figure S1: LC-DAD and MS/ES- chromatograms of crude extract from apple pomace, Figure S2: LC-DAD and MS/ES- chromatograms of purified extract from apple pomace, Figure S3: LC-DAD and MS/ES- chromatograms of fraction 1 from apple pomace, Figure S4: LC-DAD and MS/ES- chromatograms of fraction 4 from apple pomace; Figure S5: LC-DAD and MS/ES- chromatograms of fraction 5 from apple pomace, Figure S6: LC-DAD and MS/ES- chromatograms of fraction 6 from apple pomace, Figure S7: ^{1}H and ^{13}C NMR data of pinnatifidanoside D. Figure S8: ^{1}H NMR (500 MHz) spectrum of pinnatifidanoside D, in MeOH-d$_4$, 25 °C, Figure S9: ^{13}C NMR (125 MHz) spectrum of pinnatifidanoside D, in MeOH-d$_4$, 25 °C, Figure S10: ^{1}H-^{1}H 2D COSY NMR (500 MHz) spectrum of pinnatifidanoside D, in MeOH-d$_4$, 25 °C, Figure S11: ^{1}H-^{1}H 2D NOESY NMR (500 MHz) spectrum of pinnatifidanoside D, in MeOH-d$_4$, 25 °C, Figure S12: ^{1}H-^{13}C HSQC NMR (500 MHz) spectrum of pinnatifidanoside D, in MeOH-d$_4$, 25 °C, Figure S13: ^{1}H-^{13}C H2BC NMR (500 MHz) spectrum of pinnatifidanoside D, in MeOH-d$_4$, 25 °C, Figure S14: ^{1}H-^{13}C HSQC-TOCSY NMR (500 MHz) spectrum of pinnatifidanoside D, in MeOH-d$_4$, 25 °C, Figure S15: ^{1}H-^{13}C HMBC NMR (500 MHz) spectrum of pinnatifidanoside D, in MeOH-d$_4$, 25 °C. Figure S16: LH-20 chromatogram of plant specific metabolites fraction from apple pomace.

Author Contributions: Conceptualization, M.O. and M.F.; Methodology, M.O., Ł.P., K.O. and M.F.; Formal Analysis, M.O.; Investigation, M.O., Ł.P., S.K., Ż.L.-F. and K.O.; Data Curation, M.O.; Writing-Original Draft Preparation, M.O.; Writing-Review & Editing, M.O., Ł.P., S.K., K.O. and M.F.; Visualization, M.O.; Supervision, M.O. and M.F.; Funding Acquisition, M.F.

Funding: "This research was co-funded by The National Centre for Research and Development in frame of the project BIOSTRATEG, grant number BIOSTRATEG3/344433/16/NCBR/2018.

Acknowledgments: The authors wish to thank Wiesław Oleszek for numerous helpful advice and valuable comments on the manuscript. The authors would like to express their thanks also to Jerzy Żuchowski for providing us with (αS)-4′-O-β-D-glucopyranosyl-α,2′,4-trihydroxydihydrochalcone standard.

Conflicts of Interest: The authors declare no competing of interest.

References

1. Santana-Méridas, O.; González-Coloma, A.; Sánchez-Vioque, R. Agricultural Residues as a Source of Bioactive Natural Products. *Phytochem. Rev.* **2012**, *11*, 447–466. [CrossRef]

2. Heidorn, E.; Utvik, K.; Gengler, C.; Alati, K.; Collet, D.; Attivissimo, V.; Colantonio, M. *Agriculture, Forestry and Fishery Statistics 2017 Edition*; Forti, R., Ed.; European Commission: Luxemburg, Belgium, 2017. [CrossRef]

3. Górnaś, P.; Mišina, I.; Olšteine, A.; Krasnova, I.; Pugajeva, I.; Lācis, G.; Siger, A.; Michalak, M.; Soliven, A.; Seglina, A. Phenolic Compounds in Different Fruit Parts of Crab Apple: Dihydrochalcones as Promising Quality Markers of Industrial Apple Pomace By-Products. *Ind. Crops Prod.* **2015**, *74*, 607–612. [CrossRef]

4. Waldbauer, K.; McKinnon, R.; Kopp, B. Apple Pomace as Potential Source of Natural Active Compounds. *Planta Med.* **2017**, *83*, 994–1010. [CrossRef] [PubMed]

5. Bhushan, S.; Kalia, K.; Sharma, M.; Singh, B.; Ahuja, P.S. Processing of Apple Pomace for Bioactive Molecules. *Crit. Rev. Biotechnol.* **2008**, *28*, 285–296. [CrossRef] [PubMed]

6. Sánchez-Rabaneda, F.; Jáuregui, O.; Lamuela-Raventós, R.M.; Viladomat, F.; Bastida, J.; Codina, C. Qualitative Analysis of Phenolic Compounds in Apple Pomace Using Liquid Chromatography Coupled to Mass Spectrometry in Tandem Mode. *Rapid Commun. Mass Spectrom.* **2004**, *18*, 553–563. [CrossRef] [PubMed]

7. Kołodziejczyk, K.; Sójka, M.; Abadias, M.; Viñas, I.; Guyot, S.; Baron, A. Polyphenol Composition, Antioxidant Capacity, and Antimicrobial Activity of the Extracts Obtained from Industrial Sour Cherry Pomace. *Ind. Crops Prod.* **2013**, *51*, 279–288. [CrossRef]

8. Shim, S.-H.; Jo, S.-J.; Kim, J.-C.; Choi, G.-J. Control Efficacy of Phloretin Isolated from Apple Fruits Against Several Plant Diseases. *Plant Pathol. J.* **2010**, *26*, 280–285. [CrossRef]

9. Sanzani, S.M.; Schena, L.; Nigro, F.; de Girolamo, A.; Ippolito, A. Effect of Quercetin and Umbelliferone on the Transcript Level of *Penicillium expansum* Genes Involved in Patulin Biosynthesis. *Eur. J. Plant Pathol.* **2009**, *125*, 223–233. [CrossRef]

10. Frąc, M.; Gryta, A.; Oszust, K.; Kotowicz, N. Fast and Accurate Microplate Method (Biolog MT2) for Detection of *Fusarium* Fungicides Resistance/Sensitivity. *Front. Microbiol.* **2016**, *7*, 1–16. [CrossRef]

11. Becher, R.; Hettwer, U.; Karlovsky, P.; Deising, H.B.; Wirsel, S.G.R. Adaptation of *Fusarium graminearum* to Tebuconazole Yielded Descendants Diverging for Levels of Fitness, Fungicide Resistance, Virulence, and Mycotoxin Production. *Phytopathology* **2010**, *100*, 444–453. [CrossRef]

12. Lourenço, R.M.D.C.; Melo, P.; da Silva, S.; de Almeida, A.B.A. Flavonoids as Antifungal Agents. In *Antifungal Metabolites from Plants*; Razzaghi-Abyaneh, M., Rai, M., Eds.; Springer: Berlin, Germany, 2013; pp. 1–469.

13. Upasani, M.L.; Gurjar, G.S.; Kadoo, N.Y.; Gupta, V.S. Dynamics of Colonization and Expression of Pathogenicity Related Genes in *Fusarium oxysporum* f. sp. *ciceri* during Chickpea Vascular Wilt Disease Progression. *PLoS ONE* **2016**, *11*, e0156490. [CrossRef] [PubMed]

14. Wang, H.-C.; Li, L.C.; Cai, B.; Cai, L.T.; Chen, X.J.; Yu, Z.H.; Zhang, C.Q. Metabolic Phenotype Characterization of *Botrytis cinerea*, the Causal Agent of Gray Mold. *Front. Microbiol.* **2018**, *9*, 1–9. [CrossRef] [PubMed]

15. Oszust, K.; Panek, J.; Pertile, G.; Siczek, A.; Oleszek, M.; Frąc, M. Metabolic and Genetic Properties of *Petriella setifera* Precultured on Waste. *Front. Microbiol.* **2018**, *9*, 1–10. [CrossRef] [PubMed]

16. Kwaśna, H.; Łakomy, P.; Łabędzki, A. Morphological characteristics and DNA sequence analysis of Petriella setifera and Oidiodendron setiferum from twigs of diseased oak. *Acta Mycol.* **2005**, *40*, 267–275. [CrossRef]

17. Schwarze, F.W. Wood Decay Under the Microscope. *Fungal Biol. Rev.* **2007**, *21*, 133–170. [CrossRef]

18. Pertile, G.; Panek, J.; Oszust, K.; Siczek, A.; Frąc, M. Intraspecific Functional and Genetic Diversity of *Petriella setifera*. *PeerJ* **2018**, *6*, e4420. [CrossRef] [PubMed]

19. Frąc, M.; Jezierska-Tys, S.; Yaguchi, T. Occurrence, Detection, and Molecular and Metabolic Characterization of Heat-Resistant Fungi in Soils and Plants and Their Risk to Human Health. *Adv. Agron.* **2015**, *132*, 161–204. [CrossRef]

20. Sanzani, S.M.; Girolamo, A.; Schena, L.; Solfrizzo, M.; Ippolito, A.; Visconti, A. Control of *Penicillium expansum* and Patulin Accumulation on Apples by Quercetin and Umbelliferone. *Eur. Food Res. Technol.* **2009**, *228*, 381–389. [CrossRef]

21. Steinkellner, S.; Mammerler, R. Effect of Flavonoids on the Development of *Fusarium oxysporum* f. sp. *lycopersici*. *J. Plant Interact.* **2007**, *2*, 17–23. [CrossRef]

22. Naseer, R.; Sultana, B.; Khan, M.Z.; Naseer, D.; Nigam, P. Utilization of Waste Fruit-Peels to Inhibit Aflatoxins Synthesis by *Aspergillus flavus*: A Biotreatment of Rice for Safer Storage. *Bioresour. Technol.* **2014**, *172*, 423–428. [CrossRef]

23. Parvez, M.M.; Tomita-Yokotani, K.; Fujii, Y.; Konishi, T.; Iwashina, T. Effects of Quercetin and Its Seven Derivatives on the Growth of *Arabidopsis thaliana* and *Neurospora crassa*. Biochem. *Syst. Ecol.* **2004**, *32*, 631–635. [CrossRef]

24. Dick, A.J.; Redden, P.R.; DeMarco, A.C.; Lidster, P.D.; Grindley, T.B. Flavonoid Glycosides of Spartan Apple Peel. *J. Agric. Food Chem.* **1987**, *35*, 529–531. [CrossRef]

25. Lu, Y.; Foo, L.Y. Identification and Quantification of Major Polyphenols in Apple Pomace. *Food Chem.* **1997**, *59*, 187–194. [CrossRef]

26. Vasantha Rupasinghe, H.P.; Kean, C. Polyphenol Concentrations in Apple Processing By-Products Determined Using Electrospray Ionization Mass Spectrometry. *Can. J. Plant Sci.* **2008**, *88*, 759–762. [CrossRef]

27. Li, L.Z.; Gao, P.Y.; Song, S.J.; Yuan, Y.Q.; Liu, C.T.; Huang, X.X.; Liu, Q.B. Monoterpenes and Flavones from the Leaves of *Crataegus pinnatifida* with Anticoagulant Activities. *J. Funct. Foods* **2015**, *12*, 237–245. [CrossRef]

28. Oleszek, W.; Lee, C.Y.; Jaworski, A.W.; Price, K.R. Identification of Some Phenolic Compounds in Apples. *J. Agric. Food Chem.* **1988**, *36*, 430–432. [CrossRef]

29. Walker, A.A. Note on the Polyphenol Content of Ripening Apples. *N. Z. J. Sci.* **1963**, *6*, 492–494.

30. Ingle, M.; Hyde, J. The Effect of Bruising on Discoloration and Concentration of Phenolic Compounds in Apple Tissue. *Proc. Am. Soc. Hortic. Sci.* **1968**, *93*, 738–745.

31. Vamos-Vigyazo, L.; Gajzago, I. Studies on the Enzymatic Browning and the Polyphenol–polyphenol Oxidase Complex of Apple Cultivars. *Confructa* **1976**, *21*, 24–31.

32. Burda, S.; Oleszek, W.; Lee, C.Y. Phenolic Compounds and Their Changes in Apples during Maturation and Cold Storage. *J. Agric. Food Chem.* **1990**, *38*, 945–948. [CrossRef]

33. Gadkari, P.V.; Kadimi, U.S.; Balaraman, M. Catechin Concentrates of Garden Tea Leaves (*Camellia Sinensis* L.): Extraction/Isolation and Evaluation of Chemical Composition. *J. Sci. Food Agric.* **2014**, *94*, 2921–2928. [CrossRef] [PubMed]

34. Heim, K.E.; Tagliaferro, A.R.; Bobilya, D.J. Flavonoid Antioxidants: Chemistry, Metabolism and Structure-Activity Relationships. *J. Nutr. Biochem.* **2002**, *13*, 572–584. [CrossRef]

35. Brett, A.M.O.; Ghica, M.-E. Electrochemical Oxidation of Quercetin. *Electroanalysis* **2003**, *15*, 1745–1750. [CrossRef]

36. Pinzari, F.; Ceci, A.; Abu-Samra, N.; Canfora, L.; Maggi, O.; Persiani, A. Phenotype MicroArrayTM System in the Study of Fungal Functional Diversity and Catabolic Versatility. *Res. Microbiol.* **2016**, *167*, 710–722. [CrossRef]

37. Noveroske, R.L.; Kui, J.; Williams, E.B. Oxidation of Phloridzin and Phloretin Related to Resistance of Malus to Venturia Inaequalis. *Phytopathology* **1964**, *54*, 92–97.

38. Gessler, C.; Patocchi, A.; Sansavini, S.; Tartarini, S.; Gianfranceschi, L. Venturia Inaequalis Resistance in Apple. CRC. *Crit. Rev. Plant Sci.* **2006**, *25*, 473–503. [CrossRef]

39. Gosch, C.; Halbwirth, H.; Stich, K. Phloridzin: Biosynthesis, Distribution and Physiological Relevance in Plants. *Phytochemistry* **2010**, *71*, 838–843. [CrossRef] [PubMed]

40. Baldisserotto, A.; Malisardi, G.; Scalambra, E.; Andreotti, E.; Romagnoli, C.; Vicentini, C.B.; Manfredini, S.; Vertuani, S. Synthesis, Antioxidant and Antimicrobial Activity of a New Phloridzin Derivative for Dermo-Cosmetic Applications. *Molecules* **2012**, *17*, 13275–13289. [CrossRef] [PubMed]

41. Lee, J.H.; Lee, S.J.; Park, S.; Jeong, S.W.; Kim, C.Y.; Jin, J.S.; Jeong, E.D.; Kwak, Y.S.; Kim, S.T.; Bae, D.W.; et al. Determination of Flavonoid Level Variation in Onion (*Allium Cepa* L.) Infected by *Fusarium oxysporum* Using Liquid Chromatography–tandem Mass Spectrometry. *Food Chem.* **2012**, *133*, 1653–1657. [CrossRef]

42. Das, S.; Rosazza, J.P. Microbial and Enzymatic Transformations of Flavonoids. *J. Nat. Prod.* **2006**, *69*, 499–508. [CrossRef] [PubMed]

43. Gonzales, G.B.; Smagghe, G.; Wittevrongel, J.; Huynh, N.T.; Van Camp, J.; Raes, K. Metabolism of Quercetin and Naringenin by Food-Grade Fungal Inoculum, *Rhizopus azygosporus* Yuan et Jong (ATCC 48108). *J. Agric. Food Chem.* **2016**, *64*, 9263–9267. [CrossRef] [PubMed]

44. Mierziak, J.; Kostyn, K.; Kulma, A. Flavonoids as Important Molecules of Plant Interactions with the Environment. *Molecules* **2014**, *19*, 16240–16265. [CrossRef] [PubMed]

45. Gauthier, L.; Atanasova-Penichon, V.; Chéreau, S.; Richard-Forget, F. Metabolomics to Decipher the Chemical Defense of Cereals against *Fusarium graminearum* and Deoxynivalenol Accumulation. *Int. J. Mol. Sci.* **2015**, *16*, 24839–24872. [CrossRef] [PubMed]

46. Oszust, K.; Pawlik, A.; Siczek, A.; Janusz, G.; Gryta, A.; Bilińska-Wielgus, N.; Frąc, M. Efficient Cellulases Production by *Trichoderma atroviride* G79/11 in Submerged Culture Based on Soy Flour-Cellulose-Lactose. *BioResources* **2017**, *12*, 8468–8489. [CrossRef]

47. White, T.J.; Bruns, T.; Lee, S.J.W.T.; Taylor, J.L. Amplification and direct sequencing of fungal ribosomal RNA genes for phylogenetics. *PCR Protoc. A Guide Methods Appl.* **1990**, *18*, 315–322.

48. Oleszek, M.; Krzemińska, I. Enhancement of biogas production by co-digestion of maize silage with common goldenrod rich in biologically active compounds. *BioResources* **2017**, *12*, 704–714. [CrossRef]

49. Kozachok, S.; Pecio, Ł.; Kolodziejczyk-Czepas, J.; Marchyshyn, S.; Nowak, P.; Mołdoch, J.; Oleszek, W. γ-Pyrone Compounds: Flavonoids and Maltol Glucoside Derivatives from *Herniaria glabra* L. Collected in the Ternopil Region of the Ukraine. *Phytochemistry* **2018**, *152*, 213–222. [CrossRef]

50. Żuchowski, J.; Pecio, Ł.; Reszczyńska, E.; Stochmal, A. New Phenolic Compounds from the Roots of Lentil (*Lens culinaris*). *Helv. Chim. Acta* **2016**, *99*, 674–680; [CrossRef]
51. Oleszek, M.; Kozachok, S. Antioxidant Activity of Plant Extracts and Their Effect on Methane Fermentation in Bioreactors. *Int. Agrophysics* **2018**, *32*, 395–401. [CrossRef]

© 2019 by the authors. Licensee MDPI, Basel, Switzerland. This article is an open access article distributed under the terms and conditions of the Creative Commons Attribution (CC BY) license (http://creativecommons.org/licenses/by/4.0/).

Article

Detoxification of the Fumonisin Mycotoxins in Maize: An Enzymatic Approach

Johanna Alberts [1,*], **Gerd Schatzmayr** [2], **Wulf-Dieter Moll** [2], **Ibtisaam Davids** [1,3], **John Rheeder** [1], **Hester-Mari Burger** [1], **Gordon Shephard** [1] and **Wentzel Gelderblom** [1,4]

1   Mycotoxicology Research Group, Institute of Biomedical and Microbial Biotechnology, Cape Peninsula University of Technology, Bellville 7535, South Africa
2   BIOMIN Research Center, BIOMIN, Technopark 1, 3430 Tulln, Austria
3   Department of Biomedical Science, University of the Western Cape, Bellville 7535, South Africa
4   Department of Biochemistry, Stellenbosch University, Stellenbosch 7600, South Africa
*   Correspondence: albertsh@cput.ac.za

Received: 31 July 2019; Accepted: 23 August 2019; Published: 10 September 2019

Abstract: Enzymatic detoxification has become a promising approach for control of mycotoxins postharvest in grains through modification of chemical structures determining their toxicity. In the present study fumonisin esterase FumD (EC 3.1.1.87) (FUM*zyme*®; BIOMIN, Tulln, Austria), hydrolysing fumonisin (FB) mycotoxins by de-esterification, was utilised to develop an enzymatic reduction method in a maize kernel enzyme incubation mixture. Efficacy of the FumD FB reduction method in "low" and "high" FB contaminated home-grown maize was compared by monitoring FB_1 hydrolysis to the hydrolysed FB_1 (HFB_1) product utilising a validated LC-MS/MS analytical method. The method was further evaluated in terms of enzyme activity and treatment duration by assessing enzyme kinetic parameters and the relative distribution of HFB_1 between maize kernels and the residual aqueous environment. FumD treatments resulted in significant reduction (≥80%) in "low" (≥1000 U/L, $p < 0.05$) and "high" (100 U/L, $p < 0.05$; ≥1000 U/L, $p < 0.0001$) FB contaminated maize after 1 h respectively, with an approximate 1:1 µmol conversion ratio of FB_1 into the formation of HFB_1. Enzyme kinetic parameters indicated that, depending on the activity of FumD utilised, a significantly ($p < 0.05$) higher FB_1 conversion rate was noticed in "high" FB contaminated maize. The FumD FB reduction method in maize could find application in commercial maize-based practices as well as in communities utilising home-grown maize as a main dietary staple and known to be exposed above the tolerable daily intake levels.

Keywords: fumonisin; enzymatic detoxification; fumonisin esterase FumD; enzyme kinetics; maize

Key Contribution: A fumonisin B (FB) reduction method was developed for whole maize utilizing fumonisin esterase FumD (BIOMIN; Tulln; Austria). Depending on the enzyme activity and duration of treatment; reduction of FB in maize could be achieved ≥80% with an approximate 1:1 µmolar conversion ratio of FB_1 to hydrolysed FB_1 (HFB_1). The FumD FB reduction method could find application in populations exposed to maize containing FB above international regulatory levels as well as in industrial processes utilising high FB contaminated maize.

---

1. Introduction

The lack of effective and environmentally safe chemical control methods against fungal infection and mycotoxin production in maize initiated investigations into biologically safe alternatives to prevent these contaminants from entering the food chain [1]. Methods involve the application of natural resources, including plant material, microbial cultures, genetic material, clay minerals and enzymes [2–11]. Enzymatic degradation of mycotoxins in food sources, postharvest, is a new research approach, providing many opportunities for novel initiatives to improve the safety of food with respect to mycotoxin reduction [1,12]. The focus is on targeted modification of the chemical structures by enzymatic cleavage or conversion of chemical bonds/groups that play a key role during cytotoxicity [13,14]. Detoxification of the fumonisins is achieved by enzymatic deamination of the free amino group at C-2 and de-esterification of the ester bonds at C-14 and C-15. A knowledge base on reduction of mycotoxin concentrations by bacterial and fungal cultures has been established over the years [14–17]. This information was further developed by identifying microbial enzymes responsible for detoxification, characterization of genes encoding the enzymes, expression of genes in food-grade microorganisms, and development of culture and recombinant enzyme preparations for commercial application. Microorganisms and enzymes capable of degrading fumonisin B_1 (FB_1) include carboxylesterase and amino oxidase enzymes of *Exophiala spinifera* ATCC 74269, *Rhinocladiella atrovirens* ATCC 74270 as well as carboxylesterase and aminotransferase enzymes of Bacterium ATCC 55552 and *Sphingopyxis macrogoltabida* MTA144 [14–19].

Recently, a commercial fumonisin esterase FumD (EC 3.1.1.87) (FUMzyme®; BIOMIN, Tulln, Austria) capable of effectively hydrolysing the tricarballylic acid groups of FB_1 yielding hydrolysed FB_1 (HFB_1), was introduced. The enzyme activity is specific and irreversible, while HFB_1 exhibited less toxic effects when evaluated in pig intestine, as indicated by disruption of the sphinganine/sphingosine ratios in the liver and plasma, induction of an intestinal immune response, the absence of hepatotoxicity, and changes in intestinal morphology [20]. FUMzyme® has been regarded safe for humans, animals and the environment by the European Food Safety Authority (EFSA) [21]. Although enzymatic detoxification has become a promising approach and found application in the animal feed industry [21–24], a broader application to safeguard human food still has to be investigated. Such an approach could add value to commercial maize-based manufacturing processes during which the fumonisins are concentrated in certain products and co-products, as well as in subsistence farming communities where people are exposed to unacceptable levels of the fumonisins in their staple diet.

The present study describes the development of a FumD FB reduction method utilising home-grown maize batches containing "low" and "high" levels of FB. The efficacy of the reduction method was determined by monitoring the extent of FB_1 hydrolysis and the formation of HFB_1 in residual maize kernels and enzyme solutions utilising a validated LC-MS/MS analytical method. The FumD FB reduction method was optimized in terms of enzyme activity and treatment duration considering specific enzyme kinetic parameters to evaluate differences between maize samples with varying FB levels. Distribution of HFB_1 between the kernel and aqueous phase was also recorded.

2. Results

2.1. Validation of The Extraction and Chromatographic Quantification Methods

The method validation parameters are summarised in Tables 1 and 2. The lower limit of quantification (LOQ) for each fumonisin was determined from a signal to noise (s/n) ratio of at least five times the response compared to the blank response. Analyte responses were identifiable, discrete and reproducible. LOQ values for the individual fumonisins ranged between 2.8–3.5 µg/kg. Linearity (r^2) of the fumonisin calibration curve was >0.998. Selectivity of the method was confirmed by demonstrating no interfering peaks at LOQ for each of the fumonisins. Accuracy was determined by replicate analysis of samples containing known amounts of analyte, resulting in means within 15% from the theoretical values. Recoveries were calculated by comparison of the response obtained for each mycotoxin with that of known spiked concentrations in blank maize, expressed as a percentage. The recoveries for the individual fumonisins ranged from 79–84% and were consistent and reproducible. Fumonisin concentrations in the FAPAS quality control reference maize samples were always within the stipulated ranges for each fumonisin (data not shown).

Table 1. LC-MS/MS conditions for quantification of fumonisins and hydrolysed fumonisin b_1 by positive ESI at 3.5 kv capillary voltage.

Analyte	Cone Voltage	Precursor	Quantifier (Collision Energy)	Qualifier (Collision Energy)
Fumonisin B_1	50	722.3	334.3 (40)	352.3 (38)
Fumonisins B_2 and B_3	50	706.3	318.3 (40)	336.3 (40)
Hydrolysed fumonisin B_1	25	406.6	334.3 (25)	352.4 (20)

Table 2. Validation of the analytical method for fumonisin analyses in maize.

Analyte	LOQ (µg/kg)	Spike Level (µg/kg)	Recovery (%)	RSDr (%)
Fumonisin B_1	3.5	1060	84	2
Fumonisin B_2	2.8	925	66	4
Fumonisin B_3	2.8	520	79	1
Hydrolysed fumonisin B_1	2.8	800	80	2

LOQ, lower limit of quantification; RSDr, relative standard deviation for repeatability.

2.2. FB and HFB$_1$ Concentrations in Maize

FB and HFB$_1$ concentrations (µg/kg) (mean values and standard deviations in brackets) in untreated "low" FB contaminated maize: FB$_1$, 3326 (1588); FB$_2$, 1503 (749); FB$_3$, 530 (376); and HFB$_1$, 182 (106). Concentrations in "high" FB contaminated maize: FB$_1$, 9343 (1756); FB$_2$, 3559 (541); FB$_3$, 1701 (1557); and HFB$_1$, 342 (137). FB concentrations in "low" and "high" FB contaminated FumD treated maize are summarised in Figure 1 (Table insert). The HFB$_1$ levels in the "low" and "high" FB contaminated FumD treated maize were 12 (8) and 65 (59), respectively. High variation in the FB levels existed between subsamples within each of the two maize batches, which is a common phenomenon for this type of analyses mainly due to the random distribution of infected kernels throughout a specific sample [25,26].

2.3. Optimal FumD activity and Conversion Ratios (Fixed Time Incubation)

2.3.1. FB$_1$ Hydrolysis and Formation of HFB$_1$ in Residual Maize Kernels as a Function of Fumd Activity

The pH of maize-enzyme solutions before and after 1 h enzyme treatments were 5.1 ± 0.10 and 4.8 ± 0.04 respectively for "low" FB contaminated maize. For the "high" FB contaminated maize it was 5.7 ± 0.12 and 5.4 ± 0.04.

When compared to the water control sample in the absence of the enzyme, the total FB (FB$_1$, FB$_2$ and FB$_3$) concentrations were reduced as a function of an increased enzyme activity during a 1 h incubation in both the "low" and "high" FB contaminated maize (Figure 1A,B). Treatment with $\geq$1000 U/L resulted in an 80% reduction in total FB concentrations (Figure 1, table insert) when compared to the water control. Incubations of "low" FB contaminated maize with 10 and 100 U/L FumD markedly ($p > 0.05$) reduced FB$_1$ when compared to the water control incubation (Figure 1A). A significant reduction in FB$_1$ was observed with 1000 U/L and 5000 U/L when compared to the water control ($p < 0.05$), with no significant difference ($p > 0.05$) between the two incubations. The reduction in FB$_1$ concentrations coincided with a significant increase in HFB$_1$ concentrations from 10 U/L ($p < 0.01$) to 1000 U/L ($p < 0.001$) and 5000 U/L ($p < 0.001$), when compared to the water control. Only trace amounts of HFB$_1$ were detected in the water control samples of "low" FB contaminated maize.

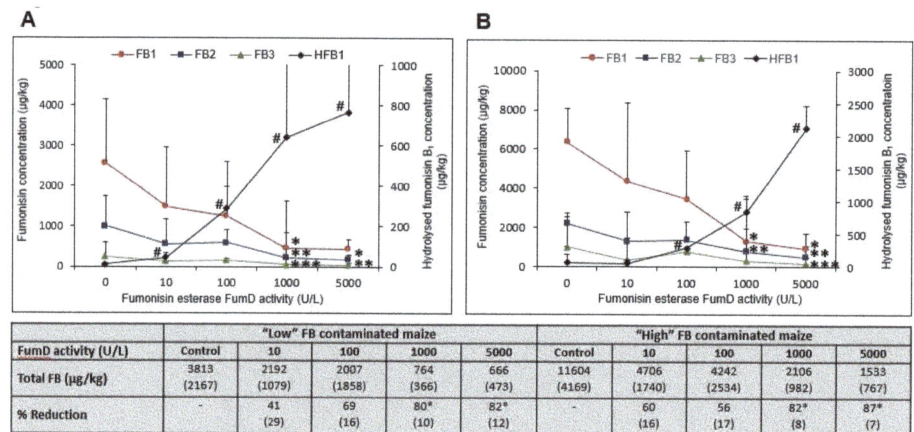

FumD activity (U/L)	"Low" FB contaminated maize					"High" FB contaminated maize				
	Control	10	100	1000	5000	Control	10	100	1000	5000
Total FB (µg/kg)	3813 (2167)	2192 (1079)	2007 (1858)	764 (366)	666 (473)	11604 (4169)	4706 (1740)	4242 (2534)	2106 (982)	1533 (767)
% Reduction	-	41 (29)	69 (16)	80* (10)	82* (12)	-	60 (16)	56 (17)	82* (8)	87* (7)

Figure 1. Fumonisin B (FB$_1$, FB$_2$, FB$_3$) and hydrolysed fumonisin B$_1$ (HFB$_1$) concentrations (µg/kg) maize as a function of fumonisin esterase FumD activity (0, 10, 100, 1000 and 5000 U/L) after 1 h treatment. (**A**) "Low" and (**B**) "High" FB contaminated maize. Values represent means of three to five replications of experiments and error bars indicate standard deviations. The statistical analyses are based on natural log (ln) transformations. The *, **, *** and # indicate significant ($p < 0.05$) differences of means from the water control (0 U/L) treatments. Table insert: % total reduction of the total FB as a function of enzyme actvity after 1 h in "low" and "high" FB contaminated maize. Standard deviations are in brackets.

In "high" FB contaminated maize, a significant reduction in FB$_1$ was observed at FumD activities of 100 U/L ($p < 0.05$), 1000 U/L ($p < 0.0001$) and 5000 U/L ($p < 0.0001$) when compared to the water control with no significant ($p > 0.05$) differences between the different incubations (Figure 1B). The total FB concentrations (Figure 1B Table insert) were again reduced by ≥80% at the two highest FumD concentrations. Incubation with 10 U/L FumD only markedly reduced the FB$_1$ concentration while incubation with 1000 U/L resulted in 79% reduction in FB$_1$, and 82% reduction in total FB concentrations, respectively. The reduction in FB$_1$ again coincided with a significant ($p < 0.0001$) increase in HFB$_1$ concentrations with treatments ≥100 U/L, while only trace amounts of HFB$_1$ were detected in the water control.

2.3.2. FB$_1$ Hydrolysis and Formation of Hfb$_1$ in the Residual Solutions as a Function of Fumd Activity

Residual water control samples obtained from the "low" FB contaminated maize contained high FB and low HFB$_1$ concentrations (Figure 2A). FumD reduced ($p < 0.05$) FB as a function of increased enzyme activity. The reduction in FB$_1$ coincided with a significant increase in HFB$_1$ from 10 U/L ($p < 0.01$) reaching a maximum at 100 U/L with no further increase >100 U/L. Treatment with 10 U/L resulted in an almost complete removal of FB$_1$, with almost no FB$_1$ detected ≥100 U/L. Complete reduction of FB$_2$ and FB$_3$ was also observed ≥10 U/L. A similar trend was observed in residual water control and enzyme solutions of the "high" FB contaminated maize, although FB$_1$ was only completely removed >1000 U/L (Figure 2B). Total reduction of FB$_2$ and FB$_3$ was observed with ≥100 U/L FumD activity. A significant ($p < 0.05$) increase in HFB$_1$ was noticed ≥10 U/L, reaching a maximum at 100 U/L.

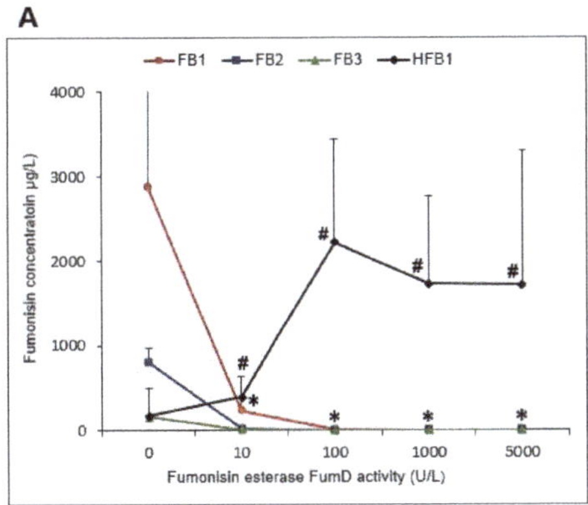
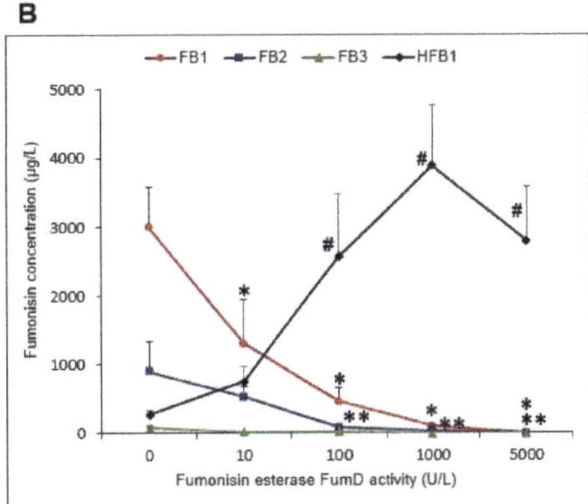

Figure 2. FB (FB$_1$, FB$_2$, FB$_3$) and hydrolysed fumonisin B$_1$ (HFB$_1$) concentrations in residual water control and enzyme solutions as a function of fumonisin esterase FumD activity (0, 10, 100, 1000 and 5000 U/L) after 1 h treatment. (**A**) "Low" and (**B**) "High" FB contaminated maize. Values represent means of three to five replications of experiments and error bars indicate standard deviations. The statistical analyses are based on natural log (ln) transformations. The *, ** and # indicate significant ($p < 0.05$) differences of means from the respective water control (0 U/L) treatments.

2.3.3. FB$_1$ Hydrolysis Relative to the Formation of HFB$_1$

When considering the total mean FB$_1$ levels in the incubation mixtures (levels in residual maize kernels plus residual solutions expressed in micromoles), it was markedly to significantly ($p < 0.05$) higher in "high" FB contaminated maize although it varied when utilizing the different enzyme incubations (Table 3). Although the mean FB$_1$ converted were similar between both samples treated with the various FumD activities, the % FB$_1$ (μmol) loss was significantly ($p < 0.05$) higher in "low" FB contaminated maize at the two lower enzyme activities with no difference when using higher FumD activities ($\geq$1000 U/L). Overall, FumD significantly ($p < 0.05$) increased the mean total HFB$_1$ levels in both samples as a function of the enzyme activity reaching a maximum level $\geq$100 U/L. When considering the FB$_1$:HFB$_1$ conversion ratio, FumD resulted in a markedly lower conversion at 10 U/L in "low" as compared to "high" FB contaminated maize. The FB$_1$:HFB$_1$ conversion ratios were similar with 100 U/L, reaching approximately a 1:1 ratio at $\geq$1000 U/L, implying maximum conversion.

2.3.4. Relative HFB$_1$ Distribution in Residual Maize Kernels and Solutions Following FumD Incubations

The total HFB$_1$ in the incubation mixture increased significantly ($p < 0.05$) in incubation $\geq$100 U/L FumD activities with no difference at higher enzyme activities and/or between the "low" and "high" FB contaminated maize samples. The % HFB$_1$ accumulation in the "low" FB contaminated maize increased as a function of increased FumD activity reaching a maximum level at the 1000 U/L (Table 4). A similar response was noticed in the "high" FB contaminated maize showing a typical dose response effect with a significant ($p < 0.05$) higher level obtained with 5000 U/L FumD activity. Maximum levels when considering the % total mean HFB$_1$ were noticed in the residual solution up to 100 U/L FumD, which markedly and significantly decreased in the "low" and "high" FB contaminated samples at higher FumD activities, respectively.

Table 3. Comparative FB_1 conversion relative to HFB_1 formation between "low" and "high" FB contaminated maize during 1 h incubation with different FumD activities.

FumD Aactivity (U/L)	Total Mean µmol FB_1 in Incubation Mixture *	Mean µmol FB_1 Converted	Mean µmol FB_1 Loss (%)	Total Mean µmol HFB_1 in Incubation Mixture	FB_1:HFB_1 µmol Conversion Ratio
"Low" FB Contaminated Maize					
0 **	1.15 ± 0.55	-	-	0.02 ± 0.01	42.75 ± 12.14
10	0.27 ± 0.14 [ab]	0.88 ± 0.14 [a]	76.58 ± 11.90 [abc]	0.13± 0.05 [a]	7.22 ± 2.40 [a]
100	0.18 ± 0.16 [ac]	0.98 ± 0.16 [a]	84.73 ± 14.14 [ab]	0.94 ± 0.62 [b]	2.52 ± 0.84 [ab]
1000	0.07 ± 0.03 [c]	1.09 ± 0.03 [ab]	94.32 ± 2.91 [a]	0.99 ± 0.60 [b]	1.30 ± 0.58 [bc]
5000	0.06 ± 0.05 [c]	1.09 ± 0.05 [ab]	94.64 ± 3.98 [ab]	1.02 ± 1.00 [b]	0.66 ± 0.32 [c]
"High" FB Contaminated Maize					
0 **	1.74 ± 1.32	-	-	0.16 ± 0.10	10.23 ± 3.96
10	0.97 ± 0.50 [b]	0.76 ± 0.50 [a]	55.04 ± 17.68 [c]	0.24 ± 0.15 [a]	3.4 ± 1.73 [abd]
100	0.71 ± 0.37 [b]	1.02 ± 0.37 [a]	58.84 ± 21.59 [c]	1.17 ± 0.53 [b]	1.38 ± 0.72 [bc]
1000	0.21 ± 0.16 [ac]	1.52 ± 0.16 [b]	87.75 ± 9.31 [ab]	1.96 ± 0.89 [b]	0.93 ± 0.46 [c]
5000	0.13 ± 0.07 [ac]	1.61 ± 0.07 [b]	92.61 ± 4.09 [ab]	1.68 ± 0.66 [b]	1.11 ± 0.54 [cd]

The statistical analyses are based on natural log (ln) transformations. Values represent means ± standard deviations of three to five replications of experiments. Statistical differences ($p < 0.05$) in a column for FumD activities within and between the "low" and "high" FB contaminated maize are indicated with different letters. * Combined FB_1 levels in maize and residual enzyme solution. ** Shaded areas represent the water control treatment (0 U/L). HFB_1 levels were corrected accordingly in the presence of the enzyme.

Table 4. Distribution of HFB_1 between residual maize and the solution following FumD treatment of "low" and "high" FB contaminated maize for 1 h.

FumD Activity (U/L)	Total Mean HFB_1 (µmol) in 100 g Residual Maize Kernels	Total Mean HFB_1 (µmol) in 200 mL Residual Solution	Mean µmol HFB_1 in Residual Maize Kernels (%)	Mean µmol HFB_1 in Residual Solution (%)
"Low" FB Contaminated Maize				
10	0.01 ± 0.01 [a]	0.13 ± 0.04 [a]	5.65 ± 3.47 [a]	94.35 ± 3.47 [a]
100	0.07 ± 0.06 [b]	0.87 ± 0.58 [b]	7.35 ± 3.28 [a]	92.65 ± 3.28 [a]
1000	0.15 ± 0.11 [b]	0.83 ± 0.50 [b]	15.82 ± 4.48 [b]	84.18 ± 4.48 [b]
5000	0.19 ± 0.22 [b]	0.83 ± 0.78 [b]	16.25 ± 3.86 [b]	83.75 ± 3.86 [b]
"High" FB Contaminated Maize				
10	0.00 [a]	0.24 ± 0.14 [a]	0.48 ± 0.58 [a]	99.52 ± 0.60 [a]
100	0.08 ± 0.06 [b]	1.09 ± 0.48 [b]	6.75 ± 2.24 [b]	93.25 ± 2.24 [b]
1000	0.19 ± 0.09 [bc]	1.77 ± 0.86 [b]	10.68 ± 4.33 [b]	89.32 ± 4.33 [b]
5000	0.50 ± 0.17 [c]	1.17 ± 0.65 [b]	32.90 ± 11.85 [c]	67.10 ± 11.90 [c]

The statistical analyses are based on natural log (ln) transformations. Values represent means ± standard deviations of three to five replications of experiments. Statistical differences ($p < 0.05$) in a column of the % HFB_1 distribution within "low" and "high" FB contaminated maize are indicated with different letters.

2.4. Comparative Enzyme Kinetics of FB_1 Conversion as a Function of Time and FumD Activity

2.4.1. FB_1 Leaching into the Aqueous Phase (Water Control Treatment)

The total mean baseline FB_1 levels, i.e., total FB_1 (µmol) in the "low" and "high" FB contaminated maize incubation mixtures as a function of the different incubation times (10 min, 1 h, 4 h and 24 h) are summarised in Figure 3. A significantly ($p < 0.05$) higher total FB_1 was noticed in the total incubation mixture utilising the "high" FB contaminated maize at each time point (Figure 3A). FB_1 leaching from "low" FB contaminated maize, reached saturation after 1 h as, despite that different samples were used, no further increase was noticed in the samples incubated for 4 and 24 h (Figure 3B). In contrast, leaching of FB_1 from the "high" FB contaminated maize, although significantly ($p < 0.05$) lower after 1 h and 4 h, steadily increased until it mimicked the level of the "low" FB contaminated maize after 24 h.

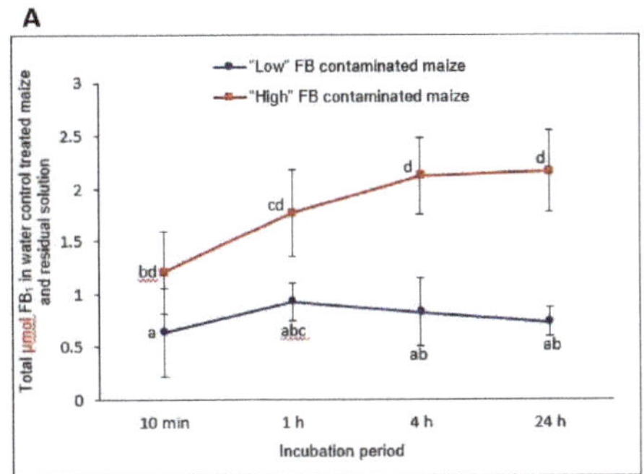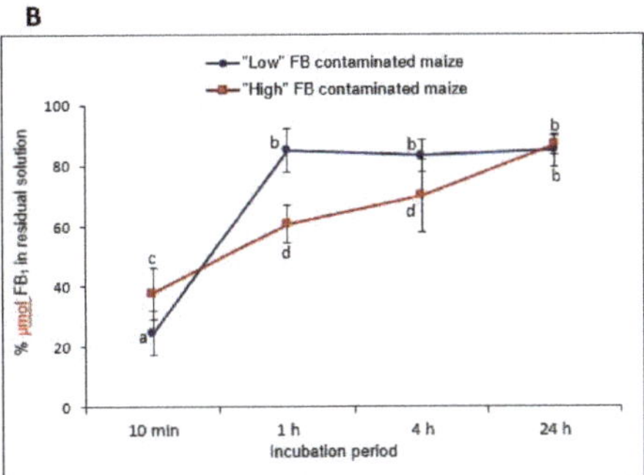

Figure 3. Total amount of FB_1 leaching into the incubation mixture in the absence of the enzyme as a function of time, of "low" and "high" FB contaminated maize (**A**). The % μmol FB_1 leaching into the residual water during a comparable time of incubation (**B**). Values represent means ± standard deviations of three to five replications of experiments. Statistical analyses are based on natural log (ln) and differences ($p < 0.05$) between treatment periods within and between "low" and "high" FB contaminated maize are indicated with different letters.

2.4.2. Comparative Enzyme Kinetics of FB_1 Hydrolysis

The conversion rates of FB_1 into HFB_1, expressed as nmol/min/mg enzyme, following the 100 and 1000 U/L FumD incubations as a function of time are summarised in Table 5. The FB_1 hydrolysis rates in the presence of 100 U/L and 1000 U/L FumD was maximum after 10 min in both maize samples and was significantly higher ($p < 0.05$) in the "high" FB contaminated maize. The hydrolysis rates decreased from 10 min to 24 h in the maize samples for both enzyme activities. The HFB_1 formation rate followed the same trend. Overall, the FB_1 hydrolyses and HFB_1 formation rates were significantly ($p < 0.05$) decreased in the 1000 U/L as compared to 100 U/L FumD incubations.

2.4.3. Comparative of FB_1 Hydrolysis to HFB_1 Formation Ratios

There was a significant ($p < 0.05$) difference between the conversion rates (FB_1 hydrolysis (nmol/min/mg enzyme)) of the 100 and 1000 U/L FumD incubations during each respective treatment period in both "low" and "high" FB contaminated maize. In both maize batches the 10 min incubation with 100 U/L exhibited a significant ($p < 0.05$) higher conversion rate when compared with the longer incubations (Table 5). The conversion ratio (FB_1 hydrolysis: HFB_1 formation) also tended to be increased in the "low" compared to the "high" FB contaminated maize with the 100 U/L incubation. A similar trend in the conversion ratios was obtained during the 1 h enzyme treatment study at the lower FumD enzyme activities (Table 3).

Table 5. Enzyme kinetic parameters regarding the conversion of FB_1 to HFB_1 by different fumonisin esterase FumD activities (U/L) as a function of treatment period (10 min, 1 h, 4 h and 24 h) in "low" and "high" FB contaminated maize.

	"Low" FB Contaminated Maize								"High" FB Contaminated Maize							
Treatment Duration	10 min	1 h	4 h	24 h	10 min	1 h	4 h	24 h	10 min	1 h	4 h	24 h	10 min	1 h	4 h	24 h
FumD Activity	FB_1 Hydrolysis (nmol/min/mg Enzyme)				HFB_1 Formation * (nmol/min/mg Enzyme)				FB_1 Hydrolysis (nmol/min/mg Enzyme)				HFB_1 Formation * (nmol/min/mg Enzyme)			
100 U/L	38.34 (1.54) a	8.23 (0.96) b	1.89 (0.04) c	0.32 (0.01) d	20.26 (10.75) A	5.15 (0.80) B	2.12 (0.56) C	0.32 (0.12) D	59.19 (12.08) i	13.18 (3.24) j	4.42 (1.03) e	0.92 (0.04) f	21.10 (8.56) A	14.55 (1.46) A	4.86 (0.82) B	1.18 (0.16) C
1000 U/L	3.06 (0.62) e	0.89 (0.11) f	0.20 (0.01) g	0.03 (0.00) h	2.78 (1.55) C	0.87 (0.16) E	0.25 (0.12) D	0.04 (0.01) F	4.53 (2.26) e	1.70 (0.19) k	0.52 (0.05) l	0.10 (0.00) m	3.64 (2.02) B	1.98 (0.46) C	0.67 (0.21) E	0.11 (0.02) D
	FB_1 Hydrolysis: HFB_1 Formation Ratio								FB_1 Hydrolysis: HFB_1 Formation Ratio							
100 U/L	3.46 (1.93) P	1.39 (0.62) Pq	0.95 (0.30) qr	0.93 (0.36) Pr	-	-	-	-	1.94 (0.16) Pr	0.91 (0.20) Pq	0.91 (0.14) qr	0.95 (0.33) qr	-	-	-	-
1000 U/L	1.45 (0.94) Pqs	1.04 (0.16) Pq	0.93 (0.38) qr	0.92 (0.35) qr	-	-	-	-	1.10 (0.29) Pqs	0.91 (0.26) qrs	0.84 (0.26) qrs	0.92 (0.24) qrs	-	-	-	-

The statistical analyses are based on natural log (ln) transformations. Values represent means of three to five replications of experiments with standard deviations indicated in brackets. Statistical differences ($p < 0.05$) of FB_1 hydrolysis rates and conversion ratios between FumD treatments (columns) as a function of time (rows) and for "low" and "high" FB contaminated maize are indicated with different lower case letters. Statistical differences ($p < 0.05$) of HFB_1 formation rates between FumD treatments (columns) as a function of time (rows) and for "low" and "high" FB contaminated maize are indicated with different uppercase letters. *Corrected for HFB_1 concentrations obtained in control (0 U/L) treatments in "low" and "high" FB contaminated maize and residual enzyme solutions. 100 U/L and 1000 U/L FumD represent 7.5 and 75 mg/L specific activity, respectively.

3. Discussion

Bioremediation, utilising enzymes for detoxification and/or degradation of environmental chemicals has become an intuitive way to reduce the risk of exposure in animals and humans. A popular approach is to use enzymes that can bind with high affinity toxic compounds and catalyze their hydrolysis, thereby abandoning or reducing their toxicity [27]. The present study established an enzymatic method for FB reduction utilizing fumonisin esterase FumD in whole maize intended for human consumption. The method involved treatment of "low" and "high" FB contaminated home-grown maize with varying FumD activities in solution as a function of time. FumD has been proved effective as a feed enzyme when incorporated into ground maize at a concentration of 40 U/kg feed [23]. In the current method, FumD was applied to whole maize, therefore the accessibility of FB would be expected to require higher FumD activities and longer incubation times for reduction than with ground maize. However, a major advantage of the current FumD FB reduction method is that the bulk of the residual enzyme and hydrolysed less toxic HFB_1 product are not associated with the treated maize kernels, which could find application when considering the reduction of FBs in food intended for human consumption.

Despite the fact FumD treatments were performed in the current reduction method at sub-optimal pH (approximately pH 5 to 5.5) and ambient temperature (±23 °C) conditions [optimal enzyme activity is obtained at pH 8 and 30 °C [14], effective hydrolysis of FB was observed in both maize batches. Treatment with 1000 U/L FumD effectively hydrolysed FB in "low" and "high" FB contaminated maize resulting in ≥80% reduction in the total FB levels after 1 h while lower enzyme activities (≤100 U/L) were less effective. At lower enzyme activities large variations in the FB levels and hydrolyses rates were noticed, which could be related to (i) sample variation within maize batches, (ii) enzyme and/or substrate availability as a function of time, and (iii) the formation of partially hydrolysed FB (PHFB) [28], which will be present as transient intermediates to the completely hydrolysed molecule at high enzyme activities. A significant ($p < 0.05$) decrease of FB_1 reduction in the "low" FB contaminated maize was noticed already at 10 U/L FumD compared to the 100 U/L in the "high" FB contaminated maize. These differences could be related to the FumD activities applied, FB_1 availability at the kernel/aqueous interphase and the formation of PHFB, and will be discussed below.

When considering differences in the formation of HFB_1 into the residual enzyme solution, maximum levels were already noticed from 100 U/L FumD activity. This suggests that some of the hydrolysed product is retained within the maize matrix as well as the possible formation of $PHFB_1$. This became apparent as, although HFB_1 was mainly associated (>90%) with the residual enzyme solution at low enzyme activities (10 and 100 U/L), it tends to accumulate in the "high" and to a lesser extent in the "low" FB contaminated maize kernels at incubations using increased FumD activities (≥1000 U/L). This could be ascribed to the formation of HFB_1 and $PHFB_1$ in the inner layers of the maize kernel matrix as compared to the "low" FB contaminated maize, where it mainly occurred on the surface.

As mentioned above, differences occurred in the extent of FB_1 hydrolysis between the two samples when considering the % FB_1 loss in the maize as well as the FB_1 hydrolysis to HFB_1 formation ratio in the incubation mixture, specifically at the lower FumD enzyme activities. These differences were also noticed when considering the enzyme kinetics of FB_1 hydrolysis and HFB_1 formation rates at 100 and 1000 U/L FumD activities, which were significantly ($p < 0.05$) increased in the "high" FB contaminated maize. The higher FB_1 conversion rate obtained with the 100 U/L FumD activity as compared to the 1000 U/L enzyme activity is related, as mentioned above, to FB availability and/or the limitation thereof when utilising excess enzyme activities.

The level of FB contamination determined the hydrolysis rate implying that fungal infiltration and FB production inside the kernel are key rate limiting factors. This became evident as *Fusarium* spp. infects maize with the fumonisins concentrated in the pericarp and embryo of the maize kernel [29], while in damaged kernels the fungus is likely to penetrate deeper into the kernel, contaminating the endosperm [30–32]. It is known that leaching of substances from the maize endosperm occurs during absorption of water while damage to the kernel pericarp increases leaching early during water absorption [33].

Differences in FB_1 hydrolysis rates between the "low" and "high" FB contaminated maize batches are therefore likely to depend on the leaching of FB_1 from the maize kernel and the interaction of FumD at the kernel outer layer/aqueous inter phase. This became apparent when considering the % FB_1 hydrolysis in the "low" FB contaminated sample that reached a maximum already after 100 U/L compared to the "high" FB contaminated maize reaching a maximum at 1000 U/L. This implies that the FB_1 became depleted as a substrate for the enzyme much earlier. This was also evident as leaching of FB_1 into the water control was faster from the "low" FB contaminated maize at 1 and 4 h of incubation compared to a more gradual effect considering the "high" FB contaminated maize. The gradual increased leaching of FB_1 from the "high" FB contaminated maize could be related to the extent of fungal damage of the kernels that will determine the diffusion rate of FB_1 from the inner layers. Therefore, the higher initial (10 min) conversion rate of FB_1 in the "high" as compared to "low" FB contaminated maize is facilitated by the accessibility of FB_1 associated with the inner as well as outer layers. The location of FB_1 in the kernel and the leaching rate into the aqueous enzyme-solution could therefore affect the efficacy of FB conversion rates and substrate to product ratios as noticed in the current study. However, the efficacy of FumD is not only determined by the concentration of FB_1 leaching into the aqueous phase but could also depend on the enzyme penetration into the endosperm which could explain the differences in FB conversion rates between "low" and "high" FB contaminated maize. This became evident as an increased % of HFB_1 was noticed in the maize kernels when utilising higher FumD activities. Therefore, depending on the extent of fungal damage and FB contamination to the maize pericarp/endosperm, a specific FumD activity seems to be required for maximum FB_1 hydrolysis, yielding a 1:1 FB_1 hydrolysis to HFB_1 formation ratio. This would become important when considering the application of the enzyme in different technological approaches utilising maize samples with varying FB contamination levels.

From a technological perspective, enzymatic detoxification of FB has been utilised effectively in the animal feed industry [21,22,24], while very little is known about possible applications to reduce exposure in humans. In this regard, commercial maize-based manufacturing processes, including dry milling and ethanol production are challenged with maintaining regulatory levels for FB in products and co-products [29,34]. During dry milling the FB mycotoxins are known to be ± 3-fold concentrated in the surface layers and are mainly confined to the total hominy feed fractions, which are mainly used in animal feed or non-food products [29]. As most micronutrients are concentrated in total hominy feed and a large portion of maize grain is lost in this fraction [35], FB decontamination could be of value with respect to human food. It could also find application during ethanol production, as the co-products wet and dry distillers' grains and solubles, containing high levels of FB, are increasingly being marketed as protein-rich and cost-saving inclusion in livestock and poultry feed [34]. Therefore, possible economic advantages of the FumD FB reduction method for commercial maize-based manufacturing processes should be further investigated.

The FumD FB reduction method in maize could also find application in communities utilising home-grown maize as a main dietary staple and known to be exposed above the tolerable daily intake levels, i.e., a PMTDI of 2 µg/kg/bw/day [36,37]. To decrease the risk of FB exposure, culturally sensitive, practical and biologically based methods of reduction are relevant and need to be implemented [1]. In the current study, the "low" FB contaminated maize contained FB levels below the regulatory maximum levels for fumonisins in maize (total FB_1 and FB_2 4000 µg/kg) set by the Codex Alimentarius Commission [38], while the high FB contaminated samples exceeded that. These maize batches were therefore ideal to evaluate the newly designed FumD FB reduction method, which could find application in maize subsistence farming communities where exposure to high levels of FB is the norm on a daily basis [37,39]. Of interest is that the bulk of HFB_1 resides into the aqueous phase, which will further minimize exposure to, not only FB_1, but also to the less toxic breakdown product, HFB_1.

However, increased amounts (up to 32%) of HFB_1 accumulated in "high" FB contaminated maize in the presence of increased FumD activities.

Recently a practical and culturally sensitive maize hand-sorting and water wash intervention method resulted in 84% reduction of FB_1 levels [40]. In the current study, the 1000 U/L FumD treatment resulted in access of 80% reduction in total FBs in both "low" and "high" FB contaminated samples. However, as only approximately 10% of FB is removed from the maize kernels during the water wash procedure [41], the newly developed FumD FB reduction method could effectively be applied to further reduce FB exposure prior to food preparation.

4. Conclusions

The present study developed an innovative FumD FB reduction method in whole maize. The method is suitable for direct application in the food chain postharvest, resulting in the reduction of FB_1 with the formation of the hydrolysed breakdown product, HFB_1, mainly associated with the aqueous phase to be discarded. Reduction of FB in contaminated maize could have a positive impact on food safety and security as well as having economic benefits during manufacturing processes.

5. Materials and Methods

5.1. Chemicals

Methanol, acetonitrile, formic acid (HPLC grade) and Whatman filter paper were obtained from Merck (Kenilworth, NJ, USA). Water for all experiments was successively purified by reverse osmosis followed by Milli-Q water purification (Millipore, Burlington, MA, USA).

5.2. Fumonisin Standard Solutions

Pure analytical standards of FB_1, FB_2, FB_3 and HFB_1 (purity > 97%) were prepared at the Institute of Biomedical and Microbial Biotechnology of the Cape Peninsula University of Technology, South Africa, according to the methods of Cawood et al. [42] and Gelderblom et al. [13]. Stock solutions of the individual purified fumonisin standards were prepared (1 mg/mL in acetonitrile-H_2O (1:1)) and aliquots used to prepare an evaporated working solution containing the fumonisin standards at individual concentrations of 5 µg/mL. For compiling matrix-matched calibration curves, five working standard dilutions were prepared with blank maize matrix extract as solvent, as described below.

5.3. Maize Sample Collection

Home-grown maize was collected in the Centane, Mnquma Local Municipality (areas of the former Transkei region) of the Eastern Cape Province, South Africa from households of subsistence maize farmers. "Good" home-grown whole maize was collected directly from visibly healthy batches destined for human consumption. "Mouldy" home-grown whole maize was collected directly from "mouldy" batches destined for chicken/livestock feed. The "good" and "mouldy" maize batches were labelled "low" and "high" FB contamination, respectively. Each maize batch was thoroughly mixed and kept at 4 °C until analysed. Control maize containing no fumonisins, was obtained from the Southern African Grain Laboratory (Pretoria, South Africa) and used for the preparation of a maize extract used for matrix-matched calibration curves.

5.4. FumD Enzyme Preparation

A fumonisin esterase, designated FumD (EC 3.1.1.87; FUM*zyme*®), was obtained from BIOMIN (Tulln, Austria) with a specific activity of 13,400 U/g. One unit is the enzymatic activity defined to release 1 µmol tricarballylic acid per minute from 100 µM FB_1 in 20 mM Tris-HCl buffer pH 8.0 containing 0.1 mg/mL bovine serum albumin at 30 °C. A stock enzyme solution (400 U/mL) was prepared in the Tris-HCl buffer and used in all experiments.

5.5. The FumD FB Reduction Method

Enzyme solutions utilised were prepared in distilled water from the FumD stock solution. Maize kernels (100 g) of the "low" and "high" FB contaminated maize batches were weighed in Erlenmeyer flasks (500 mL) and enzyme solution added (200 mL), obtaining a maize to solvent ratio of 1:2. Samples were mixed at 80 rpm on a shaker (New Brunswick Scientific, Edison Township, NJ, USA) at ambient temperature (± 23 °C) for different incubation periods. The residual enzyme solution was decanted and stored at −20 °C until analysed, while the residual maize kernels were dried onto laboratory paper towels at ambient temperature for 48 h. Residual maize kernel samples were ground in a laboratory mill (Falling Number AB, Stockholm, Sweden) to a fine meal and kept in airtight containers at −20 °C until analysed. Three to five replicates were included for each treatment.

5.6. FumD Incubation Protocols

To obtain the optimal enzyme activity required for FB hydrolysis the maize samples were incubated with different FumD activities (10, 100, 1000 and 5000 U/L) for 1 h as described above. The pH values of the maize-enzyme solutions were determined before and after the incubation treatment. In a follow-up experiment two FumD activities (100 and 1000 U/L) were used, and incubations were carried out over varying time periods including 10 min, 1 h, 4 h and 24 h to investigate the FB hydrolysis kinetics. Reference water control incubations were included in each experiment. All samples were processed and stored as described above.

5.7. Analyses of FB and HFB_1 Concentrations in Maize and Residual Solutions

FB_1, FB_2, FB_3 and HFB_1 concentrations were determined in the (i) untreated "low" and "high" FB contaminated maize batches (five replicates), (ii) the water control and enzyme treated residual maize kernel samples, and (iii) residual water control and enzyme solutions (3–5 replicate experiments).

5.7.1. Extraction Methods

FB and HFB_1 were extracted from maize according to the method of Sewram et al. [43] with minor modifications. Briefly, 100 mL of extraction solvent [methanol: acetonitrile: water (25:25:50; *v/v/v*)] was added to ground maize kernels (10 g) and placed on a shaker (80 rpm) for 20 min. The extracts were subsequently centrifuged (4000× *g*) in a refrigerated Sorvall RC-3B centrifuge (DuPont, Norwalk, CT, USA) at 4 °C for 10 min. The supernatant (20 ml) was diluted (1:1) with methanol:water (25:75), filtered (Whatman No 4 filter paper) and filtrates analyzed by direct injection into the LC-MS/MS. FAPAS (London, UK) quality control reference maize samples (Cat no T22123QC), containing the mycotoxins in known concentration ranges were included. For analyses of the residual solutions, samples were filtered (Whatman No 4 filter paper) and filtrates analysed directly by LC-MS/MS. Matrix-matched standard solutions for calibration curves were prepared utilising an extract prepared from control maize.

5.7.2. Chromatographic Quantification of FB and HFB_1

Quantification of FB and HFB_1 in maize extracts and residual enzyme solutions was performed by the Mass Spectrometry Unit of the Central Analytical Facility of Stellenbosch University, South Africa. The mycotoxins were separated on a reversed-phase BEH C_{18} column (2.1 × 100 mm; particle size 1.7 μm) (Waters, Milford, MA, USA) and analysed with positive electrospray ionisation (EI) in the multiple reaction monitoring (MRM) mode in a Waters Acquity Ultra high performance Liquid Chromatograph (UPLC) coupled to a Tandem Quadrupole Mass spectrometer (Waters Xevo TQ MS). Eluent A was water and eluent B was methanol, both containing 0.1% formic acid. The chromatographic method held the initial mobile phase composition (15% B) constant for 2 min, followed by a linear gradient to 100% B within 3 min. This final condition was held for 3.30 min, followed by 8 min of column re-equilibration at 15% B. The flow rate of the mobile phase was 0.35 ml/min. For each compound, one precursor and two product ions were monitored, one product ion for quantification and one for confirmation. A calibration curve consisting of five matrix-matched standards for each mycotoxin was used for quantification.

5.7.3. Validation of the Extraction and FB Quantification Methods

The extraction and chromatographic methods were validated by determining the LOQ, linearity (r^2) of the calibration curve, selectivity, accuracy, and % recovery according to guidelines of the United States Department of Health and Human Services, Food and Drug Administration [44].

5.8. Statistical Analyses

The NCSS Version 11 software [45] was used for statistical analysis. Data were subjected to natural log (ln) transformation of all variables and analysed within a generalised linear model ANOVA. Multiple comparisons were analysed using the Tukey-Kramer's multiple comparison procedure. This method provides joint simultaneous confidence intervals for all pairwise differences between the means; and also provides the multiple comparison p-value. Generally, $p < 0.05$ was used as statistical significance. In addition, the size of the F-ratios was used to measure relative sizes of differences.

Author Contributions: Data curation, J.A. and W.G.; Formal analysis, J.A. and I.D.; Funding acquisition, J.A.; Methodology, J.A., G.S., W.-D.M., J.R., H.-M.B., G.S. and W.G.; Resources, G.S., W.-D.M., J.R. and H.-M.B.; Supervision, J.A.; Validation, J.A.; Writing—original draft, J.A. and I.D.; Writing—review & editing, J.A., G.S., W.-D.M., J.R., H.-M.B., G.S. and W.G.

Funding: This research was funded by the South African Maize Trust (Project MTM 13/04).

Acknowledgments: The authors thank the South African Maize Trust for their financial support of research on enzymatic methods for reduction of mycotoxin concentrations in maize.

Conflicts of Interest: The authors declare no conflict of interest.

References

1. Alberts, J.F.; van Zyl, W.H.; Gelderblom, W.C.A. Biologically based methods for control of fumonisin-producing *Fusarium* species and reduction of the fumonisins. *Front. Microbiol.* **2016**, *7*, 548. [CrossRef] [PubMed]
2. Yates, I.E.; Meredith, F.; Smart, W.; Bacon, C.W.; Jaworski, A.J. *Trichoderma viride* suppresses fumonisin B_1 production by *Fusarium moniliforme*. *J. Food Prot.* **1999**, *62*, 1326–1332. [CrossRef] [PubMed]
3. Bacon, C.W.; Yates, I.E.; Hinton, D.M.; Meredith, F. Biological control of *Fusarium moniliforme* in maize. *Environ. Health Perspect.* **2001**, *109*, 325–332.
4. Duvick, J. Prospects for reducing fumonisin contamination of maize through genetic modification. *Environ. Health Perspect.* **2001**, *109*, 337–342.
5. Duvick, J.; Maddox, J.; Gilliam, J. Composition and methods for fumonisin detoxification. Patent No US6538177, 25 March 2003.
6. Cleveland, T.E.; Dowd, P.F.; Desjardins, A.E.; Bhatnagar, D.; Cotty, P.J. United States Department of Agriculture-Agricultural Research Service research on pre-harvest prevention of mycotoxins and mycotoxigenic fungi in US crops. *Pest. Manag. Sci.* **2003**, *59*, 629–642. [CrossRef] [PubMed]
7. Cavaglieri, L.; Passone, A.; Etcheverry, M. Screening procedures for selecting rhizobacteria with biocontrol effects upon *Fusarium verticillioides* growth and fumonisin B_1 production. *Res. Microbiol.* **2004**, *155*, 747–754. [CrossRef] [PubMed]
8. Samapundo, S.; De Meulenaer, B.; Osei-Nimoh, D.; Lamboni, Y.; Debevere, J.; Devlieghere, F. Can phenolic compounds be used for the protection of corn from fungal invasion and mycotoxin contamination during storage? *Food Microbiol.* **2007**, *24*, 465–473. [CrossRef]
9. Dalie, D.K.; Deschamps, A.M.; Atanasova-Penichon, V.; Richard-Forget, F. Potential of *Pediococcus pentosaceus* (L006) isolated from maize leaf to suppress fumonisin-producing fungal growth. *J. Food Prot.* **2010**, *73*, 1129–1137. [CrossRef]

10. Bacon, C.W.; Hinton, D.M. *In planta* reduction of maize seedling stalk lesions by the bacterial endophyte *Bacillus mojavensis*. *Can. J. Microbiol.* **2011**, *57*, 485–492. [CrossRef]

11. Mitchell, N.J.; Xue, K.S.; Lin, S.; Marroquin-Cardona, A.; Brown, K.A.; Elmore, S.E.; Tang, L.; Romoser, A.; Gelderblom, W.C.A.; Wang, J.; et al. Calcium montmorillonite clay reduces AFB_1 and FB_1 biomarkers in rats exposed to single and co-exposures of aflatoxin and fumonisin. *J. Appl. Toxicol.* **2014**, *34*, 795–804. [CrossRef]

12. Alberts, J.F.; Lilly, M.; Rheeder, J.P.; Burger, H.M.; Shephard, G.S.; Gelderblom, W.C.A. Technological and community-based methods to reduce mycotoxin exposure. *Food Control.* **2017**, *73*, 101–109. [CrossRef]

13. Gelderblom, W.C.A.; Cawood, M.E.; Snyman, S.D.; Vleggaar, R.; Marasas, W.F.O. Structure-activity relationships of fumonisins in short-term carcinogenesis and cytotoxicity assays. *Food Chem. Toxicol.* **1993**, *31*, 407–414. [CrossRef]

14. Heinl, S.; Hartinger, D.; Thamhesl, M.; Kunz-Vekiru, E.; Krska, R.; Schatzmayr, G.; Moll, W.; Grabherr, R. Degradation of fumonisin B_1 by the consecutive action of two bacterial enzymes. *J. Biotechnol.* **2010**, *145*, 120–129. [CrossRef] [PubMed]

15. Duvick, J.; Rood, T.; Maddox, J.; Gilliam, J. Detoxification of mycotoxins *in planta* as a strategy for improving grain quality and disease resistance: Identification of fumonisin-degrading microbes from maize. In *Molecular Genetics of Host-Specific Toxins in Plant Disease, Developments in Plant Pathology*; Kohmoto, K., Yoder, O.C., Eds.; Springer International Publishing AG: Basel, Switzerland, 1998; Volume 13, pp. 369–381.

16. Duvick, J.; Rood, T.; Wang, X. Fumonisin detoxification enzymes. Patent No US5716820, 10 February 1998.

17. Blackwell, B.A.; Gilliam, J.T.; Savard, M.E.; Miller, D.; Duvick, J.P. Oxidative deamination of hydrolysed fumonisin B_1 (AP_1) by cultures of *Exophiala spinifera*. *Nat. Toxins* **1999**, *7*, 31–38. [CrossRef]

18. Hartinger, D.; Schwartz, H.; Hametner, C.; Schatzmayr, G.; Haltrich, D.; Moll, W.D. Enzyme characteristics of aminotransferase FumI of *Sphingopyxis* sp. MTA144 for deamination of hydrolyzed fumonisin B_1. *Appl. Microbiol. Biotechnol.* **2011**, *91*, 757–768. [CrossRef]

19. Moll, D.; Hartinger, D.; Grießler, K.; Binder, E.M.; Schatzmayr, G. Method for the production of an additive for the enzymatic decomposition of mycotoxins, additive, and use thereof. Patent No US8703460B2, 22 April 2014.

20. Oswald, I.P.; Grenier, B.; Schatzmayr, G.; Moll, W. Enzymatic detoxification of mycotoxins: Hydrolysis of fumonisin B_1 strongly reduced the toxicity for piglets. In *World Nutrition Forum, Nutri Economics: Balancing Global Nutrition and Productivity*; Binder, E.M., Ed.; Anytime Publishing: Leicestershire, UK, 2012; pp. 263–271.

21. BIOMIN. Available online: https://www.BIOMIN.net/en/home/ (accessed on 12 July 2019).

22. Masching, S.; Naehrer, K.; Schwartz-Zimmermann, H.E.; Sārāndan, M.; Schaumberger, S.; Dohnal, I.; Nagl, V.; Schatzmayr, D. Gastrointestinal degradation of fumonisin B_1 by carboxylesterase FumD prevents fumonisin induced alteration of sphingolipid metabolism in turkey and swine. *Toxins* **2016**, *8*, 84. [CrossRef] [PubMed]

23. Grenier, B.; Bracarense, A.F.L.; Schwartz, H.E.; Lucioli, J.; Cossalter, A.; Moll, W.-D.; Schatzmayr, G.; Oswald, I.P. Biotransformation approaches to alleviate the effects induced by *Fusarium* mycotoxins in Swine. *J. Agric. Food Chem.* **2013**, *61*, 6711–6719. [CrossRef]

24. Grenier, B.; Schwartz, H.E.; Gruber-Dorninger, C.; Dohnal, I.; Aleschko, M.; Schatzmayr, G.; Moll, W.-D.; Applegate, T.J. Enzymatic hydrolysis of fumonisins in the gastrointestinal tract of broiler chickens. *Poult Sci.* **2017**, *96*, 4342–4351. [CrossRef]

25. Mogensen, J.M.; Sørensen, S.M.; Sulyok, M.; van der Westhuizen, L.; Shephard, G.S.; Frisvad, J.C.; Thrane, U.; Krska, R.; Nielsen, K.F. Single-kernel analysis of fumonisins and other fungal metabolites in maize from South African subsistence farmers. *Food Addit. Contam.* **2011**, *28*, 1724–1734. [CrossRef]

26. Janse van Rensburg, J.; Flett, B.C.; Mc Laren, N.W.; Mc Donald, A.H. Sampling variation in the quantification of fumonisins in maize samples. *S. Afr. J. Plant. & Soil* **2011**, *28*, 90–96.

27. Febbraio, F. Biochemical strategies for the detection and detoxification of toxic chemicals in the environment. *World J. Biol. Chem.* **2017**, *8*, 13–20. [CrossRef] [PubMed]

28. Hahn, I.; Nagl, V.; Schwartz, H.E.; Varga, E.; Schwartz, C.; Slavik, V.; Reisinger, N.; Malachová, A.; Cirlini, M.; Generotti, S.; et al. Effects of orally administered fumonisin B_1 (FB_1), partially hydrolysed FB_1, hydrolysed FB_1 and N-(1-deoxy-D-fructos-1-yl) FB_1 on the sphingolipid metabolism in rats. *Food Chem. Toxicol.* **2015**, *76*, 11–18. [CrossRef] [PubMed]

29. Burger, H.-M.; Shephard, G.S.; Louw, W.; Rheeder, J.P.; Gelderblom, W.C.A. The mycotoxin distribution in maize milling fractions under experimental conditions. *Int. J. Food Microbiol.* **2013**, *165*, 57–64. [CrossRef] [PubMed]

30. Brera, C.; Debegnach, F.; Grossi, S.; Miraglia, M. Effect of industrial processing on the distribution of fumonisin B$_1$ in dry milling corn fractions. *J. Food Prot.* **2004**, *67*, 1261–1266. [CrossRef] [PubMed]

31. Duncan, K.E.; Howard, R.J. Biology of maize kernel infection by *Fusarium verticillioides*. *Mol. Plant. Microbe Interact.* **2010**, *23*, 6–16. [CrossRef] [PubMed]

32. Kent, N.L.; Evers, D. Dry Milling. In *Kent's Technology of Cereals. An Introduction for Students of Food Science and Agriculture*, 5th ed.; Rosenstrater, K., Evers, A., Eds.; Woodhead Publishing: Cambridge, UK, 2017; pp. 421–514.

33. Bruggink, H.; Kraak, H.L.; Dijkema, M.H.G.E.; Bekendam, J. Some factors influencing electrolyte leakage from maize (*Zea mays* L.) kernels. *Seed Sci. Res.* **1991**, *1*, 15–20. [CrossRef]

34. Wu, F.; Munkvold, G.P. Mycotoxins in Ethanol Co-products: Modeling Economic Impacts on the Livestock Industry and Management Strategies. *J. Agric. Food Chem.* **2008**, *56*, 3900–3911. [CrossRef]

35. Ranum, P.; Pena-Rosas, J.P.; Garcia-Casal, M.N. Global maize production, utilization, and consumption. *Ann. N. Y. Acad. Sci.* **2014**, *1312*, 105–112. [CrossRef]

36. Burger, H.-M.; Lombard, M.J.; Shephard, G.S.; Danster-Christians, N.; Gelderblom, W.C.A. Development and evaluation of a sensitive mycotoxin risk assessment model (MYCORAM). *Toxicol. Sci.* **2014**, *141*, 387–397. [CrossRef]

37. Shephard, G.S.; Burger, H.-M.; Rheeder, J.P.; Alberts, J.F.; Gelderblom, W.C.A. The effectiveness of regulatory maximum levels for fumonisin mycotoxins in commercial and subsistence maize crops in South Africa. *Food Control.* **2019**, *97*, 77–80. [CrossRef]

38. FAO. Maximum Levels of Fumonsisin in Maize and Maize Products, Codex Alimentarius Commission, Geneva. Available online: http://www.fao.org/news/story/en/item/238558/icode/ (accessed on 30 May 2019).

39. Alberts, J.; Rheeder, J.; Gelderblom, W.; Shephard, G.; Burger, H.-M. Rural subsistence maize farming in South Africa: Risk assessment and intervention models for reduction of exposure to mycotoxins. *Toxins* **2019**, *11*, 334. [CrossRef]

40. Van der Westhuizen, L.; Shephard, G.S.; Burger, H.M.; Rheeder, J.P.; Gelderblom, W.C.; Wild, C.P.; Gong, Y.Y. Fumonisin B$_1$ as a urinary biomarker of exposure in a maize intervention study among South African subsistence farmers. *Cancer Epidemiol. Prev. Biomark.* **2011**, *20*, 483–489. [CrossRef]

41. Van der Westhuizen, L.; Shephard, G.S.; Rheeder, J.P.; Burger, H.-M.; Gelderblom, W.C.A.; Wild, C.P.; Gong, Y.Y. Optimising sorting and washing of home-grown maize to reduce fumonisin contamination under laboratory-controlled conditions. *Food Control.* **2011**, *22*, 396–400. [CrossRef]

42. Cawood, M.E.; Gelderblom, W.C.A.; Vleggaar, R.; Behrend, Y.; Thiel, P.; Marasas, W.F.O. Isolation of the fumonisins: A quantitative approach. *J. Agric. Food Chem.* **1991**, *39*, 1958–1962. [CrossRef]

43. Sewram, V.; Shephard, G.S.; Marasas, W.F.O.; De Castro, M.F.P.M. Improving extraction of fumonisin mycotoxins from Brazilian corn-based infant foods. *J. Food Prot.* **2003**, *66*, 854–859. [CrossRef]

44. United States Department of Health and Human Services, Food and Drug Administration (US FDA). USA Guidance for Industry. Bioanalytical Method Validation. Available online: https://www.fda.gov/regulatory-information/search-fda-guidance-documents/bioanalytical-method-validation-guidance-industry (accessed on 30 May 2019).

45. NCSS 11 Statistical Software. Available online: http://ncss.com/software/ncss (accessed on 30 May 2019).

© 2019 by the authors. Licensee MDPI, Basel, Switzerland. This article is an open access article distributed under the terms and conditions of the Creative Commons Attribution (CC BY) license (http://creativecommons.org/licenses/by/4.0/).

 *toxins*

Article

Biotransformation of the Mycotoxin Zearalenone to its Metabolites Hydrolyzed Zearalenone (HZEN) and Decarboxylated Hydrolyzed Zearalenone (DHZEN) Diminishes its Estrogenicity In Vitro and In Vivo

Sebastian Fruhauf [1], Barbara Novak [1], Veronika Nagl [1,*], Matthias Hackl [2], Doris Hartinger [1], Valentina Rainer [1], Silvia Labudová [1], Gerhard Adam [3], Markus Aleschko [1], Wulf-Dieter Moll [1], Michaela Thamhesl [1] and Bertrand Grenier [1]

[1] BIOMIN Research Center, Technopark 1, 3430 Tulln, Austria
[2] TAmiRNA GmbH, Muthgasse 18, 1190 Vienna, Austria
[3] Institute of Applied Genetics and Cell Biology (IAGZ), University of Natural Resources and Life Sciences, Vienna (BOKU), Konrad Lorenz-Straße 24, 3430 Tulln, Austria
* Correspondence: veronika.nagl@biomin.net

Received: 26 July 2019; Accepted: 16 August 2019; Published: 20 August 2019

Abstract: Zearalenone (ZEN)-degrading enzymes are a promising strategy to counteract the negative effects of this mycotoxin in livestock. The reaction products of such enzymes need to be thoroughly characterized before technological application as a feed additive can be envisaged. Here, we evaluated the estrogenic activity of the metabolites hydrolyzed zearalenone (HZEN) and decarboxylated hydrolyzed zearalenone (DHZEN) formed by hydrolysis of ZEN by the zearalenone-lactonase Zhd101p. ZEN, HZEN, and DHZEN were tested in two in vitro models, the MCF-7 cell proliferation assay (0.01–500 nM) and an estrogen-sensitive yeast bioassay (1–10,000 nM). In addition, we compared the impact of dietary ZEN (4.58 mg/kg) and equimolar dietary concentrations of HZEN and DHZEN on reproductive tract morphology as well as uterine mRNA and microRNA expression in female piglets (n = 6, four weeks exposure). While ZEN increased cell proliferation and reporter gene transcription, neither HZEN nor DHZEN elicited an estrogenic response, suggesting that these metabolites are at least 50–10,000 times less estrogenic than ZEN in vitro. In piglets, HZEN and DHZEN did not increase vulva size or uterus weight. Moreover, RNA transcripts altered upon ZEN treatment (EBAG9, miR-135a-5p, miR-187-3p and miR-204-5p) were unaffected by HZEN and DHZEN. Our study shows that both metabolites exhibit markedly reduced estrogenicity in vitro and in vivo, and thus provides an important basis for further evaluation of ZEN-degrading enzymes.

Keywords: zearalenone; estrogen response element; gene expression; cell proliferation; estrogen receptor; biotransformation

Key Contribution: By employing cellular and transcriptomics approaches, we demonstrated both in vitro and in vivo that the biotransformation of zearalenone by the zearalenone-lactonase Zhd101p results into two metabolites (hydrolyzed zearalenone and decarboxylated hydrolyzed zearalenone) much less estrogenic than the parent compound

---

1. Introduction

Zearalenone (ZEN), a mycotoxin produced by various *Fusarium* species, is a frequent contaminant of cereal-based foods and feeds. Although maize is regarded as the most affected commodity, ZEN can also occur in other grains, such as barley, oats, or wheat [1]. The incidence of samples tested positive for ZEN varies with factors such as region, year, commodity, or detection method [2]. A recent survey reports a ZEN occurrence of 88% in finished feed, maize, and maize silage samples collected at a global scale [3]. Although median ZEN concentrations of 0.02 mg/kg were found, individual samples reached levels of up to 11.19 mg/kg.

ZEN acts as xenoestrogen by activation of estrogen receptors α (ERα, full agonist) and β (ERβ, partial agonist) [4]. Subsequently, ERα and ERβ translocate to the nucleus, bind to estrogen response elements (ERE, 15-bp palindrome motif) in the promotor regions of target genes and induce their transcription [5]. In vitro, this leads to a dose-dependent proliferation of estrogen-dependent cells, such as the human breast cancer cell line MCF-7, upon ZEN treatment [6]. In vivo, ZEN evokes symptoms of hyperestrogenism. In pigs, the species considered to be most sensitive to this mycotoxin, clinical signs include swelling and reddening of the vulva, metaplasia of uterus, ovarian atrophy, enlargement of the mammae in both females and males, decreased testes weight, depressed spermatogenesis and libido, reduced fertility or delivery of stillborn, and weak piglets [1,4]. Alteration of hematological and biochemical measures, gene expression and subpopulations of immunocompetent cells in tissues have been observed after ZEN exposure in piglets and prepubertal gilts, albeit with limited reproducibility between studies [7]. Recently, the effect of ZEN on the expression of microRNAs has caught scientific interest [8–11]. MicroRNAs represent a class of small non-coding RNAs, that negatively regulate gene expression. They can serve as indicators for pathological processes in organs, and possess considerable potential as biomarkers [12]. Very recently it has been shown that ZEN affects uterine expression of certain microRNAs in pigs, among them ssc-miR-424-5p, ssc-miR-450a, ssc-miR-450b-5p, ssc-miR-450c-5p, ssc-miR-503, and ssc-miR-542-3p [11]. This effect was partly attributed to estrogen receptor (ER) activation, but further molecular targets of this mycotoxin, such as G protein-coupled estrogen receptor 1 [13] or the pregnane X receptor [14], need to be considered in this context. An overview of the manifold modes of action of ZEN are provided in several comprehensive reviews [1,4,6,15,16].

Due to its frequent occurrence and impairment of animal health, effective strategies for the control of ZEN are required. Many countries worldwide introduced guidance levels for ZEN in the feed. Within the European Union, recommended guidance levels for ZEN in compound feed for pigs range from 100 µg/kg (piglets, gilts) to 250 µg/kg (sows, fattening pigs) [17]. Yet, these regulations do not account for the presence of multiple (modified) mycotoxins in feed, which might potentiate the effects of ZEN [18]. Prevention of fungal growth and toxin formation by application of good agricultural practices represents a key strategy in mycotoxin reduction management [19]. Since those preharvest measures are often insufficient to completely eliminate mycotoxins in crops, different postharvest techniques for removal of ZEN have been tested. Technological possibilities include physical and chemical methods, such as UV irradiation, gamma radiation or application of ozone and peroxide, and are able to degrade ZEN to a certain extent [20]. However, their application is hampered by factors such as the formation of unknown metabolites, nutrient losses, impairment of palatability, and cost-efficient application under field conditions [21]. Feed additives, that either adsorb or biotransform mycotoxins in the gastrointestinal tract (GIT) of animals and, thus, reduce their bioavailability and/or toxicity, seem to be more applicable in practice. Certain materials (e.g. bentonites, diatomites, zeolites, or yeasts) show remarkable adsorption efficiencies for aflatoxin B_1, whereas binding of ZEN is comparatively low [22,23]. Modifications in the adsorbents' surface might lead to improved binding capacities for ZEN [20], but the limited specificity of these products and the potential binding of nutrients or veterinary drugs [24] is still a major disadvantage. Hence, the biological transformation of ZEN to non- or less toxic metabolites via microorganisms or enzymes might represent a more suitable option for the feed and livestock industry.

Several fungi and bacteria originating from soil or the GIT of animals are able to transform ZEN in vitro [1,21]. One of the first studies in this field described the reduction of ZEN to α-zearalenol (α-ZEL) and to a lesser extent to β-zearalenol (β-ZEL) by rumen protozoa [25]. Since α-ZEL has a higher binding affinity to ERs than ZEN [26], this reaction cannot be considered as detoxification. Similarly, a transformation of ZEN to zearalenone-glucosides or zearalenone-sulfate, e.g. by *Thamnidium elegans*,

Mucor bainieri or various *Rhizopus* spp. [27,28], might not be useful for further exploitation as a feed additive, as zearalenone-glucosides and zearalenone-sulfate are hydrolyzed during porcine digestion, which results in the release and subsequent systemic absorption of ZEN [29]. For many other microorganisms reported to degrade ZEN, neither the formed metabolites nor their toxicity or the responsible catabolic pathways have been elucidated in detail [1,21,30]. Two important exceptions have to be mentioned in this regard. First, *Trichosporon mycotoxinivorans* can cleave the lactone ring between C5 and C6 of the ZEN molecule, thus, producing the non-estrogenic ZOM-1 [31]. Second, *Gliocladium roseum* NRRL 1859 and *Clonostachys rosea* (synonym of *G. roseum*) IFO 7063 are capable of hydrolyzing the ester bond of the lactone ring of ZEN [32,33]. The lactonohydrolase mediating this reaction was further characterized, and the encoding gene (*zhd101*) was cloned [34]. Although the resulting metabolites had been postulated and partly confirmed before [32,33], the entire reaction mechanism was unveiled only in 2015 [30]. In this study, Vekiru et al. produced the zearalenone-lactonase Zhd101p from a codon-optimized version of the gene *zhd101*, and identified hydrolyzed zearalenone (HZEN, (E)-2,4-dihydroxy-6-(10-hydroxy-6-oxo-1-undecen-1-yl)benzoic acid) as a primary reaction product. This intermediate partly decarboxylates spontaneously to the previously described decarboxylated hydrolyzed zearalenone (DHZEN, (E)-1-(3,5-dihydroxy-phenyl)-10-hydroxy-1 -undecen-6-one, Figure 1) [30,33]. For the technological application of the zearalenone-lactonases as a feed additive, estrogenicity assessment of its metabolites HZEN and DHZEN is essential. So far, only one in vitro report on the estrogenicity of DHZEN is available, showing markedly reduced estrogenic potency in an MCF-7 proliferation assay [32]. In contrast, data on HZEN are completely lacking.

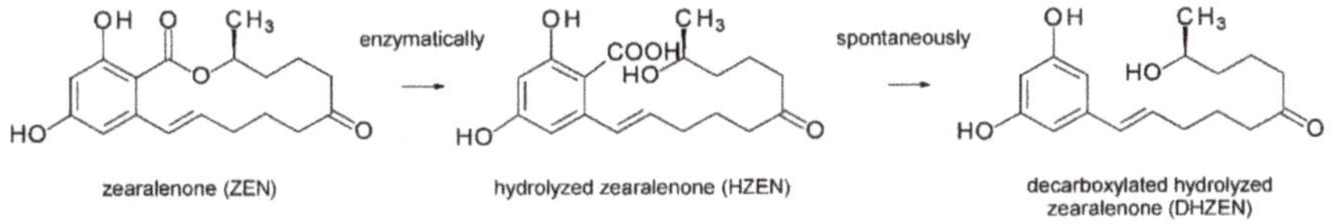

Figure 1. Structures of zearalenone (ZEN) and its metabolites hydrolyzed zearalenone (HZEN) and decarboxylated hydrolyzed zearalenone (DHZEN) produced by the zearalenone-lactonase Zhd101p (modified based on [30]).

Therefore, the aim of our study was to assess the estrogenicity of HZEN and DHZEN in vitro and in vivo, with ZEN serving as a reference. To this end, appropriate amounts of HZEN and DHZEN were produced, and subsequently tested in the MCF-7 proliferation assay as well as in a yeast bioassay. Furthermore, we evaluated the estrogenic potency of HZEN and DHZEN in prepubertal gilts based on morphological changes of the reproductive tract and alterations of mRNA and microRNA expression in uterus tissue. By generating data on the comparative estrogenicity of ZEN, HZEN, and DHZEN, our results contribute significantly to the risk assessment of this ZEN-degrading enzyme approach and its (potential) use as a feed additive.

2. Results

2.1. In Vitro Experiments

First, a proliferation assay with the human breast cancer cells MCF-7 was conducted. For that purpose, MCF-7 cells were treated with ZEN, HZEN, and DHZEN at concentrations of 0.01 to 500 nM. Cell proliferation was measured by the WST-1 assay (see Section 5.1.3). After 144 h of incubation, ZEN significantly stimulated the proliferation of human breast cancer cells at concentrations of 10, 100, and 500 nM. For comparison also ß-estradiol (E2, 10 nM) was tested, which increased cell proliferation to 226.9 ± 52% compared to the cell control. In contrast, neither HZEN nor DHZEN showed increased in cell proliferation (Figure 2).

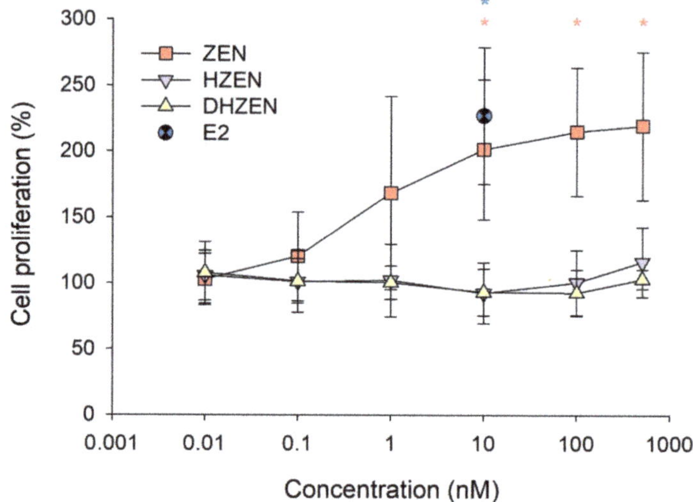

Figure 2. MCF-7 cell proliferation (%) after treatment with 0.01, 0.1, 1, 10, 100, and 500 nM ZEN (positive control), HZEN or DHZEN for 144 h. Cell control (= 0.05% DMSO) was set to 100%. 17ß-estradiol (E2, 10 nM) served as additional positive control. Each result represents the mean ± SD of five independent experiments (with three technical replicates per experiment). Significant differences are indicated with * ($p < 0.05$). The blue * only concerns the comparison between Control and E2.

To confirm the lack of estrogenicity of HZEN and DHZEN in vitro, we employed a reporter gene assay with the engineered estrogen-sensitive yeast strain YZHB817. Due to the higher tolerance of the yeast cells concentrations up to 10,000 nM of ZEN, HZEN, and DHZEN were tested. Consistent with results of the MCF-7 proliferation assay, a significant estrogenic effect was detected starting at ZEN concentrations of 10 nM, whereas no estrogenicity was seen even at the highest concentrations of HZEN and DHZEN (10,000 nM; Figure 3).

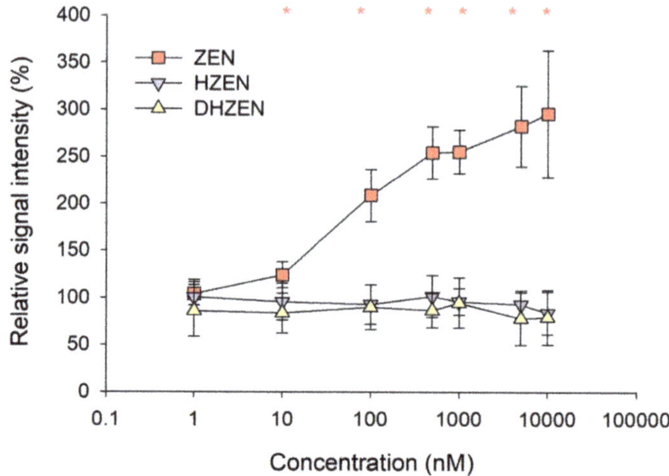

Figure 3. Estrogen dependent ß-galactosidase activity in yeast bioassay after treatment with 1, 10, 100, 500, 1000, 5000 and 10,000 nM ZEN, HZEN, or DHZEN for 18 h. Signal intensity of cell control (= 1% DMSO) was set to 100%. Each result represents the mean ± SD of four independent experiments (with four technical replicates per experiment). Significant differences are indicated with * ($p < 0.05$).

2.2. In Vivo Experiment

The effects of dietary ZEN (4.58 mg/kg), as well as equimolar concentrations of HZEN (4.84 mg/kg) and DHZEN (4.21 mg/kg) on the reproductive tract of female prepubertal pigs, were monitored in a feeding trial. None of the treatments significantly affected the body weight of animals (Table 1). While ZEN caused a time-dependent increase of the vulva size, reaching an enlargement by a factor of 3.3 compared to the control group on day 27 [11], neither HZEN nor DHZEN influenced the vulva size during the experimental period. Similarly, the reproductive tract weight at the end of the experiment was markedly increased in the ZEN group, whereas the ZEN metabolites showed no effect on this parameter (Table 1).

Table 1. Effect of dietary ZEN (4.58 mg/kg) and equimolar concentrations of HZEN (4.84 mg/kg) and DHZEN (4.21 mg/kg) on body weight, reproductive tract weight and vulva size of piglets (mean $\pm$ SD, n = 6) after 27 days of exposure. Significant differences between treatments are indicated with dissimilar superscripts [a,b] ($p < 0.05$).

Group	Body Weight (kg)	Vulva Size [1] (cm^2)	Reproductive Tract Weight [2]
Control	20.5 ± 2.8	1.47 ± 0.42^a	51.8 ± 20.6^a
ZEN	21.0 ± 2.2	4.84 ± 0.66^b	353.4 ± 45.2^b
HZEN	20.7 ± 2.5	1.70 ± 0.24^a	59.8 ± 12.0^a
DHZEN	20.5 ± 4.0	1.41 ± 0.27^a	49.9 ± 11.1^a

[1] Calculated by multiplying vulva length (cm) with vulva with (cm), [2] Expressed in g per kg body weight $\times$ 100.

In addition to investigating the effect of ZEN and its metabolites on the morphology of the reproductive tract, we also addressed potential alterations at the molecular level, by analyzing mRNA expression of estrogen-responsive genes as well as expression of ZEN responsive microRNA using qPCR.

The analysis of the porcine genomic DNA sequence for suspected estrogen-regulated genes with Dragon ERE Finder [35] resulted in the identification of fifteen potential genes with near-consensus ERE in pigs. Among them, we also found the gene EBAG9, which is one of three genes with perfect palindromic ERE identified near E2-regulated genes in the human genome. We performed mRNA expression analysis of six genes with identified EREs, and seven other genes associated with the estrogen response (see Section 5.2.2). As depicted in Table 2, ZEN exhibited only marginal effects on uterine gene expression after 28 days of exposure. The highest numerical fold change compared to the control group was observed for IL-1β (3.36-fold increase), but it lacked statistical significance due to high variability between animals. Among the genes under potential ERE activation, EBAG9 and GJA1 were significantly up-regulated upon ZEN treatment, albeit with low fold-change increase. For EBAG9, the HZEN and DHZEN groups were significantly different from ZEN group (no up-regulation). Overall, HZEN or DHZEN showed no impact on any of the other investigated genes when compared to the control group. Further statistically significant differences were only encountered between the ZEN and DHZEN groups for GAPDH and S100G, respectively. These differences can mostly be explained by high biological variability, such as seen in the control group as well.

Finally, we evaluated the impact of ZEN and its metabolites on microRNA expression in uterus. To this end, a subset of 15 microRNAs was selected for qPCR analysis based on previous data generated by Grenier et al. [11] (see Section 5.2.3 for details). Figure 4 displays the results of the hierarchical clustering analysis. While the ZEN group forms a distinct cluster, samples from the control, HZEN and DHZEN groups are heterogeneously distributed. Furthermore, individual pigs with divergent microRNA expression patterns can be deduced from this figure, e.g. pig ID 8 (Control) or ID 93 (HZEN). Analysis of data with Grubb's test substantiated this assumption for pig ID 8, for which the critical outlier Z value of 1.8 was exceeded in ten out of 15 microRNAs. Still, since not all data were normally distributed and there was no relevant evidence to exclude this sample (e.g. impaired general health of this pig during the experiment or known issue during sample preparation), we decided to use the complete dataset for expression analysis.

Table 2. Effect of dietary ZEN (4.58 mg/kg) and equimolar amounts of HZEN (4.84 mg/kg) and DHZEN (4.21 mg/kg) on gene expression in the uterus (d28, n = 6). Fold changes in treatment groups are expressed relative to the control group. Normalized Ct values (ΔCt-values) were used for statistical analysis. Significant differences between groups are indicated with dissimilar superscripts [a,b] ($p < 0.05$).

Gene Name	Protein Name	Relative Gene Expression Mean (and Range) Fold-Change Compared to Control			
		Control	ZEN	HZEN	DHZEN
EBAG9	Receptor-binding cancer antigen expressed on SiSo cells	1.00 [a] (0.61–1.65)	1.47 [b] (1.01–2.16)	1.04 [a] (0.67–1.61)	1.05 [a] (0.92–1.20)
OVGP1	Oviduct-specific glycoprotein or mucin-9	1.00 (0.54–1.85)	2.19 (1.03–4.70)	1.35 (0.68–2.67)	1.79 (1.07–3.00)
IGFBP4	Insulin-like growth factor-binding protein 4	1.00 (0.59–1.71)	0.57 (0.29–1.15)	1.16 (0.70–1.94)	1.23 (0.93–1.63)
GJA1	Gap junction alpha-1 protein or connexin 43	1.00 [a] (0.62–1.62)	1.93 [b] (1.48–2.52)	1.12 [a,b] (0.65–1.93)	1.26 [a,b] (0.98–1.62)
GAPDH	Glyceraldehyde 3phosphate dehydrogenase	1.00 [a,b] (0.46–2.15)	1.97 [a] (1.15–3.37)	1.08 [a,b] (0.48–2.43)	0.92 [b] (0.58–1.45)
C3	Complement component 3	1.00 (0.41–2.42)	0.78 (0.40–1.53)	0.86 (0.44–1.69)	1.40 (1.11–1.76)
S100G	S100 calciumbinding protein or calbindin D9K	1.00 [a,b] (0.13–7.72)	0.22 [a] (0.02–2.44)	2.48 [a,b] (1.68–3.66)	4.32 [b] (2.18–8.54)
CLU	Clusterin or apolipoprotein J	1.00 (0.53–1.90)	1.27 (0.65–2.50)	0.85 (0.53–1.35)	1.02 (0.73–1.44)
ODC	Ornithine decarboxylase	1.00 (0.55–1.82)	0.95 (0.71–1.28)	0.95 (0.62–1.45)	1.09 (0.89–1.33)
ESR1	Estrogen receptor alpha	1.00 (0.57–1.74)	1.11 (0.73–1.68)	1.18 (0.66–2.10)	1.24 (1.06–1.45)
ESR2	Estrogen receptor beta	1.00 (0.39–2.54)	1.22 (0.33–4.51)	0.40 (0.09–1.90)	1.62 (0.53–4.94)
IL-1β	Interleukin 1 beta	1.00 (0.26–3.90)	3.36 (0.47–24.12)	1.23 (0.43–3.52)	1.83 (1.12–3.01)
IL-6	Interleukin 6	1.00 (0.56–1.78)	0.89 (0.34–2.36)	1.40 (0.85–2.30)	1.90 (1.25–2.92)

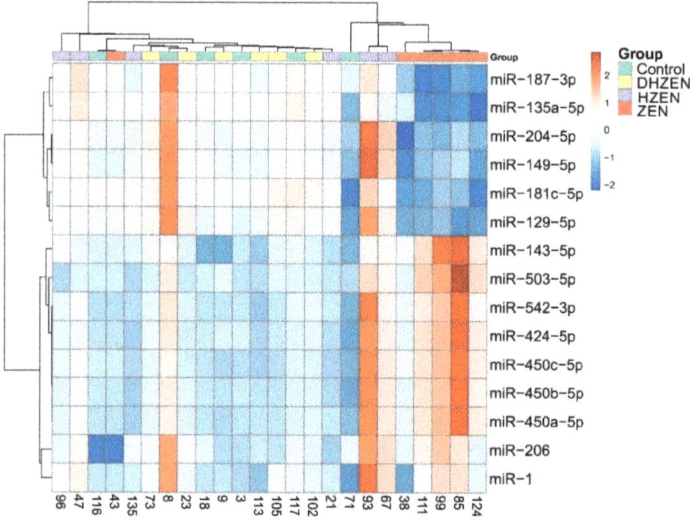

Figure 4. Hierarchical clustering analysis (average Pearson correlation coefficients) of microRNA expression in uterus of piglets exposed to uncontaminated feed (Control), ZEN (4.58 mg/kg), HZEN (4.84 mg/kg), or DHZEN (4.21 mg/kg) for 28 days (n = 6). The top row gives the treatment group, each column represents one pig. Each row represents one microRNA. Red rectangles indicate upregulation, blue rectangles indicate down regulation of the respective microRNA.

As shown in Table 3, seven of the selected microRNAs were significantly altered by the different dietary treatments. Compared to the control group, ZEN caused a significant decrease in the expression of miR-135a-5p, miR-187-3p, and miR-204-5p. In contrast, those microRNAs were unaffected in HZEN or DHZEN exposed animals. ZEN also caused numerical down-regulation of miR-129-5p and miR-149-5p, which reached statistical significance only compared to the HZEN and/or DHZEN group. None of the selected microRNAs were significantly up-regulated upon ZEN treatment, although the expression of the microRNA members of the miR-503 cluster (e.g. miR-424-5p, miR-450a-5p, miR-450b-5p, miR-450c-5p, miR-542-3p, or miR-503) showed a numerical increase of up to 4.5-fold. In the case of the miR-503 cluster, the only statistically significant difference was detected for miR-450c-5p between the ZEN and DHZEN group. In Figure 5, Ct normalized values of four miRNAs are depicted to provide a better overview of the inter-individual variance of microRNA expression, and thus the potential inter-individual difference in response to the administered substances. Similar to the mRNA analysis, neither HZEN nor DHZEN caused significant alterations of any investigated microRNAs compared to the control group.

Table 3. Effect of dietary ZEN (4.58 mg/kg) and equimolar amounts of HZEN (4.84 mg/kg) and DHZEN (4.21 mg/kg) on microRNA expression in the uterus (d 28, n = 6). Fold changes in treatment groups are expressed relative to the control group. Normalized Ct values (ΔCt-values) were used for statistical analysis. Significant differences between groups are indicated with dissimilar superscripts [a,b] ($p < 0.05$).

microRNA	Relative microRNA Expression Mean (and Range) Fold-Change Compared to Control			
	Control	ZEN	HZEN	DHZEN
miR-1	1.00 (0.47–2.11)	1.27 (0.72–2.22)	1.89 (0.63–5.66)	1.03 (0.67–1.60)
miR-129-5p	1.00 [a,b] (0.45–2.24)	0.42 [a] (0.18–0.99)	1.44 [b] (0.40–5.21)	1.19 [b] (0.78–1.82)
miR-135a-5p	1.00 [a] (0.32–3.11)	0.12 [b] (0.02–0.68)	0.94 [a] (0.21–4.27)	1.01 [a] (0.63–1.62)
miR-143-5p	1.00 [a] (0.47–2.14)	2.78 [b] (1.44–5.36)	1.45 [a,b] (0.38–5.48)	0.96 [a] (0.65–1.43)
miR-149-5p	1.00 [a,b] (0.46–2.18)	0.37 [a] (0.19–0.74)	1.77 [b] (0.59–5.31)	1.11 [a,b] (0.76–1.63)
miR-181c-5p	1.00 (0.40–2.52)	0.40 (0.17–0.96)	1.12 (0.30–4.24)	1.14 (0.70–1.87)
miR-187-3p	1.00 [a] (0.44–2.29)	0.11 [b] (0.03–0.48)	0.81 [a] (0.23–2.85)	0.70 [a] (0.42–1.16)
miR-204-5p	1.00 [a] (0.45–2.22)	0.29 [b] (0.13–0.66)	1.58 [a] (0.54–4.65)	0.99 [a] (0.66–1.47)
miR-206	1.00 (0.36–2.79)	1.34 (0.45–4.04)	1.88 (0.61–5.81)	0.88 (0.56–1.39)
miR-424-5p	1.00 (0.47–2.15)	3.69 (1.31–10.41)	1.86 (0.63–5.47)	0.95 (0.57–1.57)
miR-450a-5p	1.00 (0.50–1.98)	4.25 (1.45–12.42)	2.11 (0.72–6.21)	0.92 (0.60–1.42)
miR-450b-5p	1.00 (0.48–2.09)	3.63 (1.34–9.84)	2.00 (0.68–5.86)	0.94 (0.61–1.44)
miR-450c-5p	1.00 [a,b] (0.49–2.06)	4.52 [a] (1.60–12.75)	2.36 [a,b] (0.78–7.12)	0.84 [b] (0.54–1.30)
miR-503-5p	1.00 (0.53–1.89)	3.36 (1.29–8.74)	1.16 (0.38–3.51)	0.87 (0.49–1.54)
miR-542-3p	1.00 (0.47–2.12)	3.45 (1.20–9.89)	2.19 (0.72–6.63)	0.88 (0.54–1.42)

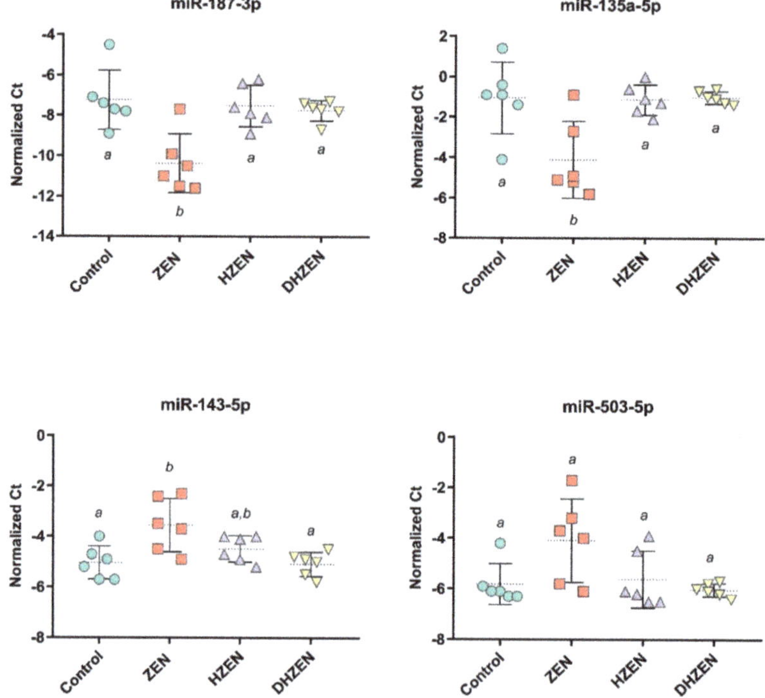

Figure 5. Normalized Ct-values of miR-187-3p, miR-135a-3p, miR-143-5p, and miR-503-5p in the uterus tissue of piglets exposed to uncontaminated feed (Control), ZEN, HZEN, or DHZEN (d28, n = 6). Individual symbols represent individual pigs. Significant differences between groups are indicated with dissimilar superscripts [a,b] ($p < 0.05$).

3. Discussion

Due to the limitations of traditional physical and chemical strategies for the detoxification of mycotoxins, biodegradation of mycotoxins by microorganisms and/or enzymes is an attractive approach to reduce the negative effects of mycotoxins. Several microorganisms capable of detoxification of ZEN and other mycotoxins have been previously reported [1,20]. However, for application as a feed additive, feed safety issues have to be sufficiently addressed. From a technological point of view, preparation of microorganisms with good stability during storage and high activity in the gastrointestinal tract is difficult. In comparison, enzymatic degradation using highly specific enzymes should be more effective. An example is the zearalenone-lactonase Zhd101p, which degrades ZEN to HZEN as the primary reaction product. HZEN is unstable and partially decarboxylates to DHZEN. To evaluate the technological potential of the zearalenone-lactonase as a feed additive, the present study aimed to investigate the estrogenic activity of its reaction products HZEN and DHZEN in vitro and in vivo.

ER-positive MCF-7 cells are widely used in endocrine research, e.g. to determine the estrogenic activity of substances (proliferation assay, also known as E-screen) or to study potential remedies for estrogen-related diseases, such as breast cancer. Various studies have demonstrated a dose-dependent pro-proliferative effect of ZEN in this cell line (e.g. [32,36–38]). Likewise, ZEN increased cell proliferation of MCF-7 cells in our experiment, which became statistically significant at a concentration of 10 nM (201.4 ± 53.0% compared to the control set to 100%). At higher ZEN concentrations, a saturation in cell proliferation at approximately 220% was observed, which is in line with previous studies, which report a plateau at around 170% [38] to 220% [32]. In contrast, neither HZEN nor DHZEN affected

cell proliferation in the tested concentration range, indicating that these metabolites are at least 50 ×
less estrogenic than ZEN (500 nM HZEN showed no effect, whereas 10 nM ZEN showed a significant
effect). Higher concentrations were not tested, since ZEN starts to impair cell viability in breast cancer
cell lines at 1000–25,000 nM [36,37,39]. So far, only one report is available on the estrogenicity of one of
those ZEN metabolites: Kakeya et al. [32] showed that DHZEN has no impact on cell proliferation of
MCF-7 cells up to a concentration of 100 nM. Thus, we could not only confirm these results but extend
them to HZEN, which was identified as the primary reaction product of the zearalenone-lactonase
only a few years ago [30].

The MCF-7 proliferation assay only provides information on whether a substance evokes the
cellular response known to be induced by estrogens, whereas definite conclusions on the underlying
mechanisms cannot be drawn [40]. Hence, we used a yeast bioassay to investigate specifically the
effect of ZEN, HZEN, and DHZEN on ER activation and subsequent transcription of the reporter gene
GAL7-lacZ. Moreover, this yeast bioassay provides several additional advantages, e.g. much easier to
perform, no cell culture equipment is needed, the concentration range of test substances is broader,
and it has a comparably low turnover time (incubation with test substances takes 18 h as compared
to 144 h for MCF-7 proliferation assay). In the present study, ZEN significantly increased the lacZ
activity at 10 nM (124.1 ± 13.8% compared to the control set to 100%). In the MCF-7 proliferation
assay, estrogenicity of ZEN was detected at the same concentration. It has been reported that yeast
bioassays can be less sensitive than the MCF-7 proliferation assay [40], but this was not observed in
our study. In line with results of the MCF-7 proliferation assay, HZEN and DHZEN did not activate
the *GAL7-lacZ* reporter gene in the yeast bioassay, even at the highest concentration tested (10,000 nM).
Given the tested concentration range, HZEN and DHZEN can be considered as at least 1000 times less
estrogenic than ZEN in this assay.

Collectively, our in vitro data showed that neither HZEN nor DHZEN have any measurable effect
on MCF-7 cell proliferation or yeast reporter gene transcription in the tested concentration range,
indicating a low or even absent intrinsic estrogenic activity of those metabolites. The highest levels
used (500/10,000 nM) correspond to 168/3364 ng/mL for HZEN, 146/2924 ng/mL for DHZEN and
159/3184 ng/mL for ZEN. To compare with in vivo exposure, these concentrations were checked against
levels of ZEN found in biological samples (ZEN was used as a reference, given no toxicokinetic data
are available for HZEN or DZHEN). For instance, one of the highest tissue levels of ZEN measured in
pigs was reported by Doell et al. [41], who found on average 5.3 ng/g ZEN in the liver after dietary
administration of 0.42 mg/kg ZEN. This tissue concentration is approximately by a factor 30/600 lower
than the highest toxin level used in our in vitro experiments. In addition, the aforementioned tissue
concentration represents a sum value of free and glucuronidated ZEN. The glucuronidation rate in liver
accounts for at least 62% [42] and ZEN-glucuronides lack substantial estrogenic activity [43], further
strengthening the assumption that HZEN and DHZEN do not exhibit estrogenic activity at practically
relevant concentrations in vivo. However, as emphasized by Andersen et al. [40], the estrogenic activity
of a substance in an intact organism cannot be directly deduced from in vitro results. Important
factors that need to be considered in this regard comprise the substance's absorption, distribution,
metabolism and excretion (ADME), its ability to enter target cells and the concentration of endogenous
estrogens. To underline this, differences in the intrinsic activity between ZEN, α-ZEL, and β-ZEL
are well documented [4], whereas variations in plasma protein binding among these substances and
among species were unraveled only recently [44]. Hence, to indirectly account for all these factors,
we evaluated the estrogenic activity of HZEN and DHZEN also in an in vivo experiment in pigs.

Prepubertal female pigs were exposed to 4.58 mg/kg ZEN and equimolar concentrations of HZEN
and DHZEN, respectively, in order to identify potential adverse health effects of ZEN metabolites at
relatively high dietary levels. This concentration clearly exceeds the EU recommendations for ZEN
in feed [17], but might still be encountered under unfavorable conditions, where up to 11.19 mg/kg
ZEN was found in an individual feed sample [3]. In our experiment, the effect of ZEN treatment on
vulva size enlargement and reproductive tract weight increase was prominent. This was expected,

as already exposure to 1.3 mg/kg ZEN for 24 d was shown to significantly affect reproductive tract morphology [45]. In contrast to ZEN, neither HZEN nor DHZEN caused alterations in any of the clinical parameters assessed, suggesting that both metabolites lack relevant estrogenic activity in vivo. To substantiate this assumption, we next focused on potential changes on the molecular level. Since previously reported data of this animal experiment showed that ZEN had no influence on the plasma metabolome [46], we examined the impact of different treatments on the target organ uterus using a transcriptomics approach.

The classical mechanism, in which ligand-bound ERs interact directly with estrogen response elements (EREs) to activate the transcription of genes, is called the genomic response. Although less extensively studied than E2, ZEN has been shown to induce an estrogenic response via this mechanism [47], and therefore, we first investigated the expression of several genes with identified or potential EREs in their DNA sequences. As in the human homolog, a perfect ERE palindrome was found in the porcine gene sequence of EBAG9. The ERE of the human homolog is recognized by ERs with high specificity in vitro [48]. Although a significant increase was observed for the transcription of this gene after ZEN ingestion, the effect seems small (1.5-fold change) given the high concentration of ZEN used. In contrast, EBAG9 expression in HZEN and DHZEN groups were not different from the control group. Similar to EBAG9, little effect was seen on the expression of GJA1, for which a potential ERE was identified. GJA1 has been already reported as highly regulated by E2 in rat endometrium [49]. In addition, the authors mentioned that the expression of the C3 complement and the ODC genes were augmented in the uterus of rodents exposed to endocrine disruptors. Based on our results, which show no significant impact of ZEN on the uterine expression of C3 and ODC, we cannot confirm these findings in pigs for this xenoestrogen. Interestingly, the uterine transcription of GAPDH, which is frequently used for qPCR normalization, was affected by ZEN treatment (however, GAPDH transcription in the liver and jejunum was not affected, data not shown). Thus, GAPDH was excluded from the list of housekeeping genes we used. This effect was already described in 1998 by Zou and Ing [50] in the endometrium of ewes exposed to E2, and it emphasizes the need to select suitable reference genes for each analysis of mRNA expression. Although our initial hypothesis on the activation of genes carrying ERE motifs upon ZEN exposure is not conclusive, activation of GREB1 containing ERE in its promoter was very recently demonstrated on MCF-7 cells treated with ZEN [51]. In this study, authors used chromatin immunoprecipitation to show that ZEN induced the recruitment of ERα DNA-binding at chromatin sites. It is plausible that the genes we selected are carrying elements that may not represent binding sites in vivo, possibly because of chromatin accessibility [48].

In line with that, we also extended our analysis to genes otherwise associated with the response to estrogens. Several studies showed that the gene S100G, better known to encode the protein Calbindin-D9k (CaBP-9k), is a potent biomarker for screening estrogen-like environmental chemicals [49,52]. In pigs, E2 treatment induced an increase in CaBP-9k mRNA levels, while exposure to progesterone caused a decline in the expression [53]. It seems that CaBP-9k expression is also fluctuating depending on the serum estrogen levels [52]. In our experiment, none of the treatments induced significant alterations in uterine CaBP-9k expression compared to the control group. Yet, the opposite effects seen between the piglets fed ZEN and DHZEN deserve further investigations. As briefly discussed in Grenier et al. [11], the lack of effects observed on the ERα is in agreement with Oliver et al. [54], who analyzed the gene and protein expression of ERs in the uterus of piglets exposed to 1.5 mg/kg of ZEN for four weeks. However, Oliver et al. reported a two-fold increase in the mRNA level of ERβ in those animals, which could not be confirmed in the present experiment.

Another transcriptomics approach we implemented in the present study is the targeted analysis of selected microRNAs. Unlike mRNA analysis focusing on the direct activation of gene transcription, this approach gives information on post-transcriptional gene regulation. Indeed, changes in microRNA levels (both in blood and tissue) have been reported for certain estrogen-associated diseases in humans, such as breast cancer [55]. In vitro experiments, e.g. in MCF-7 cells, underlined the role of microRNAs not only in estrogen signaling, but also in mediating the effects of different endocrine disruptors [55].

Since 2015, several reports on the impact of ZEN on microRNA expression have become available, addressing changes in microRNAs levels in the porcine liver and intestine [9], porcine pituitary gland and its consequences for gonadotropin regulation [10], murine Leydig cell line TM3 [8], as well as porcine uterus and serum [11]. Based on results obtained in the latter, we selected 15 microRNAs and evaluated their response to ZEN, HZEN, and DHZEN treatment by qPCR.

As shown in Table 3, seven of the selected microRNAs were significantly altered by the different dietary treatments. Compared to the control group, ZEN caused a significant decrease in the expression of miR-135a-5p, miR-187-3p, and 204-5p. In contrast, those microRNAs were unaffected in HZEN or DHZEN exposed animals. Previous literature studies identified binding sites for ERα in the microRNA gene hsa-miR-135a2, and subsequent in vitro experiments showed up-regulation of miR-135a-5p (previous name: miR-135a) in E2-treated MCF-7 and ZR-75-1 cells [56,57]. Hence, our in vivo findings on decreased miR-135a-5p levels after ZEN exposure were surprising. Yet, direct comparisons of in vivo and in vitro results on the microRNA response after exposure to ER agonists should be done with caution. First, most of the in vitro studies use cancer cell lines and among those studies, conflicting results on the microRNA response were obtained due to differences in methodology and potential time-dependent effects [55]. Also, the vast majority of in vitro studies only investigate the role of single microRNAs, which does not reflect the interplay of various microRNAs or potential regulatory feedback-loops. As a consequence, it remains challenging to determine the specific role of an individual microRNA in the complex estrogen response (which is not solely limited to ERα and ERβ activation) and related diseases. For example, miR-135a-5p was reported to promote, as well as to decrease, breast cancer cell migration [58,59]. Likewise, miR-204-5p (previous name: miR-204) has been addressed both as oncogene and tumor suppressor, and is e.g. aberrantly expressed in endometrial carcinoma [60]. In ovariectomized mice, uterine expression of miR-204 was decreased after E2 treatment [61]. This effect was counteracted by pre-treatment with ER antagonists, which indicates that microRNA deregulation was mediated via the ER pathway.

Although miR-135a-5p, miR-187-3p, and 204-5p have been associated with estrogen signaling or related diseases, they do not seem to represent the key players in estrogen signaling [5,55]. In this regard, members of the miR-503 cluster (a set of two or more microRNAs transcribed from physically adjacent microRNA genes) were suggested to have a more prominent role, with miR-503 itself being addressed as "candidate master regulator of the estrogen response" [62]. Although a numerical increase was seen in the present study for the microRNAs belonging to this cluster (e.g. miR-424-5p, miR-450a-5p, miR-450b-5p, miR-450c-5p, miR-542-3p, or miR-503), no significant ($p < 0.05$) effect was seen in the uterus of pigs exposed to ZEN, and only some of these microRNAs showed a trend ($p < 0.1$) towards increased expression. This can be explained by a high inter-individual variance in expression level, with certain pigs being very responsive (up to 8-12-fold increase). Although this increases the variability within the group, the same consistent effect was seen across these six microRNAs (similar average in expression around 3.5-4.0-fold increase). These results are yet not as pronounced as seen for these microRNAs when using an untargeted RNA sequencing approach [11]. Since samples were derived from the same feeding trial, these differences in effect intensity and significance might be due to different approaches used for microRNA quantification (qPCR and RNA sequencing). Overall, none of the investigated microRNAs were significantly altered upon HZEN or DHZEN treatment compared to the control group. A numerical increase of certain microRNAs (e.g. miR-450c-5p or miR-542-3p) was observed after HZEN treatment, which most likely is irrelevant at practically relevant metabolite concentrations. Still, follow-up studies should address this aspect, as well as the time-dependent effects of ZEN on the microRNA expression. Already now, generated results can serve for prediction and subsequent experimental verification of new mRNA targets of ZEN.

Key steps in the development of mycotoxin-degrading enzymes are the identification of enzymes with a certain degradation potential, the characterization of the produced metabolites as well as the elucidation of their toxicities [19]. All those conditions have now been met for the zearalenone-lactonase Zhd101p. For the technological application of this enzyme as a feed additive, further challenges must be met, above all the fast and efficient degradation of ZEN in vivo. In addition, the enzyme's safety (for workers, target species, consumers, and environment) and its storage stability must be demonstrated [63]. To the best of our knowledge, no ZEN-degrading enzyme has fulfilled all those criteria so far and received EU authorization [64]. Hence, further research efforts are needed to develop a safe and effective ZEN-degrading enzyme to be used as a feed additive. Furthermore, ZEN poses not only a risk for animal health, but is also of concern for humans [20]. In future, the enzyme technology might be evaluated for its application in human nutrition, e.g. during food processing [65]. Importantly, such developments should never indirectly promote the production of unsafe raw materials, but rather be considered complementary to measures for the reduction of mycotoxin formation pre- and postharvest [66].

4. Conclusions

HZEN and DHZEN exhibited markedly diminished estrogenic activity compared to ZEN in both the MCF-7 proliferation assay and the yeast bioassay. In vivo, these metabolites did not cause morphological changes in the reproductive tract in prepubertal gilts. Conclusions on their molecular effects are hampered by the limited impact observed for ZEN, especially on mRNA level. Still, for transcripts altered upon ZEN treatment (EBAG9, miR-135a-5p, miR-187-3p, and miR-204-5p), neither HZEN nor DHZEN showed an effect. Hence, our data indicate that cleavage of ZEN by the zearalenone-lactonase Zhd101p reduces its estrogenicity in vitro and in vivo. Our study represents an important step in the safety evaluation of ZEN-hydrolyzing enzymes.

5. Materials and Methods

5.1. In Vitro Experiments

5.1.1. Chemicals, Reagents and Materials

Solid ZEN was obtained from Fermentek LTD (Israel, purity 99.2%), while β-estradiol (E2) was purchased from Sigma-Aldrich (St. Louis, MO, USA). Further chemicals and reagents used for HZEN and DHZEN production comprise acetonitrile (ACN, chromasolv for HPLC, ≥ 99.9%, Sigma-Aldrich, St. Louis, MO, USA), hydrochloric acid 37% (puriss. p.a) and tris base (both Sigma-Aldrich, St. Louis, MO, USA).

For MCF-7 experiments, charcoal-stripped fetal bovine serum and dimethylsulfoxid (DMSO, both Sigma-Aldrich, St. Louis, MO, USA), fetal bovine serum (FBS) and insulin-transferrin-selenium-A supplement (ITS, both Thermo Fisher Scientific, Waltham, MA, USA), L-glutamine (Sigma-Aldrich, St. Louis, MO, USA) as well as RPMI 1640 with phenol red and RPMI 1640 without phenol red (both Sigma-Aldrich, St. Louis, MO, USA) were used. Cell proliferation reagent WST-1 was obtained from Roche (Mannheim, Germany).

For the yeast bioassay, chemicals and reagents include α-D-(+)-Glucose (water free, 96%) and 2-mercaptoethanol (both Sigma-Aldrich, St. Louis, MO, USA), 2-nitrophenyl-B-D-galactopyranosid (ONPG) (Roth, Karlsruhe, Germany), amino acid drop out mix (adenine, arginine-hydrochloride, aspartic acid, glutamic acid monosodium salt monohydrate, histidine monohydrochloride, isoleucine, leucine, lysine, methionine, phenylalanine, serine, threonine, tyrosine, uracil, valine, all amino acids in L-form and purchased from Sigma-Aldrich, St. Louis, MO, USA), magnesium sulfate, sodium carbonate anhydrous, sodium dihydrogen phosphate monohydrate, sodium phosphate (dibasic) and potassium chloride (all Sigma-Aldrich, St. Louis, MO, USA), Y-PER Yeast Protein Extraction Reagent (Sigma Aldrich, St. Louis, MO, USA), and yeast nitrogen base (Invitrogen, Carlsbad, CA, USA).

5.1.2. Production of HZEN and DHZEN

Production and purification of HZEN and DHZEN for in vitro and in vivo experiments were done according to Vekiru et al. [30] with slight modifications. Briefly, multiple batches of 50 mg ZEN each were weighed in glass bottles and dissolved in 50 mL of ACN. Afterwards, 1 L of Tris-HCl buffer (pH value 7.5, 30 °C) and 1.2 mg of a codon-optimized version of the ZEN-lactonase (Zhd101p) of *Gliocladium roseum* were added. For the production of HZEN, solutions were incubated on a magnetic stirrer for 38 h at 30 °C. For DHZEN production half of the HZEN solution was used. Decarboxylation of HZEN was achieved by adding HCl to the HZEN solution to yield a final pH value of 3, and further incubation for 10 h at 50 °C. Both solutions, HZEN and DHZEN, were concentrated by solid phase extraction and purified by preparative HPLC as described by Vekiru et al. [30]. Purity (HZEN: 95%, no residual ZEN detected, DHZEN: 98%, no residual ZEN or HZEN detected) and quantity of ZEN metabolites were assessed by LC-UV-MS/MS as described by Hahn et al. [67].

5.1.3. MCF-7 Proliferation Assay

The human breast adenocarcinoma cell line MCF-7 was purchased from the German Collection of Microorganisms and Cell Cultures (DSMZ, No. ACC 115, Braunschweig, Germany) and grown in RPMI 1640 with phenol red supplemented with 10% FBS, 1% insulin-transferrin-selenium-A supplement, 1% L-glutamine, and 1% sodium pyruvate at 37 °C and 5% CO_2. Three days prior to the experiment, cultivation medium was replaced by the hormone-free medium (RPMI 1640 without phenol red, supplemented with 10% charcoal-stripped FBS, 1% insulin-transferrin-selenium-A supplement, 1% L-glutamine, and 1% sodium pyruvate). Approximately 5000 cells/well were seeded in hormone-free medium (200 µL/well) into a 96-well microplate and cultivated at 37 °C and 5% CO_2 for 48 h.

In total, MCF-7 cells were subjected to eight different treatments. In five independent experiments, the effects of ZEN and its metabolites on cell proliferation were tested in six concentrations (0.01, 0.1, 1, 10, 100, and 500 nM) and compared to a negative (solvent) and positive (10 nM E2) control group. To this end, stock solutions of ZEN, HZEN, and DHZEN (in DMSO) were diluted with hormone-free medium to the respective concentrations. The DMSO concentration was kept constant at 0.05% in all dilutions, and thus was also used as a negative control.

After 144 h of incubation with different substances, the cell proliferation assay WST-1 (4-[3-(4-iodophenyl)-2-(4-nitrophenyl)-2H-5-tetrazolio]-1,3-benzene disulfonate) was performed according to the manufacturer's instructions. Briefly, supernatants were discarded and cells were incubated with a 10% WST-1 solution in the hormone-free medium at 37 °C and 5% CO_2 for a maximum of 4 h. Absorbance (A_{450}) was measured by a microplate reader (Synergy HT, Biotek, Winooski, VT, USA). The development of formazan dye correlates to the number of metabolically active cells in the culture.

5.1.4. Yeast Bioassay

The yeast strain YZHB817 is a derivative of the yeast two-hybrid strain PJ69-4A [68] in which the genes encoding the ABC transporter proteins Pdr5 and Snq2 were deleted to reduce efflux and increase net ZEN uptake [69]. The *GAL7*-promoter dependent *lacZ* reporter gene expression is activated by a hybrid protein consisting of amino-acids 1-848 of the yeast Gal4p (activator of galactose utilization genes, DNA binding domain) and amino-acids 282-595 of the human ERα providing the hormone dependent activation domain. This chimeric protein is expressed behind the constitutive *ADH1* promoter on the centromeric plasmid (*CEN4 TRP1*) pTK103 [70].

A single colony of YZHB817 grown on an SC-TRP agar plate was used to inoculate 25 mL SC-TRP medium (prepared according to Sherman [71]) and incubated overnight at 28 °C, 200 rpm. Dilutions of ZEN, HZEN, and DHZEN were prepared in SC-TRP medium resulting in the following concentration levels: 0, 1, 10, 100, 500, 1000, 5000, 10,000 nM. The DMSO concentration was kept constant at 1% in all dilutions. For substance testing, the yeast was diluted to an OD of 0.1 measured at 600 nm and 100 µL aliquots were transferred into U-bottom 96-well plates. After centrifugation (15 min, 2250 rcf, 12 °C) the supernatant was discarded. Next, 100 µL aliquots of the diluted samples were transferred into U-bottom 96-well plates containing the YZHB817 cells. All samples were tested in quadruplicate per plate, and the experiment was repeated four times. The plates were covered with a sterile BREATHseal™ (Greiner Bio-One, Kremsmünster, Austria) and incubated at 28 °C, 70% RH, 800 rpm for 18 h.

For detection of ß-galactosidase expression, the reaction reagent was prepared as follows: Y-Per, Z-buffer 10 × (60 mM $Na_2HPO_4 \times 7\ H_2O$, 40 mM $NaH_2PO_4 \times H_2O$, 10 mM KCl, 1 mM $MgSO_4$, 50 mM 2-mercaptoethanol, pH 7.0) and ONPG-stock (8 mg/mL 2-Nitrophenyl β-D-galactopyranoside) were mixed at a ratio of 0.45/0.275/0.275. Subsequently, 35 µL of this reagent was added to each well and mixed. The plates were incubated for 10 min at 37 °C until the samples turned slightly yellow. The reaction was stopped by the addition of 25 µL of 2 M Na_2CO_3 per well. The plates were centrifuged for 15 min at 2250 rcf, 12 °C and 60 µL of the supernatants were transferred into flat bottom 96-well plates. Finally, the absorbance was measured at 420 nm.

5.2. In Vivo Experiment

5.2.1. Animals and Study Design

The animal experiment was conducted at the Center for Animal Nutrition (Waxenecker KEG, Austria) and approved by the Institutional Ethical Committee and the Lower Austrian Region Government, Group Agriculture and Forestry, Department of Agricultural Law (LF1-TVG-39/017-2015). All experimental procedures were carried out in accordance with the European Guidelines for the Care and Use of Animals for Research Purposes [72] and the Austrian Animal Experimentation Act 2012. The approval dates for the animal experiment were 30.6.2015 (Institutional Ethical Committee) and 29.07.2015 (Lower Austrian Region Government, Group Agriculture and Forestry, Department of Agricultural Law (LF1-TVG-39/017-2015).

As described in Grenier et al. [11], female weaned piglets (sow: Landrace × Large White, boar: Pietrain, 30 ± 2 days old) were obtained from a local producer and allowed to acclimatize for seven days. Piglets (n = 24) were housed on slatted floor pens under controlled environmental conditions, and their general health status was monitored daily. To compare the effects of ZEN and equimolar concentrations of its metabolites, piglets were exposed to one of the four experimental diets (n = 6): i) uncontaminated feed (Control), ii) feed containing 4.58 mg/kg ZEN (ZEN), iii) feed containing 4.84 mg/kg HZEN (HZEN), or iv) feed containing 4.21 mg/kg DHZEN (DHZEN). Experimental diets and water were provided *ad libitum* for 28 days. Individual bodyweight, vulva length and vulva width of piglets were monitored in regular intervals during the experimental period. As presented in detail in Grenier et al. [11], animals were euthanized at day 28, the reproductive tract was dissected and weighed, and uterus samples were collected for qPCR analysis (mRNA, microRNA).

For homogenous distribution of HZEN and DHZEN within the feed, lyophilizates (Section 5.1.2) were first mixed with maltodextrin (proportion 1:8 and 1:6, respectively). Subsequently, these premixes were added to the basal feed at an inclusion rate of 0.9%. Concentrations of final diets were confirmed by LC-UV-MS/MS. The absence of any natural and significant contaminations with mycotoxins in the basal feed, as well as the procedure for artificial ZEN contamination of the diet, is described in Grenier et al. [11].

5.2.2. qPCR Analysis of mRNAs From Genes with Potential ERE or Associated with Estrogen Response

E2-liganded ERs bind with the highest affinity to EREs (15-bp palindromes) composed of PuGGTCA motifs separated by three variable bp [73]. To a lesser extent than the extensive work done by Bourdeau et al. [48] in human and mouse, we tried to identify near-consensus ERE sequences in pig based on human E2-responsive genes. Sixty genes were pulled out from the list of Bourdeau et al. [48] on the E2-upregulated genes (experimentally validated in humans), and the respective sequences of these genes in pigs were obtained (from Ensembl database, https://www.ensembl.org/index.html). Subsequently, each genomic DNA sequence was analyzed with the program Dragon ERE Finder version 2 [35] in order to identify potential EREs. In addition to this hypothesis of ZEN activating genes carrying ERE on their sequences, we also selected a few genes (Table 4) showing growing evidence in the literature that they are responsive to estrogenic compounds, such as the gene S100G (also known as CaBP-9k). We did not find any potential ERE motifs on the DNA sequences of those genes. The expression of other genes such as ERα, ERβ, and the two pro-inflammatory cytokines IL-1β and IL-6 was also evaluated via RT-qPCR.

Beforehand, isolation of total RNA was done approximately from 30 mg of uterus tissue, disrupted via bead-beating and RNA was extracted with the miRNeasy Mini Kit (Qiagen, Hilden, Germany) according to the manufacturer's recommendations. The total RNA was used for first-strand cDNA synthesis using Maxima H Minus First Strand cDNA synthesis kit including dsDNase (Thermo Fisher Scientific, Waltham, MA, USA) according to standard procedures. A few primers were used from literature but most of them were designed with the software Primer3 ([74], https://primer3plus. com/primer3web/primer3web_input.htm), and pre-experimentally validated. In addition, due to a differential effect of the ZEN treatment on the expression of housekeeping genes, we used the software GeNorm [75] to select among six candidates (ActB, GAPDH, HPRT1, RPL4, RPL32, and TBP) the most stable ones for data normalization (Table 4). All qRT-PCR reactions were conducted on the Mastercycler ep Realplex (Eppendorf, Hamburg, Germany) using SYBR green chemistry (Kapa SYBR Fast Universal, Sigma-Aldrich, St. Louis, MO, USA). The thermal cycle conditions were as follows: 1 cycle of pre-incubation at 95 °C for 3 min, 40 cycles of amplification (95 °C for 10 sec, 60 °C for 20 sec, and 72 °C for 20 sec), and melting curve program included at the end of the run. Relative gene expression was calculated using the $2^{\Delta\Delta Ct}$ method with the geometric mean of the Cts from the four housekeeping genes serving for normalization (subtracting the Ct-values of individual target genes from the Ct-value of the housekeeping genes of the same sample (ΔCt-values). Next, the mean ΔCt for each experimental group and target gene was calculated, and subsequently used for statistical evaluation and expressing the fold change (= $2^{\Delta\Delta Ct}$ value).

5.2.3. qPCR Analysis of microRNAs

To compare the effects of ZEN and its metabolites on uterine microRNA expression, microRNAs previously described to be altered by ZEN exposure [11] were chosen for targeted qPCR analysis. In total, 15 microRNAs were selected based on expression patterns (magnitude of log2 fold changes upon ZEN treatment, average tags per million count) observed after sequencing of uterus tissue. The panel included eight microRNAs up-regulated upon ZEN exposure (ssc-miR-1, ssc-miR-143-3p, ssc-miR-424-5p, ssc-miR-450a, ssc-miR-450b-5p, ssc-miR-450c-5p, ssc-miR-503, and ssc-miR-542-3p) and seven down-regulated microRNAs (ssc-miR-129a, ssc-miR-135, ssc-miR-149, ssc-miR-181c, ssc-miR-187, ssc-miR-204-5p and ssc-miR-206).

The extraction of RNA was done as described in the above section for mRNA. Synthesis of cDNA and PCR amplification was performed as described in Grenier et al. [11]. The method to calculate the fold change is the same as for the mRNA, the $2^{\Delta\Delta Ct}$ method. Normalization of the Ct values was performed with the reference U6 snRNA, subtracting the Ct-values of individual microRNAs from the Ct-value of U6 snRNA of the same sample (ΔCt-values).

Table 4. Nucleotide sequence of primers for real-time qPCR.

Gene Name	Primer Sequence	Amplicon Size	Ensembl Access #
Housekeeping Genes			
HPRT1 [1]	F (300 nM) GGACTTGAATCATGTTTGTG R (300 nM) CAGATGTTTCCAAACTCAAC	91 bp	ENSSSCG00000034896
RPL32 [1]	F (300 nM) AGTTCATCCGGCACCAGTCA R (300 nM) GAACCTTCTCCGCACCCTGT	92 bp	ENSSSCG00000035811
RPL4 [1]	F (300 nM) CAAGAGTAACTACAACCTTC R (300 nM) GAACTCTACGATGAATCTTC	122 bp	ENSSSCG00000004945
TBP [1]	F (300 nM) AACAGTTCAGTAGTTATGAGCCAGA R (300 nM) AGATGTTCTCAAACGCTTCG	153 bp	ENSSSCG00000037372
Genes with Identified ERE Motif			
EBAG9	F (300 nM) GCACAGGTTTCTCTAGTAGGCT R (300 nM) TCCCTGTCTGCTATCTTCTGC	175 bp	ENSSSCG00000006024
OVGP1	F (300 nM) GGGTCGGCTATGATGATGACA R (300 nM) CCGGTGAAGGAGTTGAGCTA	198 bp	ENSSSCG00000006791
IGFBP4	F (300 nM) CATCCCCATCCCTAACTGCG R (300 nM) CTCACTCTCGGAAGCTGTCG	185 bp	ENSSSCG00000017472
GJA1	F (300 nM) TCTGAGTGCCTGAACTTGCT R (300 nM) CAGCGGTGGAATAGGCTTGA	154 bp	ENSSSCG00000004241
GAPDH [1,2]	F (300 nM) AGGGGCTCTCCAGAACATCATCC R (300 nM) TCGCGTGCTCTTGCTGGGGTTGG	446 bp	ENSSSCG00000000694
C3 [2]	F (300 nM) GGGCAGATCTTGAGTGTCCG R (300 nM) ATGCTGGATGAACTGAGCCC	179 bp	ENSSSCG00000013551
Other Genes Associated With Estrogenic Response or Inflammation			
S100G	F (300 nM) GGAGTTGAACTTGGACGTGC R (300 nM) CGCATCCCTTCCAGTCCTTA	184 bp	ENSSSCG00000012147
CLU	F (300 nM) CCATGACATGTTCCAGCCCT R (300 nM) TCTGAGAGGAATTGCTGGCC	239 bp	ENSSSCG00000009668
ODC	F (300 nM) CGGCGATTGGATGCTCTTTG R (300 nM) AAGTCGTGGTTCCGGATCTG	144 bp	ENSSSCG00000027121
ESR1	F (300 nM) CCTGGAGAATGAGCCGAGC R (300 nM) CTTCCCTTGTCACTCGGTGCT	92 bp	ENSSSCG00000025777
ESR2	F (300 nM) TGCAGTGATTATGCGTCAGGA R (300 nM) CAGCTTTTACGCCGGTTCTT	149 bp	ENSSSCG00000005109
IL-1β	F (300 nM) CCATAGTACCTGAACCCGCC R (300 nM) GCTGGTGAGAGATTTGCAGC	165 bp	ENSSSCG00000039214
IL-6 [1]	F (300 nM) GGCAAAAGGGAAAGAATCCAG R (300 nM) CGTTCTGTGACTGCAGCTTATCC	87 bp	ENSSSCG00000020970

[1] The primers for these genes were already published in Nygard et al. [76], Grenier et al. [77], and Gessner et al. [78], and for the other genes, the primers were designed for the present study. [2] ERE motif was found in the human C3 and GAPDH sequences [48] but not in pig. In addition, we assessed GAPGH as both housekeeping gene and potential ERE gene. HPRT1, Hypoxanthine Phosphoribosyltransferase 1, RPL, Ribosomal Protein L, TBP, TATA-Box Binding Protein, GAPDH, Glyceraldehyde 3-phosphate dehydrogenase.

5.3. Statistics

Data from the MCF-7 proliferation assay and the yeast bioassay were expressed relative to the negative control, which was set to 100%. Statistical analysis was performed with SigmaPlot (Version 12.5, from Systat Software, Inc., San Jose, CA, USA). Since data were not normally distributed (Shapiro-Wilk test, $p < 0.05$), a nonparametric Kruskal-Wallis test (Dunn's multiple comparison post-hoc test) was conducted to assess the effects between different concentrations to the cell control.

Data from the feeding trial (body weight, reproductive tract weigh, vulva size, mRNA, and microRNA expression) were analyzed with IBM SPSS Statistics (Version 22, Armonk, NY, USA). If data were normally distributed (Shapiro-Wilk test, $p > 0.05$), a one-way ANOVA was conducted to assess the effects of treatment diets on dependent variables. When main effects were significant, differences between means were examined via Tukey post-hoc analysis (assumption of homogeneity of variances met, Levene's test, $p > 0.05$) or Games-Howell test (no homogeneity of variance). Data that were not normally distributed were analyzed with the non-parametric Kruskal-Wallis test (Dunn-Bonferroni post-hoc test). Differences between means were considered significant at $p < 0.05$.

GraphPad Prism version 8 for Windows (GraphPad Software, La Jolla California USA) was used for running the Grubbs' outlier test and generating figures.

Author Contributions: Conceptualization, B.G., M.T., V.N., and W.-D.M.; methodology, B.G., B.N., G.A., M.H., S.F., S.L., and V.R.; validation, B.G., B.N., G.A., M.H., S.F., S.L., and V.R.; formal analysis, B.G., B.N., M.H., S.F., and V.N.; resources, S.L. and V.R.; writing—original draft preparation, B.G., B.N., D.H., S.F., and V.N.; writing—review and editing, all; supervision, B.G., M.A., M.T., and W.-D.M.; funding acquisition, D.H., and V.N.

Funding: This research received funding from the Austrian Research Promotion Agency (Österreichische Forschungsförderungsgesellschaft FFG, grant numbers 872270 and 866384).

Acknowledgments: The GAL4-hER hybrid gene cloned into pTK103 was isolated from pHCA/GAL4(848)ER(G) was kindly provided by Prof. Didier Picard (University of Geneva, Switzerland). We would like to thank Susanna Skalicky (TAmiRNA GmbH) for technical support during microRNA qPCR analysis. We owe sincere gratitude to Elisavet Kunz-Vekiru for purification and quantification of HZEN and DHZEN. In addition, we are grateful for the practical assistance of Roger Berrios, Barbara Doupovec and Ines Taschl during the animal experiment. Finally, we want to thank Christiane Gruber for careful proofreading of the manuscript.

Conflicts of Interest: S.F., B.N., V.N., D.H., V.R., S.L., M.A., M.T., W.-D.M. and B.G. are employed by BIOMIN Holding GmbH, that operates the BIOMIN Research Center and is a producer of animal feed additives. M.H. is co-founder of TAmiRNA and employed by this company. This, however, did not influence the design of the experimental studies or bias the presentation and interpretation of results.

References

1. Zinedine, A.; Soriano, J.M.; Molto, J.C.; Mañes, J. Review on the toxicity, occurrence, metabolism, detoxification, regulations and intake of zearalenone: An oestrogenic mycotoxin. *Food Chem. Toxicol.* **2007**, *45*, 1–18. [CrossRef]
2. Rodrigues, I.; Naehrer, K. Prevalence of mycotoxins in feedstuffs and feed surveyed worldwide in 2009 and 2010. *Phytopathol. Mediterr.* **2012**, *51*, 175–192.
3. Kovalsky, P.; Kos, G.; Nährer, K.; Schwab, C.; Jenkins, T.; Schatzmayr, G.; Sulyok, M.; Krska, R. Co-Occurrence of Regulated, Masked and Emerging Mycotoxins and Secondary Metabolites in Finished Feed and Maize—An Extensive Survey. *Toxins* **2016**, *8*, 363. [CrossRef] [PubMed]
4. Fink-Gremmels, J.; Malekinejad, H. Clinical effects and biochemical mechanisms associated with exposure to the mycoestrogen zearalenone. *Anim. Feed Sci. Technol.* **2007**, *137*, 326–341. [CrossRef]
5. Vrtačnik, P.; Ostanek, B.; Mencej-Bedrač, S.; Marc, J. The many faces of estrogen signaling. *Biochem. Med.* **2014**, *24*, 329–342. [CrossRef]
6. Kowalska, K.; Habrowska-Górczyńska, D.E.; Piastowska-Ciesielska, A.W. Zearalenone as an endocrine disruptor in humans. *Environ. Toxicol. Pharmacol.* **2016**, *48*, 141–149. [CrossRef] [PubMed]
7. Hennig-Pauka, I.; Koch, F.-J.; Schaumberger, S.; Woechtl, B.; Novak, J.; Sulyok, M.; Nagl, V. Current challenges in the diagnosis of zearalenone toxicosis as illustrated by a field case of hyperestrogenism in suckling piglets. *Porc. Health Manag.* **2018**, *4*, 18. [CrossRef] [PubMed]
8. Wang, M.; Wu, W.; Li, L.; He, J.; Huang, S.; Chen, S.; Chen, J.; Long, M.; Yang, S.; Li, P. Analysis of the miRNA Expression Profiles in the Zearalenone-Exposed TM3 Leydig Cell Line. *Int. J. Mol. Sci.* **2019**, *20*, 635. [CrossRef] [PubMed]
9. Brzuzan, P.; Woźny, M.; Wolińska-Nizioł, L.; Piasecka, A.; Florczyk, M.; Jakimiuk, E.; Góra, M.; Łuczyński, M.; Gajęcki, M. MicroRNA expression profiles in liver and colon of sexually immature gilts after exposure to Fusarium mycotoxins. *Pol. J. Vet. Sci.* **2015**, *18*, 29–38. [CrossRef]
10. He, J.; Zhang, J.; Wang, Y.; Liu, W.; Gou, K.; Liu, Z.; Cui, S. MiR-7 Mediates the Zearalenone Signaling Pathway Regulating FSH Synthesis and Secretion by Targeting FOS in Female Pigs. *Endocrinology* **2018**, *159*, 2993–3006. [CrossRef]
11. Grenier, B.; Hackl, M.; Skalicky, S.; Thamhesl, M.; Moll, W.-D.; Berrios, R.; Schatzmayr, G.; Nagl, V. MicroRNAs in porcine uterus and serum are affected by zearalenone and represent a new target for mycotoxin biomarker discovery. *Sci. Rep.* **2019**, *9*, 9408. [CrossRef] [PubMed]
12. Schraml, E.; Hackl, M.; Grillari, J. Micrornas and toxicology: A love marriage micrornas in liquid biopsies are minimal-invasive biomarkers for tissue-specific toxicity. *Toxicol. Rep.* **2017**, *4*, 634–636. [CrossRef] [PubMed]

13. He, J.; Wei, C.; Li, Y.; Liu, Y.; Wang, Y.; Pan, J.; Liu, J.; Wu, Y.; Cui, S. Zearalenone and alpha-zearalenol inhibit the synthesis and secretion of pig follicle stimulating hormone via the non-classical estrogen membrane receptor GPR30. *Mol. Cell. Endocrinol.* **2018**, *461*, 43–54. [CrossRef] [PubMed]

14. Ayed-Boussema, I.; Pascussi, J.M.; Maurel, P.; Bacha, H.; Hassen, W. Zearalenone activates pregnane X receptor, constitutive androstane receptor and aryl hydrocarbon receptor and corresponding phase I target genes mRNA in primary cultures of human hepatocytes. *Environ. Toxicol. Pharmacol.* **2011**, *31*, 79–87. [CrossRef] [PubMed]

15. Gajęcka, M.; Zielonka, Ł.; Gajęcki, M. Activity of Zearalenone in the Porcine Intestinal Tract. *Molecules* **2016**, *22*, 18. [CrossRef] [PubMed]

16. Zheng, W.; Wang, B.; Li, X.; Wang, T.; Zou, H.; Gu, J.; Yuan, Y.; Liu, X.; Bai, J.; Bian, J.; et al. Zearalenone Promotes Cell Proliferation or Causes Cell Death? *Toxins* **2018**, *10*, 184. [CrossRef] [PubMed]

17. European Commission. Commission recommendation of of 17 august 2006 on the presence of deoxynivalenol, zearalenone, ochratoxin a, t-2 and ht-2 and fumonisins in products intended for animal feeding. *Off. J. Eur. Union* **2006**, *L 229*, 7–9.

18. Alassane-Kpembi, I.; Schatzmayr, G.; Taranu, I.; Marin, D.; Puel, O.; Oswald, I.P. Mycotoxins co-contamination: Methodological aspects and biological relevance of combined toxicity studies. *Crit. Rev. Food Sci. Nutr.* **2017**, *57*, 3489–3507. [CrossRef]

19. Jard, G.; Liboz, T.; Mathieu, F.; Guyonvarc'H, A.; Lebrihi, A. Review of mycotoxin reduction in food and feed: From prevention in the field to detoxification by adsorption or transformation. *Food Addit. Contam. Part A* **2011**, *28*, 1590–1609. [CrossRef]

20. Rogowska, A.; Pomastowski, P.; Sagandykova, G.; Buszewski, B. Zearalenone and its metabolites: Effect on human health, metabolism and neutralisation methods. *Toxicon* **2019**, *162*, 46–56. [CrossRef]

21. Ji, C.; Fan, Y.; Zhao, L. Review on biological degradation of mycotoxins. *Anim. Nutr.* **2016**, *2*, 127–133. [CrossRef]

22. Di Gregorio, M.C.; De Neeff, D.V.; Jager, A.V.; Corassin, C.H.; Carão, Á.C.D.P.; De Albuquerque, R.; De Azevedo, A.C.; Oliveira, C.A.F. Mineral adsorbents for prevention of mycotoxins in animal feeds. *Toxin Rev.* **2014**, *33*, 1–11. [CrossRef]

23. Fruhauf, S.; Schwartz, H.; Ottner, F.; Krska, R.; Vekiru, E. Yeast cell based feed additives: Studies on aflatoxin B1and zearalenone. *Food Addit. Contam. Part A* **2012**, *29*, 217–231. [CrossRef]

24. De Baere, S.; De Mil, T.; Antonissen, G.; Devreese, M.; Croubels, S. In vitro model to assess the adsorption of oral veterinary drugs to mycotoxin binders in a feed- and aflatoxin B1-containing buffered matrix. *Food Addit. Contam. Part A* **2018**, *35*, 1–11. [CrossRef] [PubMed]

25. Kiessling, K.H.; Pettersson, H.; Sandholm, K.; Olsen, M. Metabolism of aflatoxin, ochratoxin, zearalenone, and three trichothecenes by intact rumen fluid, rumen protozoa, and rumen bacteria. *Appl. Environ. Microbiol.* **1984**, *47*, 1070–1073. [PubMed]

26. Fitzpatrick, D.; Picken, C.; Murphy, L.C.; Buhr, M. Measurement of the relative binding affinity of zearalenone, α-zearalenol and β-zearalenol for uterine and oviduct estrogen receptors in swine, rats and chickens: An indicator of estrogenic potencies. *Comp. Biochem. Physiol. Part C Comp. Pharmacol.* **1989**, *94*, 691–694. [CrossRef]

27. El-Sharkawy, S.; Abul-Hajj, Y. Microbial Transformation of Zearalenone, I. Formation of Zearalenone-4-O-β-glucoside. *J. Nat. Prod.* **1987**, *50*, 520–521. [CrossRef]

28. Brodehl, A.; Möller, A.; Kunte, H.-J.; Koch, M.; Maul, R. Biotransformation of the mycotoxin zearalenone by fungi of the genera Rhizopus and Aspergillus. *FEMS Microbiol. Lett.* **2014**, *359*, 124–130. [CrossRef] [PubMed]

29. Binder, S.B.; Schwartz-Zimmermann, H.E.; Varga, E.; Bichl, G.; Michlmayr, H.; Adam, G.; Berthiller, F. Metabolism of Zearalenone and Its Major Modified Forms in Pigs. *Toxins* **2017**, *9*, 56. [CrossRef] [PubMed]

30. Vekiru, E.; Frühauf, S.; Hametner, C.; Schatzmayr, G.; Krska, R.; Moll, W.; Schuhmacher, R. Isolation and characterisation of enzymatic zearalenone hydrolysis reaction products. *World Mycotoxin J.* **2016**, *9*, 353–363. [CrossRef]

31. Vekiru, E.; Hametner, C.; Mitterbauer, R.; Rechthaler, J.; Adam, G.; Schatzmayr, G.; Krska, R.; Schuhmacher, R. Cleavage of Zearalenone by Trichosporon mycotoxinivorans to a Novel Nonestrogenic Metabolite. *Appl. Environ. Microbiol.* **2010**, *76*, 2353–2359. [CrossRef] [PubMed]

32. Kakeya, H.; Takahashi-Ando, N.; Kimura, M.; Onose, R.; Yamaguchi, I.; Osada, H. Biotransformation of the Mycotoxin, Zearalenone, to a Non-estrogenic Compound by a Fungal Strain of Clonostachys sp. *Biosci. Biotechnol. Biochem.* **2002**, *66*, 2723–2726. [CrossRef]

33. El-Sharkawy, S.; Abul-Hajj, Y.J. Microbial cleavage of zearalenone. *Xenobiotica* **1988**, *18*, 365–371. [CrossRef]

34. Takahashi-Ando, N.; Ohsato, S.; Shibata, T.; Hamamoto, H.; Yamaguchi, I.; Kimura, M. Metabolism of Zearalenone by Genetically Modified Organisms Expressing the Detoxification Gene from Clonostachys rosea. *Appl. Environ. Microbiol.* **2004**, *70*, 3239–3245. [CrossRef] [PubMed]

35. Bajic, V.B.; Tan, S.L.; Chong, A.; Tang, S.; Ström, A.; Gustafsson, J.-A.; Lin, C.-Y.; Liu, E.T. Dragon ERE Finder version 2: A tool for accurate detection and analysis of estrogen response elements in vertebrate genomes. *Nucleic Acids Res.* **2003**, *31*, 3605–3607. [CrossRef]

36. Tatay, E.; Espín, S.; García-Fernández, A.-J.; Ruiz, M.-J. Estrogenic activity of zearalenone, α-zearalenol and β-zearalenol assessed using the e-screen assay in mcf-7 cells. *Toxicol. Mech. Methods* **2018**, *28*, 239–242. [CrossRef]

37. LeComte, S.; Lelong, M.; Bourgine, G.; Efstathiou, T.; Saligaut, C.; Pakdel, F. Assessment of the potential activity of major dietary compounds as selective estrogen receptor modulators in two distinct cell models for proliferation and differentiation. *Toxicol. Appl. Pharmacol.* **2017**, *325*, 61–70. [CrossRef]

38. Hessenberger, S.; Botzi, K.; Degrassi, C.; Kovalsky, P.; Schwab, C.; Schatzmayr, D.; Schatzmayr, G.; Fink-Gremmels, J. Interactions between plant-derived oestrogenic substances and the mycoestrogen zearalenone in a bioassay with MCF-7 cells. *Pol. J. Vet. Sci.* **2017**, *20*, 513–520. [CrossRef] [PubMed]

39. Khosrokhavar, R.; Rahimifard, N.; Shoeibi, S.; Hamedani, M.P.; Hosseini, M.-J. Effects of zearalenone and α-Zearalenol in comparison with Raloxifene on T47D cells. *Toxicol. Mech. Methods* **2009**, *19*, 246–250. [CrossRef] [PubMed]

40. Andersen, H.R.; Andersson, A.-M.; Arnold, S.F.; Autrup, H.; Barfoed, M.; Beresford, N.A.; Bjerregaard, P.; Christiansen, L.B.; Gissel, B.; Hummel, R.; et al. Comparison of Short-Term Estrogenicity Tests for Identification of Hormone-Disrupting Chemicals. *Environ. Health Perspect.* **1999**, *107*, 89–108. [CrossRef]

41. Doll, S.; Dänicke, S.; Ueberschär, K.-H.; Valenta, H.; Schnurrbusch, U.; Ganter, M.; Klobasa, F.; Flachowsky, G. Effects of graded levels of Fusarium toxin contaminated maize in diets for female weaned piglets. *Arch. Anim. Nutr.* **2003**, *57*, 311–334. [CrossRef]

42. Zöllner, P.; Jodlbauer, J.; Kleinova, M.; Kahlbacher, H.; Kuhn, T.; Hochsteiner, W.; Lindner, W. Concentration Levels of Zearalenone and Its Metabolites in Urine, Muscle Tissue, and Liver Samples of Pigs Fed with Mycotoxin-Contaminated Oats. *J. Agric. Food Chem.* **2002**, *50*, 2494–2501. [CrossRef] [PubMed]

43. Frizzell, C.; Uhlig, S.; Miles, C.O.; Verhaegen, S.; Elliott, C.T.; Eriksen, G.S.; Sørlie, M.; Ropstad, E.; Connolly, L. Biotransformation of zearalenone and zearalenols to their major glucuronide metabolites reduces estrogenic activity. *Toxicol. In Vitro* **2015**, *29*, 575–581. [CrossRef] [PubMed]

44. Faisal, Z.; Lemli, B.; Szerencsés, D.; Kunsági-Máté, S.; Bálint, M.; Hetényi, C.; Kuzma, M.; Mayer, M.; Poór, M. Interactions of zearalenone and its reduced metabolites α-zearalenol and β-zearalenol with serum albumins: Species differences, binding sites, and thermodynamics. *Mycotoxin Res.* **2018**, *34*, 269–278. [CrossRef] [PubMed]

45. Jiang, S.; Yang, Z.; Yang, W.; Yao, B.; Zhao, H.; Liu, F.; Chen, C.; Chi, F. Effects of feeding purified zearalenone contaminated diets with or without clay enterosorbent on growth, nutrient availability, and genital organs in post-weaning female pigs. *Asian-Aust. J. Anim. Sci.* **2010**, *23*, 74–81. [CrossRef]

46. Grenier, B.; Nagl, V.; Lutz, A.; Aleschko, M.; Schatzmayr, G.; Moll, W.-D.; Thamhesl, M. Targeted approaches for zen biomarker discovery in prepubertal gilts. In Proceedings of the 10th World Mycotoxin Forum, Amsterdam, The Netherlands, 12–14 March 2018.

47. Li, Y.; Burns, K.A.; Arao, Y.; Luh, C.J.; Korach, K.S. Differential Estrogenic Actions of Endocrine-Disrupting Chemicals Bisphenol A, Bisphenol AF, and Zearalenone through Estrogen Receptor α and β in Vitro. *Environ. Health Perspect.* **2012**, *120*, 1029–1035. [CrossRef] [PubMed]

48. Bourdeau, V.; Deschênes, J.; Métivier, R.; Nagai, Y.; Nguyen, D.; Bretschneider, N.; Gannon, F.; White, J.H.; Mader, S. Genome-Wide Identification of High-Affinity Estrogen Response Elements in Human and Mouse. *Mol. Endocrinol.* **2004**, *18*, 1411–1427. [CrossRef]

49. Jung, E.-M.; An, B.-S.; Yang, H.; Choi, K.-C.; Jeung, E.-B. Biomarker Genes for Detecting Estrogenic Activity of Endocrine Disruptors via Estrogen Receptors. *Int. J. Environ. Res. Public Health* **2012**, *9*, 698–711. [CrossRef]

50. Zou, K.; Ing, N.H. Oestradiol up-regulates oestrogen receptor, cyclophilin, and glyceraldehyde phosphate dehydrogenase mRNA concentrations in endometrium, but down-regulates them in liver. *J. Steroid Biochem. Mol. Boil.* **1998**, *64*, 231–237. [CrossRef]

51. LeComte, S.; DeMay, F.; Pham, T.H.; Moulis, S.; Efstathiou, T.; Chalmel, F.; Pakdel, F. Deciphering the Molecular Mechanisms Sustaining the Estrogenic Activity of the Two Major Dietary Compounds Zearalenone and Apigenin in ER-Positive Breast Cancer Cell Lines. *Nutrients* **2019**, *11*, 237. [CrossRef]

52. Vo, T.T.B.; Jeung, E.-B. An Evaluation of Estrogenic Activity of Parabens Using Uterine Calbindin-D9k Gene in an Immature Rat Model. *Toxicol. Sci.* **2009**, *112*, 68–77. [CrossRef] [PubMed]

53. Krisinger, J. Porcine calbindin-D9k gene: Expression in endometrium, myometrium, and placenta in the absence of a functional estrogen response element in intron A. *Boil. Reprod.* **1995**, *52*, 115–123. [CrossRef] [PubMed]

54. Oliver, W.; Miles, J.; Diaz, D.; Dibner, J.; Rottinghaus, G.; Harrell, R. Zearalenone enhances reproductive tract development, but does not alter skeletal muscle signaling in prepubertal gilts. *Anim. Feed Sci. Technol.* **2012**, *174*, 79–85. [CrossRef]

55. Klinge, C.M. miRNAs regulated by estrogens, tamoxifen, and endocrine disruptors and their downstream gene targets. *Mol. Cell. Endocrinol.* **2015**, *418*, 273–297. [CrossRef] [PubMed]

56. Kim, K.; Madak-Erdogan, Z.; Ventrella, R.; Katzenellenbogen, B.S. A microrna196a2* and tp63 circuit regulated by estrogen receptor-α and erk2 that controls breast cancer proliferation and invasiveness properties. *Horm. Cancer* **2013**, *4*, 78–91. [CrossRef] [PubMed]

57. Ferraro, L.; Ravo, M.; Nassa, G.; Tarallo, R.; De Filippo, M.R.; Giurato, G.; Cirillo, F.; Stellato, C.; Silvestro, S.; Cantarella, C.; et al. Effects of Oestrogen on MicroRNA Expression in Hormone-Responsive Breast Cancer Cells. *Horm. Cancer* **2012**, *3*, 65–78. [CrossRef] [PubMed]

58. Tribollet, V.; Barenton, B.; Kroiss, A.; Vincent, S.; Zhang, L.; Forcet, C.; Cerutti, C.; Périan, S.; Allioli, N.; Samarut, J.; et al. miR-135a Inhibits the Invasion of Cancer Cells via Suppression of ERRα. *PLoS ONE* **2016**, *11*, e0156445. [CrossRef]

59. Chen, Y.; Zhang, J.; Wang, H.; Zhao, J.; Xu, C.; Du, Y.; Luo, X.; Zheng, F.; Liu, R.; Zhang, H.; et al. miRNA-135a promotes breast cancer cell migration and invasion by targeting HOXA10. *BMC Cancer* **2012**, *12*, 111. [CrossRef]

60. Guo, S.; Yang, J.; Wu, M.; Xiao, G. Clinical value screening, prognostic significance and key pathway identification of miR-204-5p in endometrial carcinoma: A study based on the Cancer Genome Atlas (TCGA), and bioinformatics analysis. *Pathol. Res. Pr.* **2019**, *215*, 1003–1011. [CrossRef]

61. Nothnick, W.B.; Healy, C. Estrogen induces distinct patterns of microRNA expression within the mouse uterus. *Reprod. Sci.* **2010**, *17*, 987–994. [CrossRef]

62. Baran-Gale, J.; Purvis, J.E.; Sethupathy, P. An integrative transcriptomics approach identifies miR-503 as a candidate master regulator of the estrogen response in MCF-7 breast cancer cells. *RNA* **2016**, *22*, 1592–1603. [CrossRef] [PubMed]

63. EFSA Panel on Additives and Products or Substances used in Animal Feed (FEEDAP). Scientific Opinion Guidance for the preparation of dossiers for technological additives. *EFSA J.* **2012**, *10*, 2528. [CrossRef]

64. Reg (EC) No 1831/2003. European Union Register of Feed Additives. Edition 6/2019 (272). Appendixes 3e, 4–26 June 2019. Available online: https://ec.europa.eu/food/safety/animal-feed/feed-additives_en (accessed on 7 August 2019).

65. Hartinger, D.; Moll, W.-D. Fumonisin elimination and prospects for detoxification by enzymatic transformation. *World Mycotoxin J.* **2011**, *4*, 271–283. [CrossRef]

66. Alberts, J.; Lilly, M.; Rheeder, J.; Burger, H.; Shephard, G.S.; Gelderblom, W. Technological and community-based methods to reduce mycotoxin exposure. *Food Control* **2017**, *73*, 101–109. [CrossRef]

67. Hahn, I.; Kunz-Vekiru, E.; Twaruzek, M.; Grajewski, J.; Krska, R.; Berthiller, F. Aerobic and anaerobic in vitro testing of feed additives claiming to detoxify deoxynivalenol and zearalenone. *Food Addit. Contam. Part A* **2015**, *32*, 922–933. [CrossRef] [PubMed]

68. James, P.; Halladay, J.; Craig, E.A. Genomic Libraries and a Host Strain Designed for Highly Efficient Two-Hybrid Selection in Yeast. *Genetics* **1996**, *144*, 1425–1436. [PubMed]

69. Mitterbauer, R.; Weindorfer, H.; Safaie, N.; Krska, R.; Lemmens, M.; Ruckenbauer, P.; Kuchler, K.; Adam, G. A Sensitive and Inexpensive Yeast Bioassay for the Mycotoxin Zearalenone and Other Compounds with Estrogenic Activity. *Appl. Environ. Microbiol.* **2003**, *69*, 805–811. [CrossRef] [PubMed]

70. Bachmann, H. Phenotypic Detection of Zearalenone in Saccharomyces Cerevisiae. Master's Thesis, University of Natural Resources and Life Sciences, Vienna (BOKU), Vienna, Austria, 2001.
71. Sherman, F. Getting started with yeast. *Methods Enzymol* **2002**, *350*, 3–41. [PubMed]
72. European Commission. Directive 2010/63/eu of the European Parliament and of the Council of of 22 September 2010 on the Protection of Animals Used for Scientific Purposes. *Off. J. Eur. Union* **2010**, *L 276*, 33–79.
73. Klinge, C.M. Estrogen receptor interaction with estrogen response elements. *Nucleic Acids Res.* **2001**, *29*, 2905–2919. [CrossRef]
74. Untergasser, A.; Cutcutache, I.; Koressaar, T.; Ye, J.; Faircloth, B.C.; Remm, M.; Rozen, S.G. Primer3—New capabilities and interfaces. *Nucleic Acids Res.* **2012**, *40*, e115. [CrossRef] [PubMed]
75. Vandesompele, J.; De Preter, K.; Pattyn, F.; Poppe, B.; Van Roy, N.; De Paepe, A.; Speleman, F. Accurate normalization of real-time quantitative RT-PCR data by geometric averaging of multiple internal control genes. *Genome Boil.* **2002**, *3*, research0034.1.
76. Nygard, A.-B.; Jorgensen, C.B.; Cirera, S.; Fredholm, M. Selection of reference genes for gene expression studies in pig tissues using SYBR green qPCR. *BMC Mol. Boil.* **2007**, *8*, 67. [CrossRef] [PubMed]
77. Grenier, B.; Loureiro-Bracarense, A.-P.; Lucioli, J.; Pacheco, G.D.; Cossalter, A.-M.; Moll, W.-D.; Schatzmayr, G.; Oswald, I.P.; Loureiro-Bracarense, A. Individual and combined effects of subclinical doses of deoxynivalenol and fumonisins in piglets. *Mol. Nutr. Food Res.* **2011**, *55*, 761–771. [CrossRef] [PubMed]
78. Gessner, D.K.; Fiesel, A.; Most, E.; Dinges, J.; Wen, G.; Ringseis, R.; Eder, K. Supplementation of a grape seed and grape marc meal extract decreases activities of the oxidative stress-responsive transcription factors nf-κb and nrf2 in the duodenal mucosa of pigs. *Acta Vet. Scand.* **2013**, *55*, 18. [CrossRef]

© 2019 by the authors. Licensee MDPI, Basel, Switzerland. This article is an open access article distributed under the terms and conditions of the Creative Commons Attribution (CC BY) license (http://creativecommons.org/licenses/by/4.0/).

Article

Photocatalytic Degradation of Deoxynivalenol over Dendritic-Like α-Fe$_2$O$_3$ under Visible Light Irradiation

Huiting Wang [1,2], Jin Mao [1,3,4], Zhaowei Zhang [1,3,4], Qi Zhang [1,2,3,5], Liangxiao Zhang [1,2,4], Wen Zhang [1,3,4,5] and Peiwu Li [1,2,3,4,5,*]

1   Oil Crops Research Institute of the Chinese Academy of Agricultural Sciences, Wuhan 430062, China; wanghuitingsdau@163.com (H.W.); maojin106@whu.edu.cn (J.M.); zwzhang@whu.edu.cn (Z.Z.); zhangqi521x@126.com (Q.Z.); zhanglx@caas.cn (L.Z.); zhangwen@oilcrops.cn (W.Z.)
2   Key Laboratory of Biology and Genetic Improvement of Oil Crops, Ministry of Agriculture, Wuhan 430062, China
3   Key Laboratory of Detection for Biotoxins, Ministry of Agriculture, Wuhan 430062, China
4   Laboratory of Risk Assessment for Oilseeds Products (Wuhan), Ministry of Agriculture, Wuhan 430062, China
5   Quality Inspection and Test Center for Oilseeds Products, Ministry of Agriculture, Wuhan 430062, China
*   Correspondence: peiwuli@oilcrops.cn; Tel.: +86-27-8681-2943

Received: 10 January 2019; Accepted: 8 February 2019; Published: 11 February 2019

Abstract: Deoxynivalenol (DON) is a secondary metabolite produced by *Fusarium*, which is a trichothecene mycotoxin. As the main mycotoxin with high toxicity, wheat, barley, corn and their products are susceptible to contamination of DON. Due to the stability of this mycotoxin, traditional methods for DON reduction often require a strong oxidant, high temperature and high pressure with more energy consumption. Therefore, exploring green, efficient and environmentally friendly ways to degrade or reduce DON is a meaningful and challenging issue. Herein, a dendritic-like α-Fe$_2$O$_3$ was successfully prepared using a facile hydrothermal synthesis method at 160 °C, which was systematically characterized by X-ray diffraction (XRD), high-resolution transmission electron microscopy (HRTEM), scanning electron microscopy (SEM), and X-ray photoelectron spectroscopy (XPS). It was found that dendritic-like α-Fe$_2$O$_3$ showed superior activity for the photocatalytic degradation of DON in aqueous solution under visible light irradiation ($\lambda > 420$ nm) and 90.3% DON (initial concentration of 4.0 μg/mL) could be reduced in 2 h. Most of all, the main possible intermediate products were proposed through high performance liquid chromatography-mass spectrometry (HPLC-MS) after the photocatalytic treatment. This work not only provides a green and promising way to mitigate mycotoxin contamination but also may present useful information for future studies.

Keywords: deoxynivalenol; degradation; photocatalysis; α-Fe$_2$O$_3$; degradation products

Key Contribution: This study provides a new, efficient, green and promising way to reduce mycotoxin contamination.

---

1. Introduction

Deoxynivalenol (DON) is a water-soluble trichothecene mycotoxin produced by *Fusarium*, which can contaminate many grains such as wheat, barley, corn and other cereal crops [1]. The data from the Food and Agriculture Organization of the United Nations (FAO, 2001) and the European Union (2002), 57% of wheat samples were contaminated by DON in approximately 22,000 tested samples [2]. The international agency for research on cancer (IARC) had classified DON as the third class of carcinogens (not classifiable) in 2002 [3,4]. The contamination of DON can cause enormous economic losses of agriculture and bring serious threats to the health of human beings and animals. It was found that exposure to excessive DON could cause many adverse reactions, including dizziness, nausea and vomiting [5,6]. Therefore, attention has been paid to reduce or mitigate DON using different strategies.

Traditional strategies including physical, chemical and biological ways have been developed to eliminate DON in agro-food and the aqueous environment. The physical ways included washing, density screening, heating, adsorption [7–9]. For instance, Pronyk et al. reported that 52% of DON could be reduced after 6 min thermal degradation at 185 °C [10]. Some chemical reagents, such as ozone (O_3), sodium carbonate (Na_2CO_3), sodium bicarbonate ($NaHCO_3$) and sodium bisulfite ($NaHSO_3$) have been used to remove DON [11]. Ozone treatment has been widely applied in agro-product processing to degrade mycotoxins. Ozone could effectively reduce 53% of DON in wheat in 4 h [12]. In addition, biological technologies have also been applied in DON detoxification [13,14]. Yang demonstrated that the *Lactobacillus plantarum* JM113 could efficiently reduce DON, which was due to the high antioxidant activity of *Lactobacillus plantarum* JM113. [15]. However, due to the stability of DON in agro-food and the environment, these methods often need a strong oxidant, high temperature and more energy consumption. Therefore, exploring the green, efficient, large-scale and environmentally friendly ways to reduce DON is a meaningful and challenging topic.

In recent years, there has been a growing interest in photocatalytic degradation technology of organic pollutants. Compared to above-mentioned ways, photocatalytic degradation has several advantages: environmental-friendly, low cost, and mild conditions [16]. Bai et al. have successfully proved that graphene/ZnO hybrids showed the ability to degrade DON under UV light irradiation [17]. However, UV light accounts for only a small percentage (4%) of solar energy compared to visible light (43%) [18]. Consequently, in order to utilize solar energy more efficiently, it is necessary to develop a visible light responding catalyst for DON degradation. It was found that microcystin-LR could be removed over magnetically N-doped TiO_2 nanocomposite under visible light irradiation [19]. Our previous studies also presented that two kinds of visible light responding catalysts show high activities for the degradation of aflatoxin B_1 under visible light irradiation [20,21]. However, to our knowledge, there are only a few reports about the photocatalytic degradation of DON under visible light irradiation.

As an earth-abundant, visible light responding and nontoxic semiconductor material, hematite α-Fe_2O_3 with a suitable band gap of ~2.4 eV has received the most attention. Recently, α-Fe_2O_3 nanomaterials with different microstructures, particular shapes and particle sizes have been applied to degrade organic pollutants under visible light irradiation, such as methylene blue dye, eosin Y, rhodamine B dyes, tetracycline [22–24]. Herein, a visible light responding catalyst, dendritic-like α-Fe_2O_3 was prepared facilely by a simple one-step hydrothermal synthesis method at 160 °C. Under visible light irradiation (>420 nm), the dendritic-like α-Fe_2O_3 showed superior performance for degradation of DON in aqueous solution compared with commercial α-Fe_2O_3. In addition, the possible degradation products after 120 min photoreaction were analyzed and proposed by HPLC-MS.

2. Results and Discussion

2.1. Crystal Phase and Morphology Analyses

To analyze the crystal phase and crystallinity of the as-prepared product, the XRD pattern of the as-prepared catalyst was presented. As shown in Figure 1a, the XRD pattern of the as-prepared α-Fe_2O_3 was consistent with JCPDS No. 01-089-0596, which was a pure phase of rhombohedral α-Fe_2O_3. In addition, the representative crystal planes of the rhombohedral α-Fe_2O_3 were labeled in the pattern. These planes were ascribed to (012), (104), (110), (113), (024), (116), (122), (214) and (300), respectively. No other peaks were found, indicating that the as-prepared α-Fe_2O_3 was pure. Furthermore, the strong and steep peaks indicated that the as-prepared α-Fe_2O_3 was of high crystallinity. The XPS was used to estimate the surface chemical compositions and valence state of the as-prepared α-Fe_2O_3.

From Figure 1b, it was found that Fe and O existed in the as-prepared α-Fe$_2$O$_3$. In Figure 1c, there are two peaks at 711.18 eV and 724.78 eV, which are consistent with Fe 2p3/2 and Fe 2p1/2 respectively. The two peaks are characteristic peaks of the Fe^{3+} state in Fe$_2$O$_3$ [25]. The O 1s spectrum showed one peak (529.85 eV in Figure 1d), corresponding to oxygen atoms of the α-Fe$_2$O$_3$. From the above XRD and XPS results, it could be concluded that the as-prepared product was a pure α-Fe$_2$O$_3$.

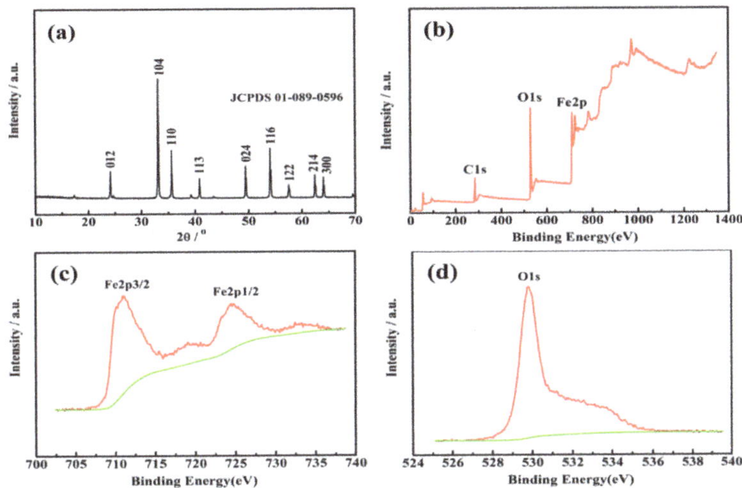

Figure 1. (**a**) XRD pattern of the as-prepared α-Fe$_2$O$_3$; (**b**) XPS spectra of the as-prepared α-Fe$_2$O$_3$; High-resolution XPS spectra of Fe (**c**) and O element (**d**).

Figure 2. SEM images (**a,b**), HRTEM images (**c,d**) of the prepared α-Fe$_2$O$_3$.

The microstructure and morphology of the as-prepared α-Fe$_2$O$_3$ were investigated by SEM and HRTEM, which are presented in Figure 2. The as-prepared α-Fe$_2$O$_3$ in the SEM image (Figure 2a,b) was a uniform dendritic-like three-dimensional microstructure. The dendritic-like microstructure had branches that stretched in different directions. As shown in Figure 2c, it was found that the α-Fe$_2$O$_3$ showed the average length of about 2 μm and the average width of the α-Fe$_2$O$_3$ was about 500 nm. In Figure 2d, the distances of the lattice in α-Fe$_2$O$_3$ were 0.27 nm, which was consistent with the (014) plane in rhombohedral α-Fe$_2$O$_3$ structure. The above SEM and HRTEM indicate that the uniform and regular dendritic-like microstructural α-Fe$_2$O$_3$ was successfully prepared through the facile hydrothermal method.

2.2. Photocatalytic Activity

The photocatalytic activities of the dendritic-like α-Fe$_2$O$_3$ were estimated by the degradation rate of DON in aqueous solution. From Figure 3a, it could be found that the intensity of DON decreased significantly with time over dendritic-like α-Fe$_2$O$_3$ under visible light irradiation. This indicated that the dendritic-like α-Fe$_2$O$_3$ had superior photocatalytic activity under the visible light irradiation. In Figure 3b, the blank line revealed that the DON could not be degraded under visible light obviously without a photocatalyst or in the presence of photocatalyst without visible light, which indicated that the catalysts and visible light were necessary for the photocatalytic degradation of DON. In addition, after the suspension was stirred with a magnetic stirring apparatus for 1 h in the dark, the content of a DON standard was reduced slightly. It was found that the dendritic-like α-Fe$_2$O$_3$ could adsorb more DON than that of commercial α-Fe$_2$O$_3$, which may enhance the photocatalytic degradation activity. As expected, the dendritic-like α-Fe$_2$O$_3$ (the blue line in Figure 3b) showed greatly a higher photocatalytic efficiency with a degradation rate of 90.3% in 2 h than that of commercial α-Fe$_2$O$_3$ (46.7%). There are two reasons why the dendritic-like α-Fe$_2$O$_3$ had better photocatalytic performance. Firstly, as the three-dimensional nanomaterial, the dendritic-like α-Fe$_2$O$_3$ is difficult to form aggregation compared with commercial α-Fe$_2$O$_3$. In addition, the unique morphology of bionic dendritic a-Fe$_2$O$_3$ tends to increase the absorption of the sunlight and provides more charge transfer pathways and more active sites to improve the efficiency of photocatalytic activities [25].

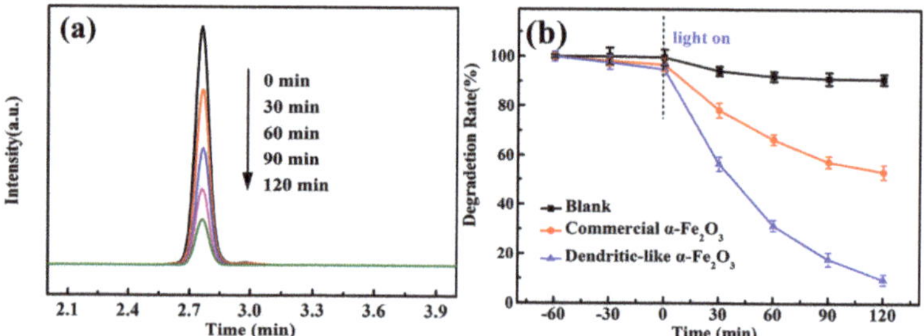

Figure 3. (**a**) HPLC chromatogram of DON photodegradation over dendritic-like α-Fe$_2$O$_3$ with different times (**b**) Photocatalytic degradation of DON over dendritic-like α-Fe$_2$O$_3$, commercial α-Fe$_2$O$_3$ under visible light and blank control.

2.3. Degradation Products Analysis

Generally, the photocatalytic degradation products of organic compounds are determined by the kind and quantity of active radicals production during the photoreaction, which may be different in the presence of different catalysts or with different irradiation time [26]. During the photoreaction over α-Fe$_2$O$_3$, the electrons could be excited and transferred from the valence band (VB) to the conduction band (CB), leaving holes in the VB, which can oxidize the hydroxide ions into hydroxyl radicals ($\bullet$OH). In addition, in the CB of the α-Fe$_2$O$_3$, the transferred electrons reacted with dissolved oxygen to form superoxide radical anions ($\bullet$O$_2^-$). Then, a number of active radicals such as $\bullet$O$_2^-$ and $\bullet$OH could react with the active site of DON and form the intermediate products [27,28]. To identify the possible intermediate products of DON, the intermediates after photocatalytic treatment were investigated by HPLC-MS. As shown in Figure 4 the total ion chromatograms (TIC) also revealed that the DON obviously reduced after 2 h photodegradation. A comparison between the initial TIC with that of after photodegradation revealed that two obvious peaks, P1 (m/z 209.17) and P2 (m/z 227.0), appeared after the photocatalysis treatment, which may be the intermediate products of DON. The m/z value of the DON in the positive ESI mode was 297.00. As shown in Figure 5, the possible structures of the two intermediate products were proposed according to previous studies [29–32]. The products P1 and P2 may have been generated from a series of reactions such as the dehydration, breakage of the carbon chain and reduction of the 12, 13-epoxy group to diene. However, the detailed reaction process needs a deep investigation. As shown in Figure 6, DON is a type B trichothecenes compound and the 12, 13-epoxy group of its structure is associated with its toxicity. In addition, the three hydroxyl groups (-OH) in the DON molecule also are related to its toxicity [33]. It was found that these toxicity sites were destroyed from the proposed structure of the two products, indicating that the toxicity of DON may decrease after efficient photocatalytic treatment over dendritic-like α-Fe₂O₃. However, the toxicity of DON products needs more evidence, systematic and long-term evaluations in our future study, which is in progress. From the above results and discussions, it could be deduced that the dendritic-like α-Fe₂O₃ have a potential application in the photocatalytic degradation of DON in aqueous solution.

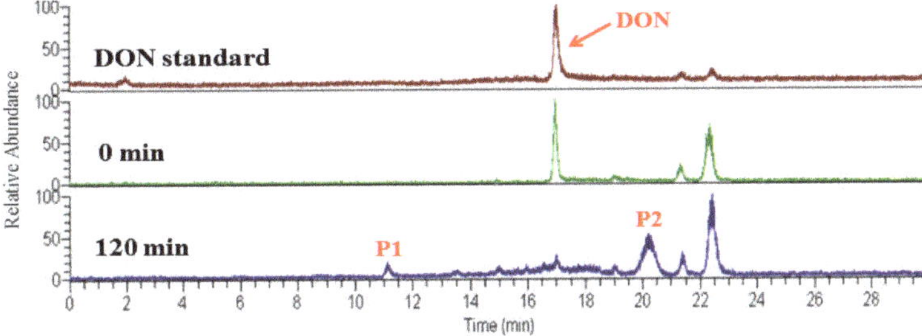

Figure 4. TIC of the sample for the DON standard, before reaction (0 min) and after 120min reaction.

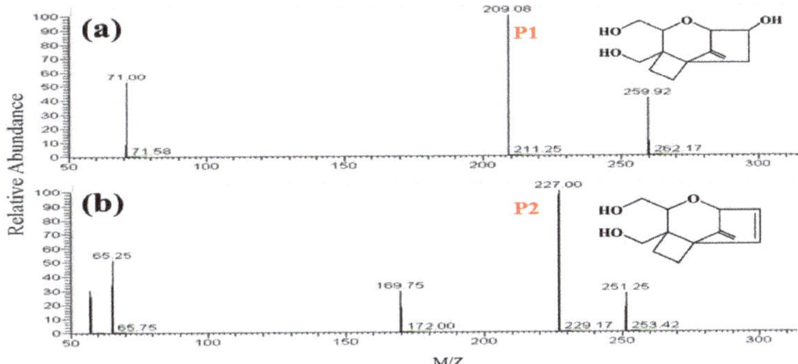

Figure 5. The full scan mass spectrums and possible structures of P1 (**a**), P2 (**b**).

Figure 6. Chemical structure of DON.

3. Conclusions

In conclusion, a light-responsive dendritic-like α-Fe₂O₃ with a length of 2 μm and the width of 500 nm was successfully synthesized through a facile hydrothermal synthesis method.

The dendritic-like α-Fe$_2$O$_3$ showed better activity for the degradation of DON in aqueous solution under visible light irradiation compared with the commercial α-Fe$_2$O$_3$, which was ascribed to the fact that the dendritic-like α-Fe$_2$O$_3$ was aggregated with difficulty and could provide more charge transfer pathways and more active sites. Most of all, two possible intermediate products, P1 (m/z 209.17) and P2 (m/z 227.00), were proposed through HPLC-MS after 2 h photocatalytic treatment. It was found that the main toxicity sites of the 12, 13-epoxy group and hydroxyl groups in DON were destroyed, indicating that the toxicity of DON may decrease after efficient photocatalytic treatment. This work not only provided a new, efficient, green and promising way to reduce mycotoxins such as DON contamination but also presents useful information for future study.

4. Materials and Methods

4.1. Materials

The deionized water was from the RIOS16/Milli-Qa10 water purification system (France). Potassium hexacyanoferrate (III) (K$_3$[Fe(CN)$_6$], >99.5% purity) was purchased from Tianjin Bodi Chemical Reagent Co. (China). For high performance liquid chromatography and high performance liquid chromatography-mass spectrometry analysis, acetonitrile and methanol (mass spectrum grade) were purchased from Fisher Chemical. Formic acid (CH$_3$COOH, >95% purity) and DON standard came from Sigma-Aldrich Co. (USA). All chemicals and reagents were used without further purification.

4.2. Synthesis of Dendritic-Like α-Fe$_2$O$_3$ Photocatalyst

In a typical hydrothermal reaction, K$_3$[Fe(CN)$_6$] (0.528 g) was ultrasonically dispersed (40 kHz, 25 °C, 5 min) in the deionized water (80 mL) and then transferred to a 100 mL Teflon-lined stainless steel autoclave. Then Teflon-lined stainless steel autoclave was put into an oven and heated for 24 h at 160 °C. After it was cooled down to room temperature, the obtained solid was washed with deionized water and methanol for 3 times. The as-prepared catalyst was dried at 60 °C for 6 h. Finally, the dried solid was ground into powder in an agate mortar for later characterization and photocatalytic measurement.

4.3. Material Characterization

The morphology and structure of the sample were characterized by X-ray diffraction (Bruker AXS, D8, Germany) with a scanning range from 10 to 70 degree. Scanning electron microscopy (Hitachi-4800, Japan) and transmission electron microscopy (FEI Tecnai G2 F30, America) were used to characterize the particle size and the morphology of the as-prepared sample. X-ray photoelectron spectroscopy analysis was performed on a Kratos XSAM 800 X-ray photoelectron spectrometer with an Mg Kα X-ray source.

4.4. Photocatalytic Measurement

The photocatalytic activity test was performed as follows: 0.01 g α-Fe$_2$O$_3$ was dispersed in 96 mL deionized water through 20 min ultrasonic treatment (40 kHz, 25 °C). Then, 4 mL of DON aqueous solution (100 µg/mL) was added to above α-Fe$_2$O$_3$ suspension. Then, an α-Fe$_2$O$_3$ suspension containing DON was magnetically stirred for 1 h in the dark. Then, the suspension was irradiated under visible light (λ > 420 nm) with magnetically stirring at room temperature. The light source was provided by a 300 W Xenon lamp (PLS-SXE 300, Beijing Trusttech Co., Beijing, China). The distance was about 20 cm from the light source to the DON suspension. The suspension after different irradiation time (30 min, 60 min, 90 min, 120 min) were collected to measure the concentration of DON using HPLC (Agilent 1200, USA) equipped with a chromatographic column (Agilent polar C18-A, 5 µm, 150 mm × 4.6 mm). The samples were filtrated through a 0.22 µm filter before detection. The mobile phases were the aqueous solution and acetonitrile in 83:17 (v/v). The flow rate of mobile phases was 1 mL/min, and the temperature of the column was 35 °C. The excitation and emission wavelengths were 220 nm and 360 nm, respectively. The injection volume was 20 µL. The control and commercial α-Fe$_2$O$_3$ were performed as the above steps.

4.5. Degraded Products Analysis

The high performance liquid chromatography-electrospray ionization mass spectrometry (HPLC-ESI/MS, Thermo Finnigan LTQ XL, USA) was used to analyze the degradation products of DON. The HPLC (Thermo, USA) equipped with a chromatographic column (Syncronis C18, 3 µm, 100×2.1 mm, Thermo). The flow rate was 200 µL/min, and the injection volume was 5 µL. The mobile phase included two components. Component A was distilled water with 0.1% formic acid and component B was pure acetonitrile. The ESI was used in positive ion mode and other parameters were as follows: capillary temperature was at 350 °C, and the flow rate of sheath gas, aux gas and sweep gas was at 28 arb, 2 arb, 0 arb, respectively. The pressure of nebulizer was at 4.50 Kv with the capillary voltage at 9 V, tube lens voltage at 95 V; multipole RF Vpp at 400 V. All other parameters in the MS were tuned for maximum signal intensity of a reference solution. The scan ranges of MS were from $m/z = 50$ to $m/z = 320$.

Author Contributions: Conceptualization, P.L.; methodology, H.W. and J.M.; software, L.Z.; validation, P.L., Q.Z. and W.Z.; formal analysis, Z.Z.; investigation, H.W.; resources, P.L.; data curation, J.M.; writing—original draft preparation, H.W.; writing—review and editing, J.M.; visualization, Z.Z.; supervision, P.L.; project administration, P.L. and J.M.; funding acquisition, J.M.

Funding: This work was supported by the Natural Science Foundation of China (31871900, 31401601), National Key Project for Agro-product Quality & Safety Risk Assessment, PRC (GJFP2018001), International Science & Technology Cooperation Program of China (2016YFE0112900).

Conflicts of Interest: The authors declare no competing financial interests.

References

1. McEvoy, J.D. Emerging food safety issues: An EU perspective. *Drug Test. Anal.* **2016**, *8*, 511–520. [CrossRef] [PubMed]
2. Schothorst, R.; Van, E.H. Critical assessment of trichothecene exposure—Report from the SCOOP project. *Toxicol. Lett.* **2004**, *153*, 133–143. [CrossRef] [PubMed]
3. De Ruyck, K.; De Boevre, M.; Huybrechts, I.; De Saeger, S. Dietary mycotoxins, co-exposure, and carcinogenesis in humans: Short review. *Mutat. Res. Rev. Mutat. Res.* **2015**, *766*, 32–41. [CrossRef] [PubMed]
4. Abdallah, M.F.; Girgin, G.; Baydar, T. Occurrence, Prevention and Limitation of Mycotoxins in Feeds. *Anim. Nutr. Feed Technol.* **2015**, *15*, 471–490. [CrossRef]
5. Peng, Z.; Chen, L.K.; Nussler, A.K.; Liu, L.G.; Yang, W. Current sights for mechanisms of deoxynivalenol-induced hepatotoxicity and prospective views for future scientific research: A mini review. *J. Appl. Toxicol.* **2017**, *37*, 518–529. [CrossRef] [PubMed]
6. Wu, W.D.; Zhou, H.R.; Pestka, J.J. Potential roles for calcium-sensing receptor (CaSR) and transient receptor potential ankyrin-1 (TRPA1) in murine anorectic response to deoxynivalenol (vomitoxin). *Arch. Toxicol.* **2017**, *91*, 495–507. [CrossRef] [PubMed]
7. Tengjaroenkul, B.; Tengjaroenku, U.; Pumipuntu, N.; Pimpukdee, K.; Wongtangtintan, S.; Saipan, P. An In Vitro Comparative Study of Aflatoxin B1 Adsorption by Thai Clay and Commercial Toxin Binders. *Thai J. Vet. Med.* **2013**, *43*, 491–495.
8. Yener, S.; Koksel, H. Effects of washing and drying applications on deoxynivalenol and zearalenone levels in wheat. *World Mycotoxin J.* **2013**, *6*, 335–341. [CrossRef]
9. Vidal, A.; Sanchis, V.; Ramos, A.J.; Marin, S. Thermal stability and kinetics of degradation of deoxynivalenol, deoxynivalenol conjugates and ochratoxin A during baking of wheat bakery products. *Food Chem.* **2015**, *178*, 276–286. [CrossRef]
10. Pronyk, C.; Cenkowski, S.; Abramson, D. Superheated steam reduction of deoxynivalenol in naturally contaminated wheat kernels. *Food Control* **2006**, *17*, 789–796. [CrossRef]
11. Moazami, F.E.; Jinap, S.; Mousa, W.; Hajeb, P. Effect of Food Additives on Deoxynivalenol (DON) Reduction and Quality Attributes in Steamed-and-Fried Instant Noodles. *Cereal Chem.* **2014**, *91*, 88–94. [CrossRef]

12. Wang, L.; Wang, Y.; Shao, H.L.; Luo, X.H.; Wang, R.; Li, Y.F.; Li, Y.N.; Luo, Y.P.; Zhang, D.J.; Chen, Z.X. In vivo toxicity assessment of deoxynivalenol-contaminated wheat after ozone degradation. *Food Addit. Contam. A* **2017**, *34*, 103–112. [CrossRef] [PubMed]

13. Niderkorn, V.; Morgavi, D.P.; Pujos, E.; Tissandier, A.; Boudra, H. Screening of fermentative bacteria for their ability to bind and biotransform deoxynivalenol, zearalenone and fumonisins in an in vitro simulated corn silage model. *Food Addit. Contam.* **2007**, *24*, 406–415. [CrossRef]

14. Vanhoutte, I.; De Mets, L.; De Boevre, M.; Uka, V.; Di Mavungu, J.D.; De Saeger, S.; De Gelder, L.; Audenaert, K. Microbial Detoxification of Deoxynivalenol (DON), Assessed via a *Lemna minor* L. Bioassay, through Biotransformation to 3-epi-DON and 3-epi-DOM-1. *Toxins* **2017**, *9*, 63. [CrossRef] [PubMed]

15. Yang, X.; Li, L.; Duan, Y.; Yang, X. Antioxidant activity of Lactobacillus plantarum JM113 in vitro and its protective effect on broiler chickens challenged with deoxynivalenol. *J. Anim. Sci.* **2017**, *95*, 837–846. [CrossRef] [PubMed]

16. Bhatkhande, D.S.; Pangarkar, V.G.; Beenackers, A.A.C.M. Photocatalytic degradation for environmental applications—A review. *J. Chem. Technol. Biotechnol.* **2002**, *77*, 102–116. [CrossRef]

17. Bai, X.J.; Sun, C.P.; Liu, D.; Luo, X.H.; Li, D.; Wang, J.; Wang, N.X.; Chang, X.J.; Zong, R.L.; Zhu, Y.F. Photocatalytic degradation of deoxynivalenol using graphene/ZnO hybrids in aqueous suspension. *Appl. Catal. B-Environ.* **2017**, *204*, 11–20. [CrossRef]

18. Yamaguchi, Y.; Usuki, S.; Yamatoya, K.; Suzuki, N.; Katsumata, K.; Terashima, C.; Fujishima, A.; Kudo, A.; Nakata, K. Efficient photocatalytic degradation of gaseous acetaldehyde over ground Rh-Sb co-doped $SrTiO_3$ under visible light irradiation. *RSC Adv.* **2018**, *8*, 5331–5337. [CrossRef]

19. Pelaez, M.; Baruwati, B.; Varma, R.S.; Luque, R.; Dionysiou, D.D. Microcystin-LR removal from aqueous solutions using a magnetically separable N-doped TiO_2 nanocomposite under visible light irradiation. *Chem. Commun.* **2013**, *49*, 10118–10120. [CrossRef] [PubMed]

20. Mao, J.; Zhang, Q.; Li, P.W.; Zhang, L.X.; Zhang, W. Geometric architecture design of ternary composites based on dispersive WO_3 nanowires for enhanced visible-light-driven activity of refractory pollutant degradation. *Chem. Eng. J.* **2018**, *334*, 2568–2578. [CrossRef]

21. Mao, J.; Zhang, L.X.; Wang, H.T.; Zhang, Q.; Zhang, W.; Li, P.W. Facile fabrication of nanosized graphitic carbon nitride sheets with efficient charge separation for mitigation of toxic pollutant. *Chem. Eng. J.* **2018**, *342*, 30–40. [CrossRef]

22. Lassoued, A.; Lassoued, M.S.; Dkhil, B.; Ammar, S.; Gadri, A. Photocatalytic degradation of methylene blue dye by iron oxide (α-Fe_2O_3) nanoparticles under visible irradiation. *J. Mater. Sci.-Mater. Electron.* **2018**, *29*, 8142–8152. [CrossRef]

23. Gobouri, A.A. Ultrasound enhanced photocatalytic properties of α-Fe_2O_3 nanoparticles for degradation of dyes used by textile industry. *Res. Chem. Intermed.* **2016**, *42*, 5099–5113. [CrossRef]

24. Li, R.; Liu, J.; Jia, Y.; Zhen, Q. Photocatalytic Degradation Mechanism of Oxytetracyclines Using Fe_2O_3-TiO_2 Nanopowders. *J. Nanosci. Nanotechnol.* **2017**, *17*, 3010–3015. [CrossRef]

25. Jiang, Z.F.; Jiang, D.L.; Wei, W.; Yan, Z.X.; Xie, J.M. Natural carbon nanodots assisted development of size-tunable metal (Pd, Ag) nanoparticles grafted on bionic dendritic α-Fe_2O_3 for cooperative catalytic applications. *J. Mater. Chem. A* **2015**, *3*, 23607–23620. [CrossRef]

26. Huang, J.H.; Cheng, W.J.; Shi, Y.H.; Zeng, G.M.; Yu, H.B.; Gu, Y.L.; Shi, L.X.; Yi, K.X. Honeycomb-like carbon nitride through supramolecular preorganization of monomers for high photocatalytic performance under visible light irradiation. *Chemosphere* **2018**, *211*, 324–334. [CrossRef] [PubMed]

27. Abd Mutalib, M.; Aziz, F.; Jamaludin, N.A.; Yahya, N.; Ismail, A.F.; Mohamed, M.A.; Yusop, M.Z.M.; Salleh, W.N.W.; Jaafar, J.; Yusof, N. Enhancement in photocatalytic degradation of methylene blue by $LaFeO_3$-GO integrated photocatalyst-adsorbents under visible light irradiation. *Korean J. Chem. Eng.* **2018**, *35*, 548–556. [CrossRef]

28. Li, H.T.; Wang, M.; Wei, Y.P.; Long, F. Noble metal-free NiS_2 with rich active sites loaded g-C_3N_4 for highly efficient photocatalytic H-2 evolution under visible light irradiation. *J. Colloid Interface Sci.* **2019**, *534*, 343–349. [CrossRef]

29. Juan-Garcia, A.; Juan, C.; Manyes, L.; Ruiz, M.J. Binary and tertiary combination of alternariol, 3-acetyl-deoxynivalenol and 15-acetyl-deoxynivalenol on HepG2 cells: Toxic effects and evaluation of degradation products. *Toxicol. In Vitro* **2016**, *34*, 264–273. [CrossRef]

30. Juan-Garcia, A.; Juan, C.; Konig, S.; Ruiz, M.J. Cytotoxic effects and degradation products of three mycotoxins: Alternariol, 3-acetyl-deoxynivalenol and 15-acetyl-deoxynivalenol in liver hepatocellular carcinoma cells. *Toxicol. Lett.* **2015**, *235*, 8–16. [CrossRef]

31. Bretz, M.; Beyer, M.; Cramer, B.; Knecht, A.; Humpf, H.U. Thermal degradation of the Fusarium mycotoxin deoxynivalenol. *J. Agric. Food Chem.* **2006**, *54*, 6445–6451. [CrossRef]

32. Sun, C.; Ji, J.; Wu, S.L.; Sun, C.P.; Pi, F.W.; Zhang, Y.Z.; Tang, L.L.; Sun, X.L. Saturated aqueous ozone degradation of deoxynivalenol and its application in contaminated grains. *Food Control* **2016**, *69*, 185–190. [CrossRef]

33. Pestka, J.J. Deoxynivalenol: Toxicity, mechanisms and animal health risks. *Anim. Feed Sci. Technol.* **2007**, *137*, 283–298. [CrossRef]

 © 2019 by the authors. Licensee MDPI, Basel, Switzerland. This article is an open access article distributed under the terms and conditions of the Creative Commons Attribution (CC BY) license (http://creativecommons.org/licenses/by/4.0/).

Article

Cold Plasma Treatment as an Alternative for Ochratoxin A Detoxification and Inhibition of Mycotoxigenic Fungi in Roasted Coffee

Paloma Patricia Casas-Junco [1], Josué Raymundo Solís-Pacheco [2], Juan Arturo Ragazzo-Sánchez [1], Blanca Rosa Aguilar-Uscanga [2], Pedro Ulises Bautista-Rosales [3] and Montserrat Calderón-Santoyo [1,*]

[1]   Laboratorio Integral de Investigación en Alimentos, Tecnológico Nacional de México/Instituto Tecnológico de Tepic, Av. Tecnológico #2595, Col. Lagos del Country, C.P. 63175 Tepic, Nayarit, Mexico; mcapalomaa@outlook.es (P.P.C.-J.); arturoragazzo@hotmail.com (J.A.R.-S.)

[2]   Laboratorio de Microbiología Industrial, Centro Universitario de Ciencias Exactas e Ingeniería, Universidad de Guadalajara, Boulevard Marcelino García Barragán #1421, Col. Olímpica, C.P. 44430 Guadalajara, Jalisco, Mexico; josuesolisp@gmail.com (J.R.S.-P.); agublanca@gmail.com (B.R.A.-U.)

[3]   Centro de Tecnología de Alimentos, Universidad Autónoma de Nayarit, Ciudad de la Cultura "Amado Nervo", C.P. 63155 Tepic, Nayarit, Mexico; u_bautista@hotmail.com

*   Correspondence: montserratcalder@gmail.com or mcalderon@ittepic.edu.mx; Tel.: +52-311-211-9400 (ext. 232)

Received: 23 May 2019; Accepted: 5 June 2019; Published: 13 June 2019

Abstract: Ochratoxin A (OTA) produced by mycotoxigenic fungi (*Aspergillus* and *Penicillium* spp.) is an extremely toxic and carcinogenic metabolite. The use of cold plasma to inhibit toxin-producing microorganisms in coffee could be an important alternative to avoid proliferation of mycotoxigenic fungi. Roasted coffee samples were artificially inoculated with *A. westerdijikiae*, *A. steynii*, *A. versicolor*, and *A. niger*, and incubated at 27 °C over 21 days for OTA production. Samples were cold plasma treated at 30 W input power and 850 V output voltage with helium at 1.5 L/min flow. OTA production in coffee was analyzed by high performance liquid chromatography coupled to a mass spectrometer (HPLC-MS). After 6 min of treatment with cold plasma, fungi were completely inhibited (4 log reduction). Cold plasma reduces 50% of OTA content after 30 min of treatment. Toxicity was estimated for extracts of artificially contaminated roasted coffee samples using the brine shrimp (*Artemia salina*) lethality assay. Toxicity for untreated roasted coffee was shown to be "toxic", while toxicity for cold plasma treated coffee was reduced to "slightly toxic". These results suggested that cold plasma may be considered as an alternative method for the degradation and reduction of toxin production by mycotoxigenic fungi in the processing of foods and feedstuffs.

Keywords: roasted coffee; mycotoxigenic fungi; ochratoxin A; cold plasma; detoxification; brine shrimp bioassay

Key Contribution: Cold plasma technology emerges as a convenient alternative to detoxify coffee products. Total inhibition of *A. niger* and 50% OTA detoxification were achieved after 6 and 30 min exposures to cold plasma, respectively.

---

1. Introduction

Ochratoxins belong to a family of structurally related, secondary fungal metabolites produced by various *Penicillium* and *Aspergillus* strains. Among them, ochratoxin A (OTA) is reputed to be important mycotoxin in coffee. OTA possesses properties such as nephrotoxicity, carcinogenicity, and teratogenicity and is classified as carcinogenic for humans (group 2B) [1,2]. According to the Commission of European Communities 2006, the maximum accepted levels for OTA are 5 µg/kg in roasted coffee beans and ground roasted coffee, and 10 µg/kg in soluble coffee. Some studies have revealed that the roasting treatment of green coffee at 200 °C for 20 min reduced the OTA levels by only 0–12% [3]. OTA binds to the plasma proteins [4], and then OTA could be bound to the coffee proteins. Changes to OTA during heat treatment have been studied; a partial isomerization of OTA in position C3 into a diastereomer occurred at high temperatures [5]. Therefore, the relative OTA thermo-stability could cause toxicological problems during the consumption of coffee products.

Several strategies are available for the detoxification of mycotoxins and inhibition of fungi growth, which include methods such as thermal inactivation, irradiation, ultrasound treatment, and biological control agents [6]. Although these are widely used methods, they are time consuming and result in reductions in the nutritional quality of food products, variations in sensorial characteristics, and higher costs can occur.

Cold plasma is an innovative technology with potential for the inactivation of various pathogenic and spoilage microorganisms and the detoxification of foods [7]. Cold plasma is the fourth state of matter, being an ionized gas containing a rich mixture of reactive neutral species, energetic charged particles, UV photons, and intense transient electric fields, which can interact simultaneously and synergistically with the food [8]. Ionization is the first important step for plasma chemistry. Plasma chemistry depends on several factors such as feed gas composition, relative humidity, the power supplied, and treatment time [9]. Various studies have shown the potential of cold plasma to inactivate toxigenic fungi, leading to the detoxification of hazelnuts, peanuts, pistachios, and nuts [6,10], date palm fruits [11], and groundnuts [9]. Solís-Pacheco et al. [12] carried out a study on the sterilization of cinnamon and camomile to eliminate fungi and yeasts by the use of cold plasma. These authors obtained reductions higher than 1.0 log CFU/g at 750 and 850 V for chamomile and 0.68 ± 0.18 log CFU/g for cinnamon at 850 V after 10 min of treatment. They concluded that the best treatment to obtain significant reductions of yeast and mold counts in chamomile and cinnamon samples with the lowest degradation of antioxidant compounds was the application of plasma energy at 750 V for 10 min.

Therefore, the objective of this study was to explore cold plasma as a technology to inhibit toxigenic fungi and to detoxify OTA in roasted coffee, as well as to demonstrate the reduction in toxicity for the cold plasma treated samples.

2. Results

2.1. Inhibition of Fungal Spores in Roasted Coffee by Cold Plasma Treatment

A complete inhibition of *A. westerdijikiae*, *A. steynii*, *A. versicolor*, and *A. niger* spores was achieved (4 log reduction) after 6 min exposure to cold plasma treatment ($p < 0.05$) (Figure 1; Table S1). Méndez-Vilas [13] reported damage in the microbial cellular membrane due to its interaction with active species and a subsequent loss of cytoplasmic material as the final cause of microorganism death. Additionally, DNA molecules—highly sensitive to ions, neutral species, and UV light—can be altered when these active species penetrate and reach the nucleus [14,15].

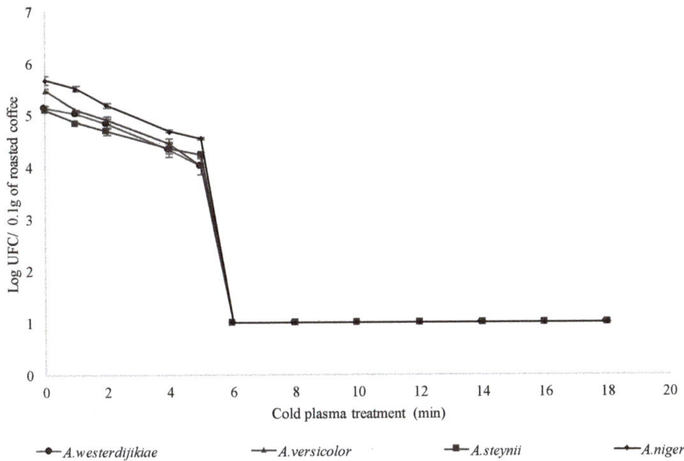

Figure 1. Logarithmic survival curves of *A. westerdijikiae, A. steynii, A. versicolor*, and *A. niger* spores on roasted coffee samples cold plasma treated (30 W input power, 850 V output voltage, He-air mixture at 1.5 L/min).

2.2. Effect of Cold Plasma on Detoxification of OTA Produced in Roasted Coffee by Mycotoxigenic Fungi

OTA production in roasted coffee artificially contaminated and incubated over 21 days was determined as follows: *A. westerdijikiae* 96.45 µg/kg, *A. steynii* 67.79 µg/kg, *A. niger* 91.03 µg/kg, and *A. versicolor* 76.74 µg/kg. After 30 min of cold plasma exposure, OTA was reduced by approximatively 50% ($p < 0.05$) (Table 1; Table S1). These results suggest the application of cold plasma may have a strong potential not only in mycotoxin degradation, but also in fungi control, and subsequently in the reduction of toxin production by mycotoxigenic fungi, thus suggesting this process can be effectively used in industrial food production. Despite an absence of OTA total reduction, the achieved reduction of 50% could be high enough to meet the requirements stablished by the European Union legislation which established the maximum accepted levels for OTA at 5 µg/kg for roasted coffee beans and ground roasted coffee and 10 µg/kg for soluble coffee [16]. Therefore, this technique can be promising for mycotoxin detoxification in the coffee industry as an alternative to preventing or eliminating OTA and toxigenic fungi.

Table 1. Effect of cold plasma applied for different periods of time on ochratoxin A (OTA) detoxification of roasted coffee.

Exposure Time (min)	OTA (µg/kg)							
	A. westerdijikiae	Reduction (%)	*A. steynii*	Reduction (%)	*A. niger*	Reduction (%)	*A. versicolor*	Reduction (%)
0	96.45 ± 3.08 [a]		67.69 ± 3.66 [a]		91.03 ± 12.06 [a]		76.741 ± 10.12 [a]	
1	89.86 ± 4.43 [a]	6.83	64.57 ± 1.92 [ab]	4.61	87.31 ± 3.56 [a]	4.09	76.61 ± 1.91 [a]	0.17
4	81.34 ± 16.91 [ab]	15.67	59.07 ± 4.97 [bc]	12.73	72.81 ± 1.99 [b]	20.02	75.44 ± 10.96 [a]	1.69
8	77.17 ± 14.79 [bac]	19.99	55.17 ± 1.63 [dc]	18.50	71.77 ± 1.28 [b]	21.16	69.19 ± 22.76 [a]	9.84
12	73.62 ± 8.43 [bac]	23.67	49.48 ± 5.49 [de]	26.90	69.50 ± 2.86 [b]	23.65	68.36 ± 2.38 [a]	10.91
16	62.89 ± 22.09 [bdc]	34.80	43.48 ± 1.75 [ef]	35.37	66.4 ± 1.99 [cb]	27.06	67.08 ± 8.85 [a]	12.59
20	51 ± 5.26 [bdc]	47.12	41.42 ± 0.69 [f]	38.31	65.37 ± 9.97 [cb]	28.19	63.67 ± 1.29 [a]	17.02
24	54.39 ± 1.08 [dc]	43.61	38.28 ± 1.59 [fg]	43.45	63.76 ± 3.40 [cb]	29.96	63.28 ± 13.36 [a]	17.53
30	38 ± 4.20 [d]	60.60	31.39 ± 0.03 [g]	52.86	54.34 ± 0.85 [c]	40.31	51.18 ± 6.64 [a]	33.31

Values followed by the same letter in the columns are not significantly different.

The effect of cold plasma on OTA detoxification could be attributed to the reactive gas species generation of such ions as (H_3O^+, O^+, O^-, OH^-, N_2^+), molecular species (O_3, H_2O_2), and reactive radicals (O•, OH•, NO•), as well as to UV irradiation and etching [7,17]. These species could promote the opening of the chlorophenolic group containing a dihydro-isocoumarin and/or amide-linked to L-phenylalanine.

Some studies have reported the efficacy of cold plasma on detoxification of mycotoxins in food products. Basaran et al. [6] indicated a 50% reduction in aflatoxins (AFB1, AFB2, AFG1, and AFG2) after 20 min exposure to air plasma, while only a 20% reduction was obtained after 20 min of exposure to sulfur hexafluoride plasma. Ouf et al. [11] demonstrated synthesis inhibition of fumonisin B2 and ochratoxin A by *A. niger* after treatment for 6 and 7.5 min with argon plasma jet. Devi et al. [9] showed 70% and 90% reductions on groundnuts samples treated with air plasma at 60 W (12 min) and 40 W (15 min), respectively.

Atalla et al. [16] explained the aflatoxin B1 reduction by UV and fluorescent light in terms of degradation of aflatoxin crystals mediated by their high sensitivity to UV radiation and changes in the structure of the terminal furan ring. During cold plasma generation, UV radiation is an important component for the detoxification of mycotoxins.

2.3. Sensitivity of Brine Shrimp Larvae to OTA

According to the obtained results for OTA detoxification by cold plasma, it was desirable to study the toxicity of structures resulting from the application of this technology. Brine shrimp incubated with untreated OTA extracts showed a toxicity level classified as "toxic". On the other hand, larval mortality for cold plasma treated OTA extracts was significantly reduced, and a toxicity level of "slightly toxic" was obtained for all samples (Table 2). Mycotoxin detoxification is mediated by the union of free radicals to the heterocyclic rings in their molecule [17]. In this way, when mycotoxins are irradiated, three possible results can be obtained: first, the resulting structures are more toxigenic than the original molecule; second, the resulting molecules are equally toxigenic as the original toxin; and third, the resulting fragments present lower toxicity compared with the original toxin molecule [18]. Ultimately, in this study, a lower degree of toxicity was obtained when OTA was cold plasma treated.

Table 2. Toxic activity of untreated and 30 min cold plasma treated coffee extracts against brine shrimp.

Mycotoxigenic Fungi in Coffee	OTA in Solvent Extract	Treatment	Mortality (%)	Degree of Toxicity
A. westerdijikiae	1.19 ng/mL	Untreated	55%	T
		Cold plasma treatment	21.66%	ST
A. steynii	1.23 ng/mL	Untreated	76.66%	T
		Cold plasma treatment	33.33%	ST
A. niger	1.57 ng/mL	Untreated	88.33%	T
		Cold plasma treatment	10%	ST
A. versicolor	1.27 ng/mL	Untreated	50%	T
		Cold plasma treatment	16%	ST

OTA toxicity was rated either as nontoxic (NT), slightly toxic (ST), toxic (T), very toxic (VT).

3. Conclusions

Cold plasma is a novel non-thermal technology that has shown significant potential for applications in food industries for food safety and shelf life extension. The results presented here suggest this method could be a viable option for commercial application in the food industry. The effect of cold plasma treatment on ochratoxin A contained in roasted coffee can be considered promising in terms of safety and process efficiency. However, the application of cold plasma needs further extensive research to reveal the mechanisms of detoxification.

4. Materials and Methods

4.1. Biological Materials

Roasted ground coffee was procured from a local market in Nayarit, Mexico. Coffee samples were autoclaved for 20 min at 121 °C and 15 psi, and dried in a laminar flow hood (CFLV-120, NOVATECH) for 1 h.

OTA producing strains of *A. westerdijikiae* (KP329736.1), *A. steynii* (FM956458.1), *A. niger* (JN226991.1), and *A. versicolor* (NR131277.1) were isolated from roasted coffee (*Coffea arabica* L.) from a coffee-growing region of Nayarit, Mexico [19]. Individual fungal cultures were grown on potato dextrose agar (PDA) for 7 days at 28 °C. Spore suspension was prepared by flooding the mycelia surface with 10 mL of sterile distilled water. The suspension was filtered and the spores counted using a Neubauer camera. The suspension was adjusted to a final value of 5×10^5 spores mL^{-1}.

4.2. Plasma Generator

The plasma apparatus was designed and built at the Laboratory of Plasma Physics of the National Nuclear Research Institute in Toluca, Mexico. The plasma reactor was described by Solis-Pacheco et al. [12].

4.3. Inhibition of Toxigenic Fungi Using Cold Plasma Treatment on Roasted Coffee

Samples of 0.5 g of sterilized roasted coffee were inoculated with 5 µL of a spore's suspension (5×10^5 spores g^{-1}) and rested for 30 min to facilitate proper attachment of spores on coffee particles. Samples were allowed to dry in a laminar flow hood. Afterwards, the samples were exposed to plasma using commercial helium gas (Praxair, Mexico) at a constant flow of 1.5 L/min, applied at 850 volts for different periods of time (0, 1, 2, 4, 5, 6, 8, 10, 12, 14, 16, and 18 min). Before and after plasma treatment, the enumeration of fungi (log CFU/g) were determined. The samples were diluted in 1 mL of 0.1% peptone water and homogenized for 10 s. Decimal dilutions were performed and pour plated with potato dextrose agar (PDA, Bioxon, Mexico) containing 2% ampicillin (Bayer, México) and 0.6% Bengal rose (Analytyka, Mexico). All plates were incubated at 28 °C for 48 h. Fungi colonies were enumerated and results reported in log CFU/g of sample. Each experiment was conducted in triplicate, and the whole analysis was performed in duplicate.

4.4. OTA Production by Mycotoxigenic Fungi and Detoxification by Cold Plasma

Isolated fungi were inoculated in the center of a plate containing PDA medium and incubated at 27 °C for 5 days. One gram of roasted coffee was added to 3 mL of distilled water and autoclaved for 15 min at 121 °C and 15 psi. After cooling, samples were inoculated with 3 agar disks taken from the periphery of the colony [20]. Once inoculated, the roasted coffee samples were incubated at 27 °C for 21 days. The exposure times of cold plasma for detoxification of OTA in the samples of coffee were 0, 1, 4, 8, 12, 16, 20, 24, and 30 min, using commercial helium gas at 1.5 L/min, 30 W input power, and a voltage of 850 V.

4.5. OTA Detection by HPLC-MS

At the end of the plasma treatment, extraction was carried out in an alkaline solvent (pH 9.4) consisting of 3% sodium bicarbonate in methanol/water (20/80, *v/v*), and the suspension was homogenized for 50 min at 60 °C. The extract was centrifuged (model EBA 21, Heittich) at 1036.39× *g* for 5 min. Ten milliliters of filtered extract were diluted in 40 mL phosphate buffer (pH 7.0, 0.1 M). The mixture was purified using an immunoaffinity column (Ochrastar R) purchased from Romer Labs (Tulin, Austria). OTA was eluted by 6 mL of methanol and evaporated to dryness in a rotary evaporator [21]. The residue was re-suspended in 1 mL of the mobile phase (water/acetonitrile, 50:50). Quantification was done by HPLC-MS (Agilent Technologies). The operating conditions were as follows: mobile phase of 5 mM ammonium acetate/acetonitrile (65/35, *v/v*) at 40 °C with a flow rate of 0.2 mL/min for 4 min run. A column Agilent Technologies Zorbax SB-C18 (2.1 × 50 mm id: 1.8 µm) was used. The detection of OTA was performed in a HPLC-MS system model 6120 series LC quadrupole equipped with an electrospray ion (ESI) MS (Agilent Technologies) in ionization mode and positive mode (capillary 3500 V, nebulizer 25 psi (nitrogen), under dry gas nitrogen at 9 L/min, at 350 °C, and the fragmentor voltage set at 95 V; ion monitoring SIM selected, *m/z* 404 (OTA). A standard curve prepared with different OTA concentrations (1, 3, 5, 7, and 21 µg/mL) in acetonitrile was used to determine OTA in coffee samples (Figure 2).

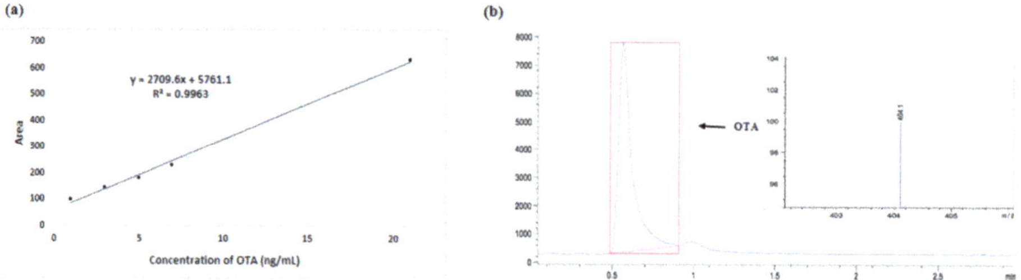

Figure 2. OTA standard curve (1–21 ng/mL) (**a**), OTA chromatogram (0.021 ng/mL) by HPLC-MS (**b**).

4.6. Brine Shrimp Bioassay

Brine shrimp eggs (*Artemia salina*) were incubated in artificial seawater illuminated by artificial light (60 W lamp) and gently aerated for 24 h at 28 °C. Afterwards, the nauplii (larvae) were collected by pipette from the lighted side, whereas their shells were left in the dark side. The test samples (extract from roasted coffee inoculated with toxigenic fungi) were prepared in acetonitrile-water (50:50). The actual concentrations of the whole solutions were determined by HPLC-MS analyses (quantitation based on OTA as internal standard) and considered with the data evaluation. Filter paper disks (d = 0.5 cm) were submerged in microtubes containing 30 µL of OTA extracts and air-dried. Then, each disc was transferred into a 6-well microtitrate plate containing artificial sea water (2 mL) and 10 nauplii; plates were maintained at room temperature for 48 h under light and dead larvae were counted by examination under a binocular stereoscope. Discs submerged in acetonitrile-water solvent (50:50) were used as controls. Mortality was determined according to Equation (1).

$$\% \text{ deaths} = \frac{\text{Dead nauplii in the biossay}}{\text{Nauplii used in the biossay} - \text{Dead nauplii in the control}} \times 100 \tag{1}$$

OTA toxicity was rated as follows: 0–9% mortality, nontoxic (NT); 10–49% mortality, slightly toxic (ST); 50–89% mortality, toxic (T); 90–100% mortality, very toxic (VT) [22].

4.7. Statistical Analysis

Statistical analysis was performed using STATISTICA version 9.0. Analysis of variance (ANOVA) and multiple comparison procedures (least significant difference–LSD) test were conducted to determine whether there were significant differences ($p < 0.05$) among treatments for spores' inhibition or OTA detoxification.

Supplementary Materials: The following are available online at http://www.mdpi.com/2072-6651/11/6/337/s1, Table S1: raw data.

Author Contributions: P.P.C.-J. performed the experiment and wrote the manuscript. J.R.S.-P. contributed to supervision and experiments design. J.A.R.-S. contributed to data analysis and the manuscript writing. B.R.A.-U. contributed to supervision and manuscript review. P.U.B.-R. contributed to supervision. M.C.-S. contributed to research design, supervision, and manuscript revision and final editing.

Funding: This research was funded by Tecnologico Nacional de Mexico through the project 5851.16-P and CONAHCYT through the fellowship 374701 granted to P. P. Casas Junco.

Conflicts of Interest: The authors declare no conflict of interest. The funders had no role in the design of the study; in the collection, analyses, or interpretation of data; in the writing of the manuscript, or in the decision to publish the results.

References

1. Nganou Donkeng, N.; Durand, N.; Tatsadjieu, N.L.; Metayer, I.; Montet, D.; Mbofung, C.M. Fungal llora and ochratoxin a associated with coffee in Cameroon. *Br. Microbiol. Res. J.* **2014**, *4*, 1–17. [CrossRef]

2. Group 2B: Possibly carcinogenic to humans. In *Monographs on the Evaluation of Carcinogenic Risks of Chemicals to Humans*; International Agency for Research on Cancer (IARC): Lyon, France, 1993; Volume 56, pp. 245–395.

3. Tsubouchi, H.; Yamamoto, K.; Hisada, K.; Sakabe, Y.; Udagawa, S.-i. Effect of roasting on ochratoxin A level in green coffee beans inoculated with Aspergillus ochraceus. *Mycopathologia* **1987**, *97*, 111–115. [CrossRef]

4. Il'ichev, Y.V.; Perry, J.L.; Simon, J.D. Interaction of ochratoxin A with human serum albumin. Preferential binding of the dianion and pH effects. *J. Phys. Chem. B* **2002**, *106*, 452–459. [CrossRef]

5. Suárez-Quiroz, M.; Louise, B.D.; Gonzalez-Rios, O.; Barel, M.; Guyot, B.; Schorr-Galindo, S.; Guiraud, J.P. The impact of roasting on the ochratoxin A content of coffee. *Int. J. Food Sci. Technol.* **2005**, *40*, 605–611. [CrossRef]

6. Basaran, P.; Basaran-Akgul, N.; Oksuz, L. Elimination of *Aspergillus parasiticus* from nut surface with low pressure cold plasma (LPCP) treatment. *Food Microbiol.* **2008**, *25*, 626–632. [CrossRef] [PubMed]

7. Pankaj, S.; Keener, K.M. Cold plasma: Background, applications and current trends. *Curr. Opin. Food Sci.* **2017**, *16*, 49–52. [CrossRef]

8. Guo, J.; Huang, K.; Wang, J. Bactericidal effect of various non-thermal plasma agents and the influence of experimental conditions in microbial inactivation: A review. *Food Control* **2015**, *50*, 482–490. [CrossRef]

9. Devi, Y.; Thirumdas, R.; Sarangapani, C.; Deshmukh, R.; Annapure, U. Influence of cold plasma on fungal growth and aflatoxins production on groundnuts. *Food Control* **2017**, *77*, 187–191. [CrossRef]

10. Siciliano, I.; Spadaro, D.; Prelle, A.; Vallauri, D.; Cavallero, M.C.; Garibaldi, A.; Gullino, M.L. Use of cold atmospheric plasma to detoxify hazelnuts from aflatoxins. *Toxins* **2016**, *8*, 125. [CrossRef] [PubMed]

11. Ouf, S.A.; Basher, A.H.; Mohamed, A.A.H. Inhibitory effect of double atmospheric pressure argon cold plasma on spores and mycotoxin production of Aspergillus niger contaminating date palm fruits. *J. Sci. Food Agric.* **2015**, *95*, 3204–3210. [CrossRef] [PubMed]

12. Solís-Pacheco, J.R.; Aguilar-Uscanga, B.R.; Villanueva-Tiburcio, J.E.; Macías-Rodríguez, M.E.; Viveros-Paredes, J.M.; González-Reynoso, O.; Peña-Eguiluz, R. Effect and mechanism of action of non-thermal plasma in the survival of Escherichia coli, *Staphylococcus aureus* and *Saccharomyces cerevisiae*. *J. Microbiol. Biotechnol. Food Sci.* **2017**, *7*, 143–144. [CrossRef]

13. Méndez-Vilas, A. *Microbial Pathogens and Strategies for Combating Them: Science, Technology and Education*; Formatex Research Center: Badajoz, Spain, 2013.

14. Jalili, M.; Jinap, S.; Noranizan, M. Aflatoxins and ochratoxin a reduction in black and white pepper by gamma radiation. *Radiat. Phys. Chem.* **2012**, *81*, 1786–1788. [CrossRef]

15. Moreau, M.; Lescure, G.; Agoulon, A.; Svinareff, P.; Orange, N.; Feuilloley, M. Application of the pulsed light technology to mycotoxin degradation and inactivation. *J. Appl. Toxicol.* **2013**, *33*, 357–363. [CrossRef] [PubMed]

16. Commission Regulation (EC) No 1881/2006 of 19 December 2006 Setting Maximum Levels for Certain Contaminants in Foodstuffs. Available online: https://www.ecolex.org/es/details/legislation/commission-regulation-ec-no-18812006-setting-maximum-levels-for-certain-contaminants-in-foodstuffs-lex-faoc068134/? (accessed on 7 July 2018).

17. Park, B.J.; Takatori, K.; Sugita-Konishi, Y.; Kim, I.-H.; Lee, M.-H.; Han, D.-W.; Chung, K.-H.; Hyun, S.O.; Park, J.-C. Degradation of mycotoxins using microwave-induced argon plasma at atmospheric pressure. *Surf. Coat. technol.* **2007**, *201*, 5733–5737. [CrossRef]

18. Atalla, M.; Hassanein, N.; El-Beih, A.; Youssef, Y. Effect of fluorescent and UV light on mycotoxin production under different relative humidities in wheat grains. *Int. J. Agric. Biol.* **2004**, *6*, 1006.

19. Casas-Junco, P.P.; Ragazzo-Sánchez, J.A.; de Jesus Ascencio-Valle, F.; Calderón-Santoyo, M. Determination of potentially mycotoxigenic fungi in coffee (Coffea arabica L.) from Nayarit. *Food Sci. Biotechnol.* **2018**, *27*, 891–898. [CrossRef] [PubMed]

20. Tournas, V.; Stack, M.; Mislivec, P.; Koch, H.; Bandler, R. Yeasts, Molds and Mycotoxins, Chap. 18. US Food and Drug Administration, Bacteriological Analytical Manual (BAM). pp. 259–268. Available online: http://www.fda.gov/Food/FoodScienceResearch/LaboratoryMethods/ucm071435.htm (accessed on 2 May 2018).

21. Mounjouenpou, P.; Durand, N.; Guyot, B.; Guiraud, J.P. Effect of operating conditions on ochratoxin A extraction from roasted coffee. *Food Addit. Contam.* **2007**, *24*, 730–734. [CrossRef] [PubMed]

22. Harwing, J.; Scott, P.M. Brine shrimp (Artemia salina L.) larvae as a screening system for fungal toxins. *Appl. Environ. Microbiol.* **1971**, *21*, 1011–1016.

 © 2019 by the authors. Licensee MDPI, Basel, Switzerland. This article is an open access article distributed under the terms and conditions of the Creative Commons Attribution (CC BY) license (http://creativecommons.org/licenses/by/4.0/).

Article

Reduction of Deoxynivalenol in Wheat with Superheated Steam and Its Effects on Wheat Quality

Yuanxiao Liu, Mengmeng Li, Ke Bian *, Erqi Guan, Yuanfang Liu and Ying Lu

National Engineering Laboratory for Wheat and Corn Further Processing, Henan University of Technology, Zhengzhou 450001, China
* Correspondence: kebian@haut.edu.cn

Received: 16 May 2019; Accepted: 15 July 2019; Published: 16 July 2019

Abstract: Deoxynivalenol (DON) is the most commonly found mycotoxin in scabbed wheat. In order to reduce the DON concentration in scabbed wheat with superheated steam (SS) and explore the feasibility to use the processed wheat as crisp biscuit materials, wheat kernels were treated with SS to study the effects of SS processing on DON concentration and the quality of wheat. Furthermore, the wheat treated with SS were used to make crisp biscuits and the texture qualities of biscuits were measured. The results showed that DON in wheat kernels could be reduced by SS effectively. Besides, the reduction rate raised significantly with the increase of steam temperature and processing time and it was also affected significantly by steam velocity. The reduction rate in wheat kernels and wheat flour could reach 77.4% and 60.5% respectively. In addition, SS processing might lead to partial denaturation of protein and partial gelatinization of starch, thus affecting the rheological properties of dough and pasting properties of wheat flour. Furthermore, the qualities of crisp biscuits were improved at certain conditions of SS processing.

Keywords: deoxynivalenol; wheat; superheated steam; wheat quality; crisp biscuit

Key Contribution: Deoxynivalenol in wheat could be reduced by SS treatment effectively although the qualities of wheat may be damaged for some degree. The wheat treated with SS could still be used as the materials of crisp biscuits with good quality.

---

1. Introduction

Head blight is a kind of fungal disease affecting wheat plant throughout the world. Both the yield and the quality of wheat are seriously affected by this disease. What's worse, *Fusarium* fungi in wheat grains can produce many kinds of mycotoxins, among which deoxynivalenol (DON) is the most commonly found [1]. In 2017, the DON levels in 48 wheat samples harvested in Brazil range from 1329 to 3937 µg/kg and the average level is 2398 µg/kg [2]. Survey of 54 wheat samples in Southeast Romania in 2009 indicated that 66% of the samples were positive with DON with the average level of 4772 µg/kg [3]. In 2014, a three-year survey of the occurrence of DON in wheat in some regions of Jiangsu Province, China indicated that DON was found in 74.4% of samples at levels ranging from 14.52 to 41,157.13 µg/kg [4]. Besides, a survey of the global occurrence of deoxynivalenol in food commodities and exposure risk assessment in humans in the last decade showed that the problem of DON contamination in wheat is becoming more and more terrible [5]. Since wheat grains are common materials of foods and feeds, DON in wheat may be toxic to humans and animals. High concentration of DON has acute toxic effects, such as vomiting and diarrhea, on humans and animals. Meanwhile, taking in DON for a long period may do harm to the immune system, hematopoietic system, stomach, bone marrow and lymphoid tissues [6]. Scientists all over the world have done much research into the degradation and reduction methods of DON, such as heat treatment [7], activated carbon adsorption [8], ozone fumigation [9], ray irradiation [10], alkali steeping [11] and biodegradation [12]. However, many of these methods are hard to apply in wheat processing up to now. Therefore, it is meaningful to explore new means for reducing DON in wheat and their application in food processing.

Superheated steam (SS) is a kind of steam with temperature above the saturation point of water at given pressure. Compared to hot air, SS has some advantages, including higher drying rate, oxygen-free environment, free of pollution and energy conservation [13]. For these features, SS has been used in the drying of wood [14], paper [15], coal [16], sludge [17], etc. In recent years, researchers have tried to apply SS to the processing of food products. They mainly focus on the drying of foodstuffs, such as potato slices [18], oat groats [19], carrot [20], pork [21], beet, etc. What's more, drying of beet with SS has been applied into some factories in the world [13]. Meanwhile, researchers have also studied the application of SS in food sterilization [22], toxin degradation [23] and enzymatic inactivation [24].

DON can be damaged when temperature is high enough and many studies on the stability of DON during different kinds of heat treatment have been carried out [25–28]. Therefore, SS may be an available way to reduce the content of DON in scabbed wheat. Besides, much of DON can be reduced when wheat kernels are milled into wheat flour. With the combination of wheat milling and SS treatment, a lot of DON in wheat flour may be reduced. Wheat protein may be partially denatured at high temperature, thus leading to the weakening of gluten strength. As a result, the scabbed wheat processed with SS can be milled into low-gluten wheat flour, which is fit to make crisp biscuits and cakes. The objectives of this study were to: (i) investigate the application of SS (with temperature above 185 °C) to reduce the content of DON in scabbed wheat; (ii) explore the effect of SS on wheat quality; (iii) understand the effect of SS on the quality of crisp biscuits.

2. Results and Discussion

2.1. Reduction of Deoxynivalenol

2.1.1. Steam Temperature

In order to study the effect of steam temperature on DON reduction, the other two factors were set as follows: processing time, 6 min; steam velocity, 3 m/s. Both the concentration of DON in wheat kernels and wheat flour reduced significantly with the increase of steam temperature (Figure 1). This was because that the increase of temperature led to the stability weakening of DON. Since the surrounding environment of wheat kernels was oxygen-free during SS processing, oxidation could hardly occur. Therefore, large quantities of heat might be the primary factor leading to DON degradation. Charlene, Wolf and Wolf-Hall has drawn the similar conclusion [29–31]. In all the experiments, the reduction rate of DON in wheat kernels was always higher than that of wheat flour. The highest reduction rate of DON in wheat kernels was 77.4%, while it was only 60.5% for wheat flour. This phenomenon might due to the fact that DON was mainly distributed in the pericarp of wheat grain and heat was transferred from the outer surface to the inner endosperm. As a result, the DON in the endosperm couldn't absorb as much heat as that in the bran of wheat. Since high temperature might do much harm to grain quality, the steam temperature used in practical application should be determined based on the real content of DON in wheat and the changes of wheat quality.

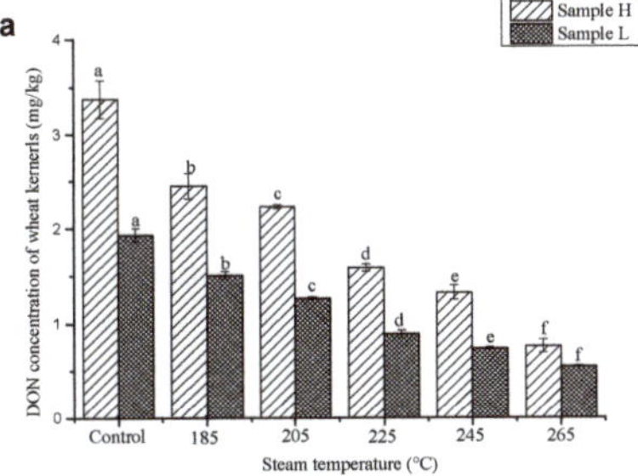
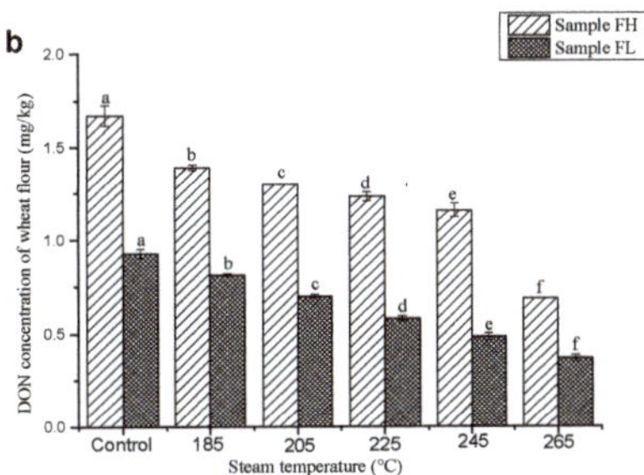

Figure 1. Effect of steam temperature on the contents of DON in wheat kernels (**a**) and wheat flour (**b**) * Sample H: Scabbed wheat sample with relatively higher content of DON; Sample L: Scabbed wheat sample with relatively lower content of DON; FH: Wheat flour milled from Sample H; FL: Wheat flour milled from Sample L. Same designations were used in Figures 2 and 3. Processing time and steam velocity are fixed at 6 min and 3 m/s respectively. Data are given as means of triplicate assays ± SD (standard deviation). Values with different letters on the same kind of bars are significantly different ($p < 0.05$). The bars are labeled by the letters a to f from the highest to the lowest. The same expressing methods are used in Figures 2 and 3.

2.1.2. Processing Time

In order to study the effect of processing time on the reduction of DON, the other two factors were set as follows: steam temperature, 225 °C; steam velocity, 3 m/s. In the initial stage of SS processing, heated steam condensed on the cool surface of wheat kernels. During this stage, some DON on the surface of wheat kernels might be washed away as well as thermally degraded [7]. At different temperature, the quantity of condensed water was different. As a result, the time needed to evaporate the condensation was also different [32,33]. After the above stage, the temperature of wheat kernels began to rise and condensed water began to be evaporated. During this course, little heat was transferred into the inner of wheat kernels. As a result, when processed with SS for 2 min, some DON in the pericarp could be degraded, while little DON in wheat flour was reduced (Figure 2). When processing time was increased, considerable heat was transferred into the inner of wheat kernels. As a result, the concentration of DON in wheat and wheat flour decreased significantly and the reduction rate increased notably. Because of the harmful effect of SS on wheat quality, the processing time should be chosen according to the changes of DON content and wheat quality. In most experiments, the reduction of DON in wheat kernels was higher than that of wheat flour. However, when processing time was above 8 min, the gap was narrowed. This result suggested that when processing time was long enough, enough heat could be transferred into the endosperm of wheat.

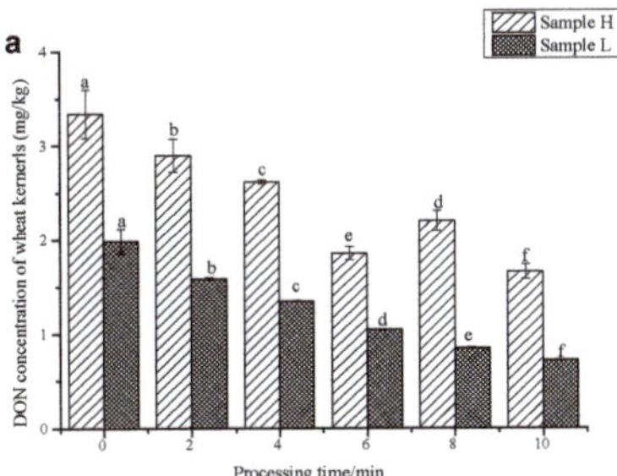
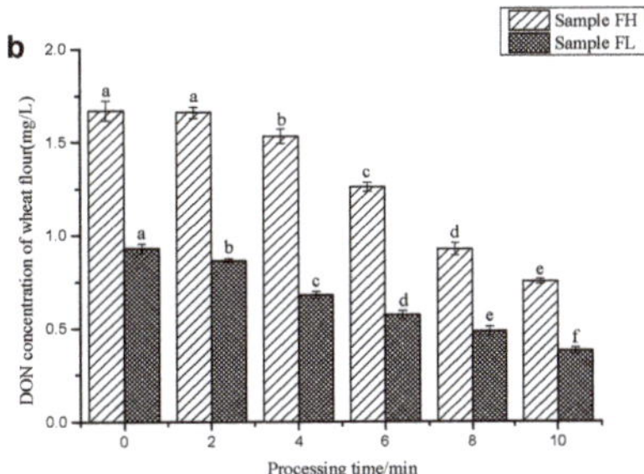

Figure 2. Effect of processing time on the contents of DON in wheat kernels (**a**) and wheat flour (**b**). * Steam temperature and steam velocity are fixed at 225 °C and 3 m/s respectively.

2.1.3. Steam Velocity

In order to study the effect of steam velocity on the reduction of DON, the other two factors were set as follows: steam temperature, 225 °C; processing time, 6 min. Steam velocity had great effects on the heat transfer efficiency of SS, thus affecting the reduction of DON significantly. Different variation trends of DON concentration in wheat kernels and wheat flour were observed when steam velocity was increased. For wheat kernels, when steam velocity was increased from 1 m/s to 3 m/s, the concentration of DON in wheat kernels decreased significantly (Figure 3); when steam velocity was increased from 3 m/s to 5 m/s, the DON concentration increased. However, for wheat flour, DON concentration decreased significantly when steam velocity was increased from 1 m/s to 5 m/s. These results indicated that high velocity of SS might enhance the transfer of heat from the surface of wheat to the inner endosperm. However, when steam velocity was increased to 5 m/s, the reduction rate of DON in wheat flour was much higher than that of wheat kernels. Since more energy will be consumed when steam velocity was increased, both energy conservation and DON content should be taken into consideration when choosing the optimum steam velocity.

The above results showed that steam temperature, processing time and steam velocity all had significant influence on the reduction of DON. Meanwhile, wheat quality may also be affected negatively. Therefore, under the condition that safety is insured, lower steam temperature, less time and suitable steam velocity should be used during SS processing. Compared to the research by Pronky [7], the reduction rate of DON in wheat was increased from 52% to 77.4% by improving the conditions of SS processing in this research. Besides, SS processing might be a more effective method to reduce DON compared to other methods of treatment, such as cooking, baking and extrusion [25–28]. However, no instrument exists now for the processing of wheat with SS in the industry.

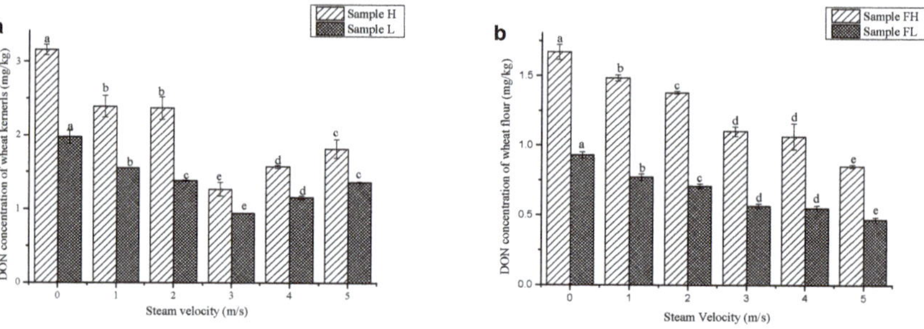

Figure 3. Effect of steam velocity on the contents of DON in wheat kernels (**a**) and wheat flour (**b**). * Steam temperature and processing time are fixed at 225 °C and 6 min respectively.

2.2. Rheological Properties of Dough

Wheat flour milled with Sample L was used to study the effects of SS processing on rheological properties of dough. Rheological properties of dough are closely related to the quality of food made with wheat flour. Rheological properties obtained from Chopin Mixolab are positively correlated with those obtained from Brabender Farinograph [34]. Therefore, rheological qualities determined by Chopin Mixolab can reflect the quality of wheat flour very well. Water absorption increased significantly with the increase of steam temperature, processing time and steam velocity (Table 1). This might due to the increased content of damaged starch in wheat caused by heat during SS processing [35]. For all the samples, the development time and stability decreased significantly when wheat kernels were processed with SS. This phenomenon resulted from the denaturation of gliadin and glutenin in wheat at high temperature, which made them unable to form gluten network with enough strength. Besides, SS processing also led to the aggregation of gluten proteins, which decreased the strength of dough obviously [36]. Sulfydryl and disulfide bond in wheat protein are the two most important groups for the formation of gluten matrix. However, they were easy to be changed during SS processing. Firstly,

SS treatment led to the reduction of free sulfydryl and the cross-linking of glutenin [37]. Besides, SS processing affected the hydratability of wheat protein, thus changing the water absorption and the development of dough [38]. In a word, the destructive effect of SS processing on wheat protein led to the shortening of development time and stability time of dough. As a result, the qualities of crisp biscuits might be affected. The detailed contents will be discussed latterly.

Table 1. Effect of steam temperature on rheological properties of dough.

Processing Condition		Water Absorption/%	Development Time/min	Stability /min
	Control	57.5 ± 0.1^f	3.01 ± 0.04^a	6.29 ± 0.11^a
Steam temperature (°C)	185	61.5 ± 0.1^e	2.17 ± 0.06^b	2.24 ± 0.11^b
	205	65.2 ± 0.2^d	2.06 ± 0.06^b	2.18 ± 0.06^b
	225	66.7 ± 0.1^c	1.87 ± 0.03^c	1.95 ± 0.08^c
	245	71.9 ± 0.1^b	1.90 ± 0.03^c	1.86 ± 0.03^c
	265	74.3 ± 0.1^a	2.08 ± 0.04^b	1.83 ± 0.03^c
Processing time (min)	2	62.2 ± 0.1^e	2.50 ± 0.04^b	2.28 ± 0.04^b
	4	65.5 ± 0.1^d	2.10 ± 0.05^c	2.16 ± 0.04^{bc}
	6	68.1 ± 0.1^c	1.94 ± 0.04^d	2.15 ± 0.07^{bc}
	8	70.4 ± 0.1^b	2.01 ± 0.01^{cd}	2.02 ± 0.01^c
	10	72.3 ± 0.1^a	2.03 ± 0.02^c	2.01 ± 0.03^c
Steam velocity (m/s)	1	62.2 ± 0.1^e	2.21 ± 0.04^b	2.02 ± 0.03^d
	2	63.6 ± 0.2^d	2.09 ± 0.03^c	2.18 ± 0.08^c
	3	67.7 ± 0.1^c	1.95 ± 0.04^d	2.30 ± 0.04^{bc}
	4	70.7 ± 0.1^b	2.08 ± 0.08^c	2.33 ± 0.02^{bc}
	5	75.6 ± 0.2^a	2.25 ± 0.03^b	2.38 ± 0.04^b

Data are given as means of triplicate assays ± SD. The values are labeled by the letters a to f from the maximum to the minimum. Values with different letters in the same column are significantly different ($p < 0.05$).

2.3. Pasting Properties of Wheat Flour

Wheat flour milled with Sample L was used to study the effects of SS processing on pasting properties of wheat flour. Starch, which accounts for about 75%–80% of wheat endosperm, is the most abundant component in wheat flour. Pasting properties can reflect the characteristics of starch in wheat flour. The SS treatment of wheat kernels might lead to starch gelatinization, retrogradation and the generation of damaged starch, thus changing the pasting properties of wheat flour. Analyzing Table 2, the following results could be found. All the pasting properties of flour treated with SS were higher than those of the control sample. With the increase of steam temperature, peak viscosity (PV), hold viscosity (HV), breakdown (BD), final viscosity (FV), setback (SB) and time to peak (TP) showed trends of decreasing. However, pasting temperature (PT) didn't change significantly. When processing time was increased from 2 min to 10 min, PV, HV, BD, FV, SB and TP showed trends of decreasing while PT didn't change significantly. Besides, the PT at every processing time was higher than that of the untreated sample. With the increase of steam velocity all the parameters showed trends of decreasing except PT while PT didn't change significantly.

The changes of pasting properties mainly resulted from the gelatinization and retrogradation of wheat starch and damaged starch caused by SS treatment [35,39]. Besides, the hydrophobicity of wheat albumin, globulin and starch were modified during SS processing [40,41]. The above changes might finally increase the water absorbing capacity of starch granules, which led to the increase of PV, TP and PT [42]. The damaged starch in wheat flour might lead to the decrease of PV, HV, BD, FV, SB and PT [43–45]. In addition, the activity of α-amylase were damaged for some degree during SS processing, thus bringing about the decrease of BD [46].

Table 2. Effect of steam temperature on pasting properties of wheat flour.

Processing Condition		PV/mPa·s	HV/mPa·s	BD/mPa·s	FV/mPa·s	SB/mPa·s	TP/min	PT/°C
	Control	2707 ± 11 [c]	1797 ± 23 [d]	910 ± 23 [a]	3026 ± 34 [c]	1229 ± 11 [b]	6.34 ± 0.09 [a]	67.7 ± 0.1 [b]
Steam temperature (°C)	185	3281 ± 64 [a]	2390 ± 69 [a]	891 ± 4 [a]	3874 ± 49 [a]	1484 ± 20 [a]	6.20 ± 0.10 [ab]	84.4 ± 0.6 [a]
	205	3201 ± 13 [ab]	2284 ± 30 [ab]	917 ± 17 [a]	3857 ± 21 [a]	1574 ± 9 [a]	6.13 ± 0.02 [bc]	84.4 ± 0.5 [a]
	225	3032 ± 41 [b]	2247 ± 17 [b]	785 ± 25 [b]	3818 ± 13 [a]	1571 ± 30 [a]	6.00 ± 0.10 [bc]	84.0 ± 1.2 [a]
	245	2835 ± 112 [c]	2203 ± 48 [b]	632 ± 64 [c]	3663 ± 122 [a]	1460 ± 74 [a]	5.97 ± 0.05 [c]	83.6 ± 0.6 [a]
	265	2486 ± 134 [d]	1974 ± 81 [c]	512 ± 53 [d]	3275 ± 165 [b]	1301 ± 84 [b]	5.94 ± 0.09 [c]	84.0 ± 0.1 [a]
Processing time (min)	2	3214 ± 2 [a]	2339 ± 21 [a]	875 ± 23 [ab]	3880 ± 4 [a]	1541 ± 24 [a]	6.17 ± 0.05 [ab]	83.9 ± 1.2 [a]
	4	3217 ± 26 [a]	2377 ± 18 [a]	840 ± 8 [b]	3924 ± 43 [a]	1547 ± 25 [a]	6.17 ± 0.05 [ab]	84.4 ± 0.5 [a]
	6	2961 ± 30 [b]	2244 ± 9 [b]	717 ± 40 [c]	3652 ± 88 [b]	1408 ± 79 [b]	6.07 ± 0.09 [b]	83.9 ± 1.2 [a]
	8	2567 ± 21 [d]	2005 ± 23 [c]	563 ± 2 [d]	3414 ± 25 [c]	1410 ± 49 [b]	5.80 ± 0.10 [c]	83.5 ± 0.5 [a]
	10	1944 ± 10 [e]	1533 ± 1 [e]	412 ± 11 [e]	2576 ± 4 [e]	1044 ± 5 [d]	5.57 ± 0.05 [d]	84.0 ± 0.1 [a]
Steam velocity (m/s)	1	3255 ± 38 [a]	2345 ± 17 [a]	910 ± 34 [a]	3842 ± 64 [a]	1497 ± 62 [a]	6.11 ± 0.03 [b]	84.2 ± 0.5 [a]
	2	3162 ± 12 [b]	2286 ± 34 [ab]	876 ± 40 [a]	3811 ± 72 [ab]	1526 ± 97 [a]	6.09 ± 0.08 [bc]	84.2 ± 0.9 [a]
	3	2966 ± 6 [c]	2225 ± 51 [b]	741 ± 57 [b]	3723 ± 26 [b]	1498 ± 77 [a]	6.07 ± 0.09 [bc]	84.4 ± 0.5 [a]
	4	2707 ± 17 [d]	2115 ± 31 [c]	592 ± 52 [c]	3506 ± 3 [c]	1391 ± 32 [ab]	5.97 ± 0.14 [bc]	84.4 ± 0.6 [a]
	5	2493 ± 23 [e]	2008 ± 10 [d]	485 ± 13 [d]	3282 ± 10 [d]	1274 ± 1 [b]	5.90 ± 0.04 [c]	84.0 ± 0.1 [a]

PV, peak viscosity: the highest viscosity of sample after heating up and before cooling down; HV, hold viscosity: the minimum viscosity of sample during cooling down; BD, breakdown: the value of PV minus HV; FV, final viscosity: the viscosity of sample at the ending of test; SB, setback: the value of FV minus HV; TP, time to peak: the time when the viscosity reaches PV; PT, pasting temperature: the time when the viscosity of sample begins to increase. Data are given as means of triplicate assays ± SD. The values are labeled by the letters a to f from the maximum to the minimum. Values with different letters in the same column are significantly different ($p < 0.05$).

2.4. Quality of Crisp Biscuits

Crisp biscuits made with Sample L were used to study the effects of SS processing on the qualities of crisp biscuits. Although the qualities of wheat were affected by SS greatly, the wheat processed by SS was still suitable for making some kinds of food products, such as biscuits, cakes and pastries. In this study, crisp biscuits were chosen as the final products to study whether the wheats processed by SS were suitable for making food products or not.

The effects of steam temperature on the qualities of biscuits were as follows (Tables 3 and 4). The hardness and working value of biscuits showed trends of decreasing with the increase of steam temperature, although they decreased non-significantly during some ranges. The above results indicated that the increase of steam temperature improved the crisping of crisp biscuits. However, when the temperature excessed 225 °C, the color of biscuit surfaces began to darken and the number of cracks increased.

The effects of processing time on the qualities of biscuits were as follows (Tables 3 and 5). The hardness and working value of biscuits showed trends of decreasing with the increase of processing time. However, when processing time was too long (≥8 min), the color of biscuit surfaces began to darken and the number of cracks increased.

The effects of steam velocity on the qualities of biscuits were as follows (Tables 3 and 6). The hardness and working value of biscuits showed trends of decreasing with the increase of steam velocity, although they decreased non-significantly during some ranges. However, when steam velocity reached 4 m/s, the color of biscuit surfaces began to darken and the number of cracks increased.

The variations of biscuit quality were closely related to the changes of dough rheology and flour pasting properties. To some degree, the shortening of dough stability time made the wheat flour more suitable for making crisp biscuits. However, short time of stability was still needed. Besides, SS processing of wheat kernels at high temperature or for long time might result in the color change of wheat flour, which had negative effects on biscuit color. Therefore, suitable conditions for processing wheat kernels should be considered to avoid the deterioration of biscuit quality to the greatest extent.

Table 3. Effect of steam temperature on texture properties of biscuits.

Processing Condition		Hardness/g	Working Value /g·s
	Control	2101 ± 186 [a]	3896 ± 152 [a]
Steam temperature (°C)	185	2093 ± 90 [a]	3859 ± 22 [a]
	205	1823 ± 95 [b]	1942 ± 253 [b]
	225	1552 ± 94 [c]	1516 ± 117 [bc]
	245	1466 ± 112 [cd]	1213 ± 88 [c]
	265	1253 ± 78 [d]	1175 ± 69 [c]
Processing time (min)	2	1708 ± 26 [b]	2657 ± 178 [b]
	4	1563 ± 38 [bc]	2325 ± 60 [bc]
	6	1396 ± 105 [cd]	1750 ± 20 [cd]
	8	1295 ± 98 [d]	1623 ± 14 [d]
	10	1497 ± 87 [bcd]	1425 ± 30 [d]
Steam velocity (m/s)	1	1898 ± 166 [a]	2556 ± 129 [b]
	2	1879 ± 3 [ab]	2009 ± 111b [cd]
	3	1656 ± 68 [bc]	2377 ± 6 [bc]
	4	1482 ± 76 [cd]	1744 ± 1 [cd]
	5	1254 ± 103 [d]	1505 ± 6 [d]

Data are given as means of triplicate assays ± SD. The values are labeled by the letters a to f from the maximum to the minimum. Values with different letters in the same column are significantly different ($p < 0.05$).

Table 4. Effect of steam temperature on apparent condition of biscuits.

Steam Temperature/°C	Control	185	205	225	245	265
Apparent condition						

Table 5. Effect of processing time on apparent condition of biscuits.

Processing Time/min	Control	2	4	6	8	10
Apparent condition						

Table 6. Effect of steam velocity on apparent condition of biscuits.

Steam Velocity/(m/s)	Control	1	2	3	4	5
Apparent condition						

3. Conclusions

SS had proven to be an effective method to reduce the content of DON in wheat and wheat flour. DON on the surface of wheat kernels could be washed as well as thermally degraded, while DON in the inner of wheat kernels could be reduced by thermal degradation. With the increase of steam temperature and processing time, the reduction rate increased significantly. Stem velocity also had significant influence on the reduction rate. In this research, the reduction rate of DON could reach 77.4% for wheat kernels, while it was 60.4% for wheat flour. Besides reducing DON in wheat, superheated steam also had great effects on wheat quality, thus affecting the qualities of food made from wheat flour. During SS processing, protein denaturation and starch gelatinization occurred in wheat. The degree of protein denaturation and starch gelatinization depended on the conditions of SS processing. As a result, the water absorption of dough was increased, while the development time and stability were decreased. In addition, the pasting properties of wheat flour were affected significantly due to the increase of damaged starch, the gelatinization of starch and the passivation of α-amylase. What's more, the qualities of crisp biscuits made with SS-processed wheat could be improved under certain conditions. Last but not least, if we want to use SS treatment in the industry, we just need to combine a steam generator with a heater to produce SS and the residual can be recycled by a centrifuge fan. The device is quite easy to design and manufacture. Therefore, for scabbed wheats, superheated steam processing could be an effect way to reduce DON as well as improving the qualities of crisp biscuits made from processed wheat and it is likely to be used in practical production. In further research, we will aim at the analysis of DON degradation products and the toxicity of degradation products. Besides, the mechanism of the quality change of wheat will also be studied further.

4. Materials and Methods

4.1. Materials and Reagents

Scabbed wheat with about 3.8 mg/kg of DON (Sample H) was harvested on a farmland of Xinyang City, Henan Province. Scabbed wheat with about 2.3 mg/kg of DON (Sample L) was harvested on a farmland of Nanyang City, Henan Province. All samples were cleaned and stored at 10 °C and no more than 50% of relative humidity before experiments.

4.2. Determination of Moisture Content

Moisture content of wheat and wheat flour were tested according to AACC (American Association of Cereal Chemists) Method 44-15A.

4.3. Tempering of Wheat

Tempering of wheat was conducted according to AACC Method 26-10A.

4.4. Superheated Steam Processing

The power was connected and the device was turned on in advance. When beginning the experiment, 200 g of wheat kernels were placed on the screen mesh of a sieve with 1.98 mm of aperture. Steam temperature was changed from 185 °C to 265 °C (185 °C, 205 °C, 225 °C, 245 °C and 265 °C) to study the effect of steam temperature on experimental results. Processing time was changed from 2 min to 10 min (2 min, 4 min, 6 min, 8 min and 10 min) to study the effect of processing time on experimental results. Steam velocity was changed from 1 m/s to 5 m/s (1 m/s, 2 m/s, 3m/s, 4 m/s and 5 m/s) to study the effect of steam velocity on experimental results. Put the screen mesh with wheat kernels into the processing chamber of SS device (Figure 4). During processing, steam passed through the wheat kernels in the processing chamber. At the beginning of processing, steam condensed on the surface of wheat kernels and then the condensed water and the water in wheat kernels was evaporated. When the processing procedure was finished, wheat kernels were moved out of the processing chamber as quickly as possible and weighed immediately. For each condition, about 3000 g of sample is needed which equaled to 15 times of SS processing. The processed wheat kernels were cooled at room temperature on the screen mesh. After cooling, mix the samples processed at the same condition. The processed samples were moved into airtight bags and stored at room temperature for no more than 24 h before smashing and milling. Then, 100g of processed wheat kernels were smashed (4.5) to determine the moisture and DON concentration and others were milled into wheat flour (4.6).

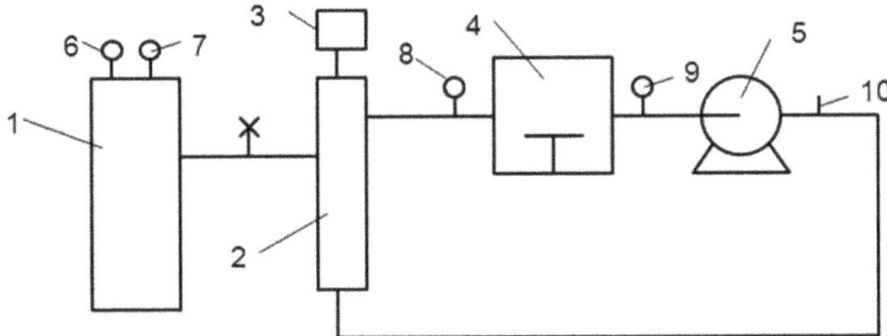

Figure 4. Schematic diagram of superheated steam device (1, Steam generator; 2, Heater; 3, Thermostat; 4, Processing chamber; 5, Centrifugal fan; 6, Pressure gauge; 7, 8, 9, Thermometer; 10, Flowmeter.).

4.5. Wheat Smashing

200 g of wheat kernels were smashed in the high-speed universal grinder for about 45 s. Fractured samples were sifted with a sieve with mesh number of 40 and the undersized parts were collected. The oversized parts were put in the grinder and smashed for 20 s again. Sift the fractured materials with a sieve with mesh number of 40 and collect the undersized parts. Then the undersized parts were mixed and stored at room temperature.

4.6. Wheat Milling

Wheat kernels were milled according to AACC Method 26–31 and Bühler model MLU-202 was used as the milling instrument. The feed rate was adjusted to 80 g/min. The sifters of the flour mill were cleaned thoroughly between of milling of two different samples. After milling, the wheat flour was moved into airtight bags and stored in the refrigerator at 4 °C.

4.7. Extraction and Purification of DON

The extraction and purification of DON were conducted according to the method as follows [9].

Extraction of DON: Weigh 25 g of smashed wheat (or 20 g of wheat flour) in a beaker. 100 mL of extracting solution ($V_{acetonitrile}/V_{water}$ = 84/16) was added into the beaker. After then, the sample and solution were stirred with a magnetic stirrer for 20 min.

Purification of DON: After the above steps, keep the beaker still for at least 5 min, and then 5 mL of the supernatant was transferred into a SPE (Solid Phase Extraction) column to wipe off some of the impurities in the DON solution. 4 mL of the purified solution was transferred into a tube and dried with nitrogen. The residue was re-dissolved with 2 mL of mobile phase ($V_{acetonitrile}/V_{water}$ = 16/84). All the re-dissolved solution was transferred into a centrifuge tube and centrifuged at the speed of 12,000 r/min. Then, the supernatant was transferred into a vial through a filter membrane with 0.22 μm of pore size. All the samples were stored at 4 °C in a dark environment.

4.8. Preparation of Standard Solution of DON

Stock solution of DON (100 mg/L, 1 mL) was transferred into a test tube (10 mL). The solution was dried with nitrogen gas and the remained DON was re-dissolved with mobile phase ($V_{acetonitrile}/V_{water}$ = 16/84). The DON solution was diluted successively to get DON solution with the following concentration: 0.1 mg/L, 0.2 mg/L, 0.5 mg/L, 1 mg/L, 2 mg/L, 5 mg/L and 10 mg/L.

4.9. HPLC Analysis

DON contents were determined by high performance liquid chromatography (HPLC) with Symmetry C18 column (250 mm × 4.6 mm i.d., 5-µm particle size) and ultraviolet (UV) detector. The HPLC conditions were optimized based on the method of Cui [47]. The mobile phase was prepared from acetonitrile and water ($V_{acetonitrile}/V_{water}$ = 16/84). The flow rate was 600 µL/min. The absorbance of DON at 218 nm was monitored and the full running time was 10 min. The retention time (RT) of DON was about 7.6 min and the limit of quantification (LOD) for DON was 0.03 µg/mL.

4.10. Determination of Dough Rheological Properties

Rheological properties were analyzed with Chopin Mixolab and AACC Method 54–60 was used as reference.

4.11. Analysis of Pasting Properties

Pasting properties were determined with Rapid Viscosity Analyzer (RVA) according to AACC Method 76-21.

4.12. Manufacturing of Crisp Biscuits

The method of making crisp biscuits was modified according to LS/T 3206-1993. The detailed steps were as follows: Powdered sugar (28.5 g), sodium bicarbonate (0.6 g), and salt (0.3 g) were mixed in 5 mL of water and stirred to accelerate dissolution. 33 g of shortening was melted by hot-water bath and added into the above solution. The above ingredients were stirred until they formed emulsion. 100 g of wheat flour was added into the emulsion. Then all the ingredients were mixed by dough mixer to accelerate the formation of dough. The dough was kept still for nearly 10 min until it was not sticky. After this, the dough was tableted until the thickness reached about 3.5 mm–4 mm. Then it was molded and baked for 10 min.

4.13. Texture Analysis

Texture properties of biscuits were determined with TA-XT2i texture analyzer (STABLE MICRO SYSTEM). The main parameters were set as follows: 5 mm/s of pre-test speed, 1 mm/s of test speed, 4 mm of target displacement and 5 g of trigger force.

4.14. Apparent Conditions of Biscuits

Crisp biscuits were taken photos of to observe the apparent conditions of crisp biscuits.

4.15. Statistical Analysis

Data are shown as the mean ± SD of at least three parallel experiments. One-way ANOVA (analyses of variance) approach was used to compare values among more than two different experimental groups. Duncan multiple comparison method was used and p values less than 0.05 were considered statistically significant. Statistics were analyzed with SPSS 16.0 (SPSS Institute, CHI, USA). All the diagrams were drawn with Origin 8.5 (OriginLab Corporation, MA, USA).

Author Contributions: Author contributions: conceptualization, K.B. and E.G.; methodology, K.B., E.G. and Y.L.; software, Y.L.; validation, K.B.; formal analysis: Y.L.; investigation: Y.L. and M.L. data curation, Y.L. and M.L.; supervision, K.B.; funding acquisition, K.B.; writing—original draft preparation, Y.L.; writing—review and editing, M.L., K.B. and Y.L.; funding acquisition, K.B.

Funding: This research was funded by the National Modern Agricultural Industry Technology System Construction Program under Grant [grant numbers: CARS-03], National Natural Science Foundation of China (NSFC) grant numbers: 31801654], Scientific Research Fund of Henan University of Technology [grant numbers: 2017QNJH15] and Major Science and Technology Project of Henan province [grant numbers: 141100110900].

Conflicts of Interest: The authors declare no conflicts of interest. The funders had no role in the design of the study; in the collection, analyses, or interpretation of data; in the writing of the manuscript, or in the decision to publish the results.

References

1. Marin, S.; Ramos, A.J.; Cano-Sancho, G.; Sanchis, V. Mycotoxins: Occurrence, toxicology, and exposure assessment. *Food Chem. Toxicol.* **2013**, *60*, 218–237. [CrossRef] [PubMed]
2. Rocha, D.F.D.L.; Oliveira, M.D.S.; Furlong, E.B.; Junges, A.; Paroul, N.; Valduga, E.; Toniazzo, G.B.; Zeni, J.; Cansian, R.L. Evaluation of the TLC quantification method and occurrence of deoxynivalenol in wheat flour of Southern Brazil. *Food Addit. Contam. Part A* **2017**. [CrossRef] [PubMed]
3. Tabuc, C.; Marin, D.; Guerre, P.; Sesan, T.; Bailly, J.D. Moldsand mycotoxin content of cereals in southeastern Romania. *J. Food Prot.* **2009**, *72*, 662–665. [CrossRef] [PubMed]
4. Ji, F.; Xu, J.H.; Liu, X.; Yin, X.C.; Shi, J.R. Natural occurrence of deoxynivalenol and zearalenone in wheat from Jiangsu province, China. *Food Chem.* **2014**, *157*, 393–397. [CrossRef] [PubMed]
5. Mishra, S.; Srivastava, S.; Dewangan, J.; Dikakar, A.; Rath, S.K. Global occurrence of deoxynivalenol in food commodities and exposure risk assessment in humans in the last decade: A survey. *Crit. Rev. Food Sci. Nutr.* **2019**, *14*, 1–29. [CrossRef] [PubMed]
6. Larsen, J.C.; Hunt, J.; Perrin, I.; Ruckenbauer, P. Workshop on trichothecenes with a focus on DON: Summary report. *Toxicol. Lett.* **2004**, *153*, 1–22. [CrossRef]
7. Pronyk, C.; Cenkowski, S.; Abramson, D. Superheated steam reduction of deoxynivalenol in naturally contaminated wheat kernels. *Food Control* **2006**, *17*, 789–796. [CrossRef]
8. Huwig, A.; Freimund, S.; Käppeli, O.; Dulter, H. Mycotoxin detoxication of animal feed by different adsorbents. *Toxicol. Lett.* **2001**, *122*, 179–188. [CrossRef]
9. Li, M.M.; Guan, E.Q.; Bian, K. Effect of ozone treatment on deoxynivalenol and quality evaluation of ozonised wheat. *Food Addit. Contam. Part A* **2015**, *32*, 544–553. [CrossRef]
10. O'Neill, K.; Damoglou, A.P.; Patterson, M.F. The stability of deoxynivalenol and 3-acetyl deoxynivalenol to gamma irradiation. *Food Addit. Contam.* **1993**, *10*, 209–215. [CrossRef]
11. Young, J.C. Reduction in levels of deoxynivalenol in contaminated corn by chemical and physical treatment. *J. Agric. Food Chem.* **1986**, *34*, 465–467. [CrossRef]
12. Shima, J.; Takase, S.; Takahashi, Y.; Iwai, Y.; Fujimoto, H.; Yamazaki, M.; Ochi, K. Novel detoxification of the trichothecene mycotoxin deoxynivalenol by a soil bacterium isolated by enrichment culture. *Appl. Environ. Microbiol.* **1997**, *63*, 3825–3830. [PubMed]
13. Sehrawat, R.; Nema, P.K.; Kaur, B.P. Effect of superheated steam drying on properties of foodstuffs and kinetic modeling. *Innov. Food Sci. Emerg. Technol.* **2016**, *34*, 285–301. [CrossRef]
14. Johansson, A.; Fyhr, C.; Rasmuson, A. High temperature convective drying of wood chips with air and superheated steam. *Int. J. Heat Mass Transfer* **1997**, *40*, 2843–2858. [CrossRef]
15. Douglas, W.J.M. Drying Paper in Superheated Steam. *Dry. Technol.* **1994**, *12*, 1341–1355. [CrossRef]
16. Stokie, D.; Meng, W.W.; Bhattacharya, S. Comparison of Superheated Steam and Air Fluidized-Bed Drying Characteristics of Victorian Brown Coals. *Energy Fuels* **2013**, *27*, 6598–6606. [CrossRef]
17. Hoadley, A.F.; Qi, Y.; Nguyen, T.; Hapgood, K.; Desai, D.; Pinches, D. A field study of lignite as a drying aid in the superheated steam drying of anaerobically digested sludge. *Water Res.* **2015**, *82*, 58–65. [CrossRef]
18. Iyota, H.; Nishimura, N.; Onuma, T.; Nomura, T. Drying of Sliced Raw Potatoes in Superheated Steam and Hot Air. *Dry. Technol.* **2001**, *19*, 1411–1424. [CrossRef]
19. Head, D.S.; Cenkowski, S.; Arntfield, S.; Henderson, K. Superheated steam processing of oat groats. *Food Sci. Technol.* **2010**, *43*, 690–694. [CrossRef]
20. Tatemoto, Y.; Michikoshi, T. Drying Characteristics of Carrots Immersed in a Fluidized Bed of Fluidizing Particles Under Reduced Pressure. *Dry. Technol.* **2014**, *32*, 1082–1090. [CrossRef]
21. Sa-Adchom, P.; Swasdisevi, T.; Nathakaranakule, A.; Soponronnarit, S. Mathematical model of pork slice drying using superheated steam. *J. Food Eng.* **2011**, *104*, 499–507. [CrossRef]
22. Hu, Y.M.; Nie, W.; Hu, X.Z.; Li, Z.G. Microbial decontamination of wheat grain with superheated steam. *Food Control* **2016**, *62*, 264–269. [CrossRef]

23. Cenkowski, S.; Pronyk, C.; Zmidzinska, D.; Muir, W.E. Decontamination of food products with superheated steam. *J. Food Eng.* **2007**, *83*, 68–75. [CrossRef]

24. Prachayawarakorn, S.; Prachayawasin, P.; Soponronnarit, S. Heating process of soybean using hot-air and superheated-steam fluidized-bed dryers. *Food Sci. Technol.* **2006**, *39*, 770–778. [CrossRef]

25. Kushiro, M. Effects of milling and cooking processes on the deoxynivalenol content in wheat. *Int. J. Mol. Sci.* **2008**, *9*, 2127–2145. [CrossRef] [PubMed]

26. Scott, P.M.; Kanhere, S.R.; Lau, P.Y.; Dexer, J.E.; Greenhalgh, R. Effects of experimental flour milling and breadbaking on retention of deoxynivalenol (vomitoxin) in hard red spring wheat. *Cereal Chem.* **1984**, *60*, 421–424.

27. Yumbe-Guevara, B.E.; Imoto, T.; Yoshizawa, T. Effects of heating procedures on deoxynivalenol, nivalenol and zearalenone levels in naturally contaminated barley and wheat. *Food Addit. Contam. Part A* **2003**, *20*, 1132–1140. [CrossRef]

28. Accerbi, M.; Rinaldi, V.E.; Ng, P.K. Utilization of Highly Deoxynivalenol-Contaminated Wheat via Extrusion Processing. *J. Food Prot.* **1999**, *62*, 1485. [CrossRef]

29. Wolf, C.E.; Bullerman, L.B. Heat and pH alter the concentration of deoxynivalenol in an aqueous environment. *J. Food Prot.* **1998**, *61*, 365–367. [CrossRef]

30. Vidal, A.; Sanchis, V.; Ramos, A.J. Thermal stability and kinetics of degradation of deoxynivalenol, deoxynivalenol conjugates and ochratoxin A during baking of wheat bakery products. *Food Chem.* **2015**, *178*, 276–286. [CrossRef]

31. Wolf-Hall, C.E.; Hanna, M.A.; Bullerman, L.B. Stability of deoxynivalenol in heat-treated foods. *J. Food Prot.* **1999**, *62*, 962. [CrossRef] [PubMed]

32. Nathakaranakule, A.; Kraiwanichkul, W.; Soponronnarit, S. Comparative study of different combined superheated-steam drying techniques for chicken meat. *J. Food Eng.* **2007**, *80*, 1023–1030. [CrossRef]

33. Taechapairoj, C.; Dhuchakallaya, I.; Soponronnarit, S.; Wetchacama, S.; Prachayawarakorn, S. Superheated steam fluidised bed paddy drying. *J. Food Eng.* **2003**, *58*, 67–73. [CrossRef]

34. Xhabiri, G.Q.; Durmishi, N.; Idrizi, X.; Ferati, I.; Hoxha, I. Rheological qualities of dough from mixture of flour and wheat bran and possible correlation between Brabender and Mixolab Chopin equiments. *MOJ Food Process. Technol.* **2016**, *2*, 42. [CrossRef]

35. Jovanovich, G.; Lupano, C.E. Correlation between Starch Damage, Alveograph Parameters, Water Absorption and Gelatinization Enthalpy in Flours Obtained by Industrial Milling of Argentinian Wheats. *J. Food Technol.* **2003**, *1*, 167–171.

36. Mann, J.; Schiedt, B.; Baumann, A.; Conde-Petit, B.; Vilgis, T.A. Effect of heat treatment on wheat dough rheology and wheat protein solubility. *Food Sci. Technol. Int.* **2014**, *20*, 341–351. [CrossRef]

37. Wang, J.S.; Wei, Z.Y.; Li, L.; Bian, K.; Zhao, M.M. Characteristics of enzymatic hydrolysis of thermal-treated wheat gluten. *J. Cereal Sci.* **2009**, *50*, 205–209. [CrossRef]

38. Bucsella, B.; Takács, Á.; Vizer, V.; Schwendener, U.; Tömösközi, S. Comparison of the effects of different heat treatment processes on rheological properties of cake and bread wheat flours. *Food Chem.* **2016**, *190*, 990–996. [CrossRef]

39. Kim, W.; Choi, S.G.; Kerr, W.L.; Johonson, J.W.; Gainse, C.S. Effect of heating temperature on particle size distribution in hard and soft wheat flour. *J. Cereal Sci.* **2004**, *40*, 9–16. [CrossRef]

40. Barlow, K.K.; Buttrose, M.S.; Simmonds, D.H.; Vesk, M. The nature of the starch-protein interface in wheat endosperm. *Cereal Chem.* **1973**, *50*, 443–454.

41. Seguchi, M. Oil-binding ability of heat-treated wheat starch. *Cereal Chem.* **1984**, *61*, 248–250.

42. Neill, G.; Almuhtaseb, A.H.; Tra, M. Optimisation of time/temperature treatment, for heat treated soft wheat flour. *J. Food Eng.* **2012**, *113*, 422–426. [CrossRef]

43. Barrera, G.N.; Bustos, M.C.; Iturriaga, L.; Flores, S.K.; León, A.E.; Ribotta, P.D. Effect of damaged starch on the rheological properties of wheat starch suspensions. *J. Food Eng.* **2013**, *116*, 233–239. [CrossRef]

44. Wu, F.F.; Li, J.; Yang, N.; Chen, Y.S.; Jin, Y.M.; Xu, X.M. The roles of starch structures in the pasting properties of wheat starch with different degrees of damage. *Starch* **2017**. [CrossRef]

45. Saxena, D.C.; Rao, P.H. Effect of damaged starch on the pasting characteristics and tandoori roti making quality of whole wheat flour. *Sci. Aliments* **2000**, *20*, 591–602. [CrossRef]

46. Noda, T.; Ichinose, Y.; Takigawa, S.; Matsuura-Endo, C.; Abe, H.; Saito, K.; Hashimoto, N.; Yamauchi, H. The pasting properties of flour and starch in wheat grain damaged by α-amylase. *Food Sci. Technol. Int. Tokyo* **2003**, *9*, 387–391. [CrossRef]
47. Cui, L.; Selvaraj, J.N.; Xing, F.G.; Zhao, Y.J.; Zhou, L.; Liu, Y. A minor survey of deoxynivalenol in Fusarium, infected wheat from Yangtze-Huaihe river basin region in China. *Food Control* **2013**, *30*, 469–473. [CrossRef]

 © 2019 by the authors. Licensee MDPI, Basel, Switzerland. This article is an open access article distributed under the terms and conditions of the Creative Commons Attribution (CC BY) license (http://creativecommons.org/licenses/by/4.0/).

Article

A Novel Niosome-Encapsulated Essential Oil Formulation to Prevent *Aspergillus flavus* Growth and Aflatoxin Contamination of Maize Grains During Storage

Marta García-Díaz, Belén Patiño, Covadonga Vázquez and Jessica Gil-Serna *

Department of Genetics, Physiology and Microbiology, Faculty of Biology, University Complutense of Madrid, Jose Antonio Novais 12, 28040 Madrid, Spain; martga43@ucm.es (M.G.-D.); belenp@ucm.es (B.P.); covi@ucm.es (C.V.)
* Correspondence: jgilsern@ucm.es

Received: 18 October 2019; Accepted: 5 November 2019; Published: 6 November 2019

Abstract: Aflatoxin (AF) contamination of maize is a major concern for food safety. The use of chemical fungicides is controversial, and it is necessary to develop new effective methods to control *Aspergillus flavus* growth and, therefore, to avoid the presence of AFs in grains. In this work, we tested in vitro the effect of six essential oils (EOs) extracted from aromatic plants. We selected those from *Satureja montana* and *Origanum virens* because they show high levels of antifungal and antitoxigenic activity at low concentrations against *A. flavus*. EOs are highly volatile compounds and we have developed a new niosome-based encapsulation method to extend their shelf life and activity. These new formulations have been successfully applied to reduce fungal growth and AF accumulation in maize grains in a small-scale test, as well as placing the maize into polypropylene woven bags to simulate common storage conditions. In this latter case, the antifungal properties lasted up to 75 days after the first application.

Keywords: essential oils; *Satureja montana*; *Origanum virens*; *Aspergillus flavus*; aflatoxin; corn; nanoparticles

Key Contribution: A safe, ecofriendly, novel strategy was developed to prevent aflatoxin contamination of maize during storage. This method uses niosome-encapsulated EOs extracted from *Satureja montana* and *Origanum virens* and is able to control *Aspergillus flavus* growth for long periods.

---

1. Introduction

Aflatoxins (AFs) are secondary metabolites produced primarily by *Aspergillus flavus* and *Aspergillus parasiticus*. Aflatoxin B_1, B_2, G_1, and G_2 (AFB_1, AFB_2, AFG_1, and AFG_2) are the most important ones, with AFB_1 being the most toxic naturally occurring human carcinogen [1,2]. The International Agency for Research on Cancer (IARC) has classified the "naturally occurring mixes of aflatoxins" as Group 1 carcinogens in humans [3].

AFs contaminate a variety of staple crops including cereals (maize, sorghum, barley, oat, rye, rice, and wheat), soya, dry nuts (nuts, pistachios, almonds, and hazelnuts), cottonseed, coffee, cacao, and spices [4].

According to the Food and Agriculture Organization of the United Nations (FAO) [5], maize is one of the most important cereals, with an annual worldwide production of 1134 million tons in 2017, and most of it is intended for direct human and animal consumption. Moreover, maize and its derivatives are considered the main source of AFs worldwide [6]. For all these reasons, the European Union established strict regulations regarding maximum permitted AF levels for maize [7].

The impact of AF contamination of agri-food products is significant. It causes important economic losses because infected products cannot be sold and the contamination also raises veterinary and health costs. Establishing adequate controls to avoid AFs in the food chain is thus essential [8].

Different strategies to prevent AF contamination have been proposed to reduce fungal development in the field or during storage. Applying good agricultural practices and maintaining adequate humidity and temperature in silos are indispensable in reducing fungal growth and, therefore, mycotoxin contamination [9]. Chemical compounds are useful in preventing fungal growth and, for a long time, have been widely used both in the field and during storage to prevent mycotoxin contamination [10]. However, synthetic fungicides are in the spotlight and consumers are now demanding safer foodstuffs that are produced using sustainable and ecofriendly methods. The indiscriminate use of chemical fungicides has important drawbacks, including residue on grain that threatens human and animal health or causes extensive environmental contamination [11]. Moreover, the indiscriminate use of fungicides has caused an increased number of resistant isolates, which makes it very difficult to effectively control fungal growth [12].

The risks of using synthetic chemicals have increased public awareness and demand for safer and ecofriendly products and, in this context, natural plant extracts are now considered good alternatives. Essential oils (EOs) extracted from aromatic plants have demonstrated strong antibacterial, antifungal, and food preservative properties, together with low toxicity, fewer environmental effects, and wider public acceptance [13]. Many EO-based formulations are listed on the generally recognized as safe (GRAS) list, fully approved by the Food and Drug Administration, and are currently commercially available as food preservatives and/or agricultural supplements [14]. Several EOs have been reported to not only reduce growth in toxigenic fungal species, but also to interfere to some extent in mycotoxin biosynthesis. Da Silva et al. [15] reported that *Rosmarinus officinalis* EO has a strong effect against *Fusarium verticillioides*, as it showed the ability to rupture the cell wall and inhibit the production of fumonisins. *Aspergillus flavus* growth and its ability to produce AFs were also significantly affected by treatment with *Origanum virens* and *Ageratum conyzoides* EOs in corn and soybeans [16], and using *Mentha spicata* EO in chickpeas [17].

Therefore, the use of EOs to prevent fungal growth during maize storage could be a sustainable solution to minimize food losses owing to mycotoxin contamination. However, their direct application in food products seems to be limited because of their high volatility, low water solubility, and susceptibility to oxidation [18]. To solve these problems, various encapsulation techniques have been developed that can preserve EOs through a physical or chemical interaction with a matrix that maintains the compounds for a longer time [19]. Encapsulation of EOs also increases stability against oxidation, which helps to prolong their antimicrobial activity [18,20]. Different methods of encapsulation have also been demonstrated to enhance the antifungal and antiaflatoxigenic properties when applied to control *A. flavus* [21].

These encapsulation particles form a protective film that isolates the nucleus that contains the active agent. The composition of the particles should be carefully chosen depending on the encapsulated compound. To date, several natural and synthetic matrices have been successfully used to encapsulate EOs including polyethylene, carbohydrates, proteins, lipids, and gum [22]. The choice of the encapsulation material is a crucial step in developing an appropriate application method for EOs. Different parameters should be taken into account such as the polarity, solubility, and volatility of the active compounds, as well as the composition of the food matrix [23]. Niosomes are lipid-based systems, composed by non-toxic self-assembly vesicles, with a single or multiple layered structure, which are able to encapsulate hydrophobic and hydrophilic compounds [24]. Niosomes are biodegradable, easily stored and handled, and present low toxicity, which are important advantages for their application in the food industry [24].

The aim of this work was to evaluate the in vitro antifungal and antitoxigenic effects of different aromatic plant EOs and to design an effective niosome-based encapsulation protocol to avoid AF contamination during maize storage.

2. Results

2.1. The Efficacy of Plant Essential Oils Against Fungal Growth and Mycotoxin Production

Figures 1 and 2 show the results for the *Aspergillus flavus* growth rate (μ, mm/day) and the lag phase prior to growth (λ, h), respectively, in CYA (Czapek Yeast Autolysate Agar) plates supplemented with different essential oils (EOs) (*Rosmarinus officinalis*, *Thymus vulgaris*, *Satureja montana*, *Origanum virens*, *O. majoricum*, and *O. vulgare*) at several different concentrations (0, 10, 100, 500, and 1000 µg/mL).

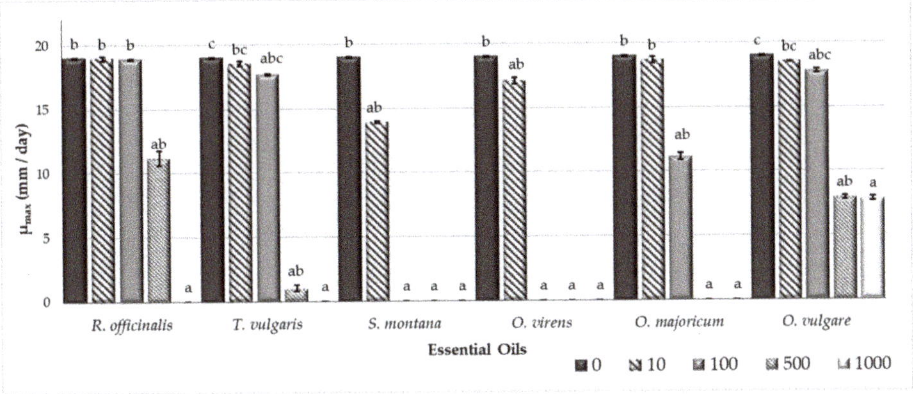

Figure 1. *Aspergillus flavus* S.44-1 growth rate (mm/day) at different concentrations (0, 10, 100, 500, and 1000 µg/mL) of essential oils (*R. officinalis*, *T. vulgaris*, *S. montana*, *O. virens*, *O. majoricum*, and *O. vulgare*). Each value is the mean of three replications and the thin vertical bars represent the standard error of the corresponding data. Groups with the same letter are not significantly different ($p > 0.05$).

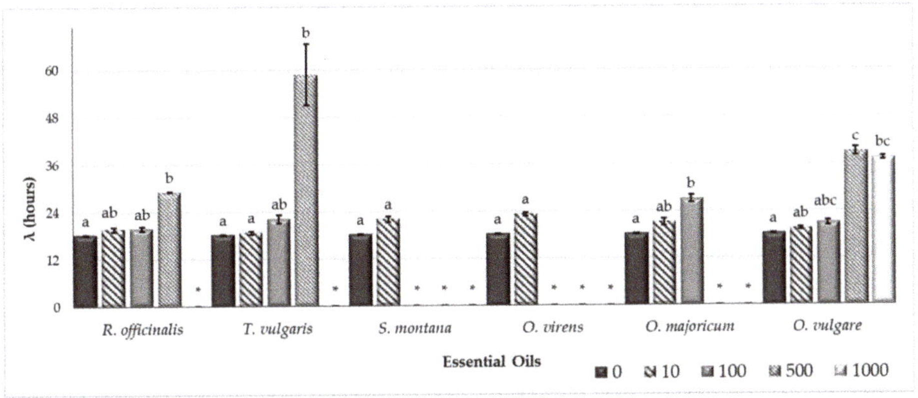

Figure 2. *Aspergillus flavus* S.44-1 lag phase (h) at different concentrations (0, 10, 100, 500, and 1,000 µg/mL) of essential oils (*R. officinalis*, *T. vulgaris*, *S. montana*, *O. virens*, *O. majoricum*, and *O. vulgare*). Each value is the mean of three replications and the thin vertical bars represent the standard error of the corresponding data. Groups with the same letter are not significantly different ($p > 0.05$). * No data.

All EOs tested had a significant effect on the *A. flavus* growth rate at the maximum concentration (Figure 1). Fungal growth was completely inhibited, except for the *O. vulgare* EO. For the EOs of *O. virens* and *S. montana*, total inhibition was reached at 100 µg/mL. *Thymus vulgaris* and *O. majoricum* EOs also showed reductions of at least 95% at 500 µg/mL.

The lag phase got longer as the EO concentration increased (Figure 2). It was not possible to calculate the lag phase when the EO treatment completely inhibited growth in the plates.

Aflatoxin production (AFB_1, AFB_2, AFG_1 y AFG_2) was significantly reduced compared with the control group in all treatments at the highest concentration tested (1000 µg/mL). No AFs were detected at these concentrations (Table 1) in any case, except for the *O. vulgare* EO treatment, which achieved reductions of more than 97% in all toxins.

Table 1. Aflatoxin (AF) concentration (B_1, B_2, G_1, and G_2) in Czapek Yeast Autolysate Agar (CYA) plates supplemented with different concentrations (0, 10, 100, 500, and 1000 µg/mL) of essential oils (EOs) (*R. officinalis*, *T. vulgaris*, *S. montana*, *O. virens*, *O. majoricum*, and *O. vulgare*). Each value is the mean of three replications ± standard error. Groups with the same letter are not significantly different ($p > 0.05$). ND: non detected.

EOs	µg/mL	AFB_1 (µg/g agar)	AFB_2 (µg/g agar)	AFG_1 (µg/g agar)	AFG_2 (µg/g agar)
R. officinalis	0	10.754 ± 0.925 [c]	0.201 ± 0.021 [c]	0.485 ± 0.055 [c]	0.088 ± 0.014 [b]
	10	5.205 ± 1.033 [bc]	0.1 ± 0.022 [bc]	0.213 ± 0.045 [bc]	ND [a]
	100	5.223 ± 0.171 [abc]	0.11 ± 0.006 [bc]	0.216 ± 0.012 [abc]	ND [a]
	500	1.09 ± 0.152 [ab]	0.017 ± 0.002 [ab]	0.058 ± 0.007 [ab]	ND [a]
	1000	ND [a]	ND [a]	ND [a]	ND [a]
T. vulgaris	0	10.754 ± 0.925 [b]	0.201 ± 0.021 [b]	0.485 ± 0.055 [b]	0.088 ± 0.014 [b]
	10	7.04 ± 0.977 [b]	0.122 ± 0.019 [ab]	0.256 ± 0.059 [ab]	ND [a]
	100	5.994 ± 0.554 [ab]	0.117 ± 0.019 [ab]	0.264 ± 0.069 [ab]	ND [a]
	500	ND [a]	ND [a]	ND [a]	ND [a]
	1000	ND [a]	ND [a]	ND [a]	ND [a]
S. montana	0	10.754 ± 0.925 [b]	0.201 ± 0.021 [b]	0.485 ± 0.055 [b]	0.088 ± 0.014 [b]
	10	8.27 ± 0.686 [ab]	0.151 ± 0.011 [ab]	0.314 ± 0.029 [ab]	ND [a]
	100	ND [a]	ND [a]	ND [a]	ND [a]
	500	ND [a]	ND [a]	ND [a]	ND [a]
	1000	ND [a]	ND [a]	ND [a]	ND [a]
O. virens	0	10.754 ± 0.925 [b]	0.201 ± 0.021 [bc]	0.485 ± 0.055 [b]	0.088 ± 0.014 [b]
	10	10.52 ± 1.334 [b]	0.245 ± 0.039 [c]	0.508 ± 0.065 [b]	ND [a]
	100	0.033 ± 0.044 [ab]	0.003 ± 0 [ab]	0.004 ± 0.002 [ab]	ND [a]
	500	ND [a]	ND [a]	ND [a]	ND [a]
	1000	ND [a]	ND [a]	ND a	ND [a]
O. majoricum	0	10.754 ± 0.925 [b]	0.201 ± 0.021 [ab]	0.485 ± 0.055 [b]	0.088 ± 0.014 [b]
	10	10.939 ± 0.21 [b]	0.234 ± 0.008 [b]	0.611 ± 0.057 [b]	ND [a]
	100	7.999 ± 0.628 [ab]	0.186 ± 0.022 [ab]	0.375 ± 0.028 [ab]	ND [a]
	500	0.003 ± 0 [a]	ND [a]	ND [a]	ND [a]
	1000	0.008 ± 0.008 [a]	ND [a]	ND [a]	ND [a]
O. vulgare	0	10.754 ± 0.925 [c]	0.201 ± 0.021 [b]	0.485 ± 0.055 [b]	0.088 ± 0.014 [b]
	10	10.143 ± 0.86 [bc]	0.192 ± 0.015 [b]	0.475 ± 0.056 [b]	ND [a]
	100	7.867 ± 0.409 [abc]	0.168 ± 0.015 [ab]	0.466 ± 0.08	ND
	500	0.738 ± 0.08 [ab]	0.012 ± 0.002 [a]	0.043 ± 0.003	ND
	1000	0.298 ± 0.068 [a]	0.004 ± 0.003 [a]	0.014 ± 0.004	ND

AFB_1 production was significantly affected at 500 µg/mL, with reductions of nearly 100% in all of the EOs tested. The same results were obtained when *O. virens* and *S. montana* EOs were applied at 100 µg/mL.

AFB_2 production was also significantly reduced by least 90% compared with the control plates after treatment with *R. officinalis*, *T. vulgaris*, *O. vulgare*, and *O. majoricum* EOs at 500 µg/mL. The most effective treatments, reaching complete inhibition of AFB_2 production at 100 µg/mL, were *O. virens* and *S. montana* EOs.

Aflatoxin G_1 was not detected on *S. montana* and *O. virens* at 100 µg/mL. The rest of the EOs showed a reduction greater than 80% at 500 µg/mL. In all cases, AFG_2 concentration was below the detection limit (<0.0025 µg/g agar), except in the case of control plates without treatment.

2.2. Techniques for the Application of Essential Oils to Prevent Fungal Growth and Mycotoxin Production

The main nanocapsule characteristics of the *O. virens* and *S. montana* niosomes can be found in Table 2. According to these data, both encapsulation processes yield high quality niosomes with low aggregation of nanoparticles. The particle size of both samples was approximately 140 nm with a polydipersity index (PDI) of 0.251 and a ζ-potential of −14 mV.

Table 2. Characterization of *O. virens* and *S. montana* essential oil particles encapsulated in niosomes.

	ZETASICER			NANOSIGHT	
	PDI	**Z-AVERAGE (nm)**	**POTENCIAL-ζ (mV)**	**Size (nm)**	**CONCENTRATION (Particle/mL)**
O. virens	0.251 ± 0.019	156.2 ± 3.9	-14.5 ± 0.5	142.4 ± 1.0	$(2.96 \pm 0.12) \times 10^{14}$
S. montana	0.251 ± 0.011	153.3 ± 2.8	-14.6 ± 2.3	140.6 ± 3.8	$(1.86 \pm 0.07) \times 10^{14}$

PDI: polydipersity index.

2.2.1. Small-Scale Assay

Satureja montana and *O. virens* EOs applied by direct contact to artificially inoculated maize grains and incubated at 28 °C reduced *A. flavus* growth compared with the control group without EOs after seven days of incubation, although no significant differences were observed (Figure 3a). When these EOs were applied in their niosome-encapsulated form and incubated for seven days, colony forming units (CFU) per gram values were slightly increased, although no significant differences were found. When incubation time was extended to 21 days, the opposite effect was observed (Figure 3b). EOs applied by direct contact seemed to lose their effect over time. However, when these compounds were encapsulated in niosomes, *S. montana* EO reduced fungal growth by 58% and, in the case of *O. virens*, this reduction was 32% with respect to the control group.

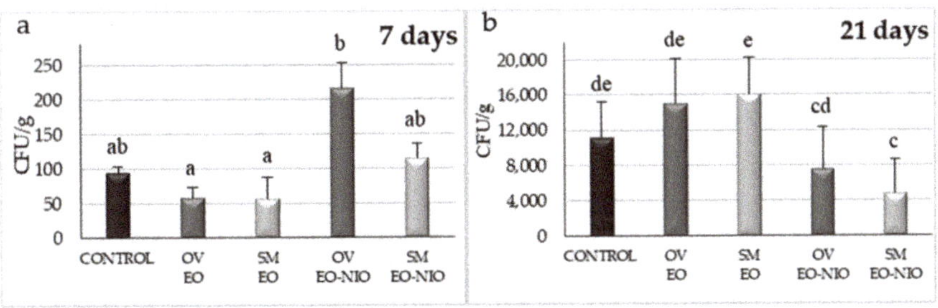

Figure 3. Effect of *S. montana* (SM) and *O. virens* (OV) by direct contact (essential oil, EO) and immobilized in niosomes (EO-NIO) on corn grains inoculated with *A. flavus*, incubated for 7 (**a**) and 21 days (**b**). Each values is the mean of three replications and the thin vertical bars represent the standard error of the corresponding data. Groups with the same letter are not significantly different ($p > 0.05$). CFU, colony forming units.

AFB_1 production was low at the short incubation time (7 days) and very significant for longer periods (21 days) (Figure 4). In the thin layer chromatography (TLC) analysis, the intensity and thickness of the fluorescent band is related to AFB_1 concentration. Therefore, in the seven-day long assay, AFB_1 concentration was reduced in all cases with respect to control. When plates were incubated over 21 days, a reduction in AFB_1 concentration was observed in plates treated directly with *O. virens*

EO as well as both EO niosome-encapsulated. However, no differences were found after *S. montana* EO application.

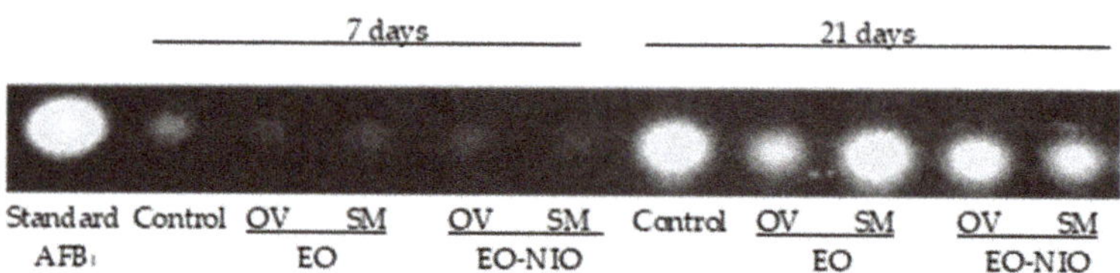

Figure 4. Effect of *S. montana* (SM) and *O. virens* (OV) by direct contact (EO) and encapsulated in niosomes (EO-NIO) on aflatoxin (AF) B_1 concentration of corn grains inoculated with *A. flavus*, incubated for 7 and 21 days. The standard corresponds to the application of purified AFB_1 (0.05 mg/mL).

2.2.2. Polypropylene Woven Bags Assays

Figure 5 shows the results in CFU/g of *A. flavus* inoculated corn stored in polypropylene woven bags at all incubation times and after treatment with EOs encapsulated in niosomes.

After 45, 60, and 75 days of incubation, both of the EOs encapsulated in niosomes were able to control fungal growth, with a maximum reduction of up to 79% and 69% for *S. montana* and *O. virens* EOs, respectively. However, this effect seems to be lost over time and no significant differences with respect to the control group were found after 90 days of incubation.

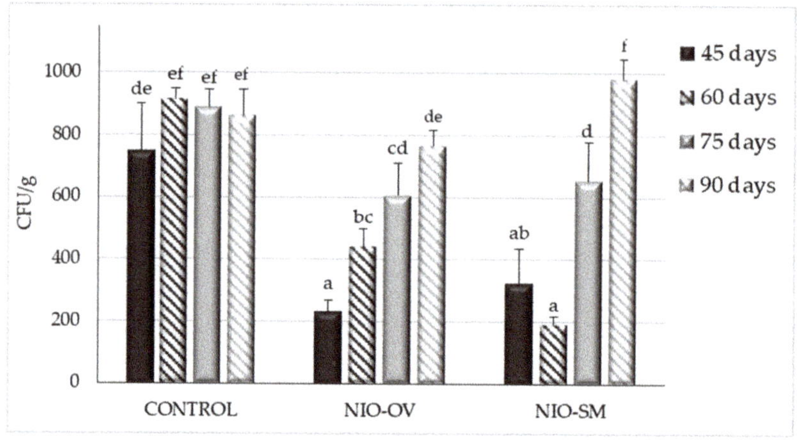

Figure 5. Effect of *S. montana* (NIO-SM) and *O. virens* (NIO-OV) EO encapsulated in niosomes on *A. flavus* growth in corn grains incubated for 45, 60, 75, and 90 days. Each value is the mean of three replications and the thin vertical bars represent the standard error of the corresponding data. Groups with the same letter are not significantly different ($p > 0.05$).

The results regarding AFB_1 detected on inoculated control bags or after niosome-encapsulated EO treatments are shown in Figure 6. The band intensity of each treatment was apparently lower with respect to their corresponding control in all cases except for NIO-OV after 90 days of incubation when a slight increase in production was detected. Treatment using niosome-encapsulated *S. montana* EO was the most effective to control AFB_1 production by *A. flavus*.

Temperature and humidity were recorded during the experiment with ranges of 25–27 °C and 75–85%, respectively.

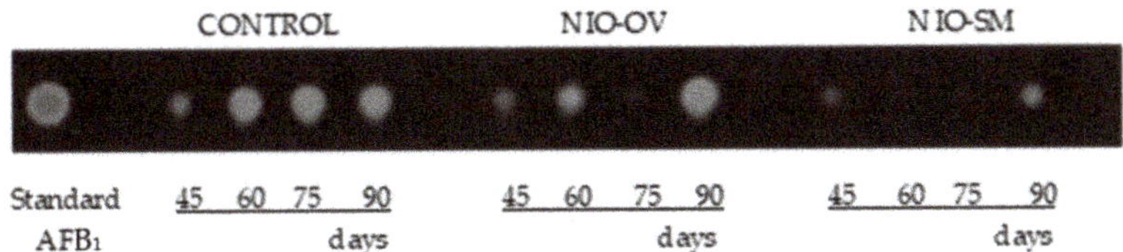

Figure 6. AFB$_1$ detection by thin layer chromatography (TLC) in polypropylene woven bags inoculated with *A. flavus* after *S. montana* (NIO-SM) and *O. virens* (NIO-OV) niosome treatment of corn incubated for 45, 60, 75, and 90 days. The intensity and thickness of the fluorescent band are related to the concentration of toxin. The standard corresponds to the application of purified AFB$_1$ (0.05 mg/mL).

3. Discussion

Aflatoxin (AF) presence poses a high risk to food security and most countries have established maximum levels of these contaminants allowed in food products [3]. Appropriate control mechanisms are needed to keep these toxins from entering the food chain. In recent years, essential oils (EOs) have come to be considered as a safe, ecofriendly, renewable, and easily biodegradable option to be used as a food supplement [13]. Moreover, many EOs have been described as potent antifungal compounds that are able to interfere in mycotoxin synthesis [15,25–27]. In this work, we selected EOs extracted from *Rosmarinus officinalis*, *Thymus vulgaris*, *Satureja montana*, *Origanum virens*, *O. majoricum*, and *O. vulgare* to determine if they were able to control *Aspergillus flavus* growth and if they reduced AF production by this fungus. To some extent, all of the EOs tested modified the fungal growth rate and extended the lag phase at high concentrations. However, at lower doses, the EOs extracted from *S. montana* and *O. virens* were the most effective at reducing both fungal growth and AF production, and they were selected to perform subsequent studies. Chromatographic characterization of the EOs used in this study were carried out in the Agricultural Research Centre of Albaladejito (data not shown) and the results revealed that *S. montana* and *O. virens* EOs are highly rich in carvacrol and thymol, respectively. These compounds have been reported to be able to interact with the cell membrane, disrupt cell permeability, and produce cell death [28]. Pure extracts of both carvacrol and thymol have been reported to inhibit the growth of important mycotoxin-producing species such as *A. niger*, *A. flavus*, *A. ochraceus*, and *F. graminearum* [29,30].

Different authors consider that EO-based formulations could be safe, ecofriendly preservatives to avoid post-harvest losses due to mycotoxin contamination [14]. Therefore, taking into account the potent antifungal properties of EOs, many studies have focused on developing successful application protocols to minimize their drawbacks, which limit their direct use in food products [18] and, therefore, it is essential to protect them to extend their shelf life and activity [19]. The controlled liberation of EOs and their encapsulation in nanoparticles made of different materials are considered a good option [19,23]. These technologies attempt to reduce the rapid loss of their active principles. In general, EOs are a complex mix of lipophilic compounds and, therefore, lipid nanoparticle systems such as liposomes are the most appropriate [23,31]. In our study, the small-scale experiments showed that the effectiveness of EOs to control *A. flavus* growth diminished over time. However, when both EOs were encapsulated in niosomes, a significant reduction in fungal viable counts with respect to the untreated control group were found even after 21 days of incubation. Hence, it seems that the niosome-based nanoparticles were able to reduce the loss of the EOs' active principles and produce a controlled release of their compounds. Similar results were obtained in the larger-scale experiments using maize stored in polypropylene woven bags. This type of storage is very common in African countries because it offers low-cost protection for grain from pests. However, a higher contamination by toxigenic fungi and mycotoxins in the grains has been reported owing to the change in moisture content, which increases the relative humidity inside the bags [32]. Even under the worst conditions for maize storage, both of the EOs encapsulated in niosomes were able to control fungal development, significantly reducing aflatoxin levels, and their effect was extended for up to 75 days. Hence, these promising approaches might be useful to prevent AF contamination under more appropriate storage conditions such as PICS (Purdue improved crop storage) bags or directly in silos.

The use of EOs in the agri-food industry is not a safety concern because several studies have ensured that they are safe as food additives and many of them are included in the GRAS list [14]. However, data are scarce regarding the effects that vesicle materials might have on human health and it is essential to carry out ecotoxicity studies to assess the impact of encapsulation matrices [19]. The niosome vesicles used in this work are commercially available and their non-toxic properties have been fully demonstrated. Moreover, they have been approved as a good option for the development of nanoparticles to improve medical therapies, including the controlled delivery of drugs or even vaccine antigens [24].

EO-based formulations need to overcome several tests before their application in food systems, as some active components can interact with food matrix components [14,33]. AFs often occur in maize, one of the most important basic cereal products worldwide for food and feeds [6]. In the present work, *S. montana* and *O. virens* EOs encapsulated in niosomes were directly applied to control *A. flavus* growth and its mycotoxigenic potential in artificially contaminated maize as a preliminary step to optimizing their application. EOs' release was effective over time in both small-scale tests and simulated storage conditions and, therefore, no interactions with the components of the maize seemed to occur. It would be interesting to apply this newly developed technology to other food matrices often contaminated with AFs to confirm that this effect could be extrapolated to other products.

4. Conclusions

In this work, we proposed a novel niosome-based EO product that was successfully applied in polypropylene woven bags simulating common storage conditions of maize. The involvement of the company Nanovex Biotechnologies S.L. guarantees a correct and standardized encapsulation protocol and the reduction of problems that might arise during product formulation. The presence of encapsulated EOs in the bags significantly reduced *A. flavus* development and the effect was observed until 75 days after inoculation. The effect of this formulation could be easily maximized by applying the products regularly during maize storage, that is, every 45 days. Regularly scheduled application, together with good agricultural practices and the maintenance of adequate storage conditions, may be a sustainable way to avoid the occurrence of aflatoxins in stored maize.

5. Materials and Methods

5.1. Fungal Strains and Essential Oils

All *Aspergillus flavus* strains used in this study were able to produce aflatoxins from group B (AFB$_1$ and AFB$_2$) and G (AFG$_1$ and AFG$_2$) and were isolated from wheat from Morocco (S.44-1) and maize from Spain (A7). The correct identification of these isolates was confirmed using species-specific PCR protocols [34].

The strains were stored as a spore suspension at −80 °C in 15% glycerol (Panreac, Barcelona, Spain) until required. They were subcultured on potato dextrose agar (PDA, Pronadisa, Spain) and incubated at 28 °C for four days. The spore suspensions were prepared in sterile saline solution (9 g/L sodium chloride) (Merck, Darmstadt, Germany) supplemented by Tween 80 0.5% (Panreac, Spain). The spore concentration was determined using a Thoma counting chamber (Marienfeld, Lauda-Königshofen, Germany) and adjusted to a final concentration of 10^2 or 10^6 spores/mL depending on the assay.

The EOs tested were extracted from rosemary (*Rosmarinus officinalis* L.), thyme (*Thymus vulgaris* L.), savory (*Satureja montana* L.), and three species of oregano (*Origanum virens* Hoffmanns. & Link, *O. majoricum* Camb., and *O. vulgare* L.). The EOs were provided by The Agricultural Research Centre of Albaladejito (Cuenca, Spain). Each species was processed in batches of 100–150 g of plant aerial parts, following the methodology proposed by the European Pharmacopoeia by hydrodistillation, in a Clevenger-type apparatus for 2 h. These EOs were analyzed in gas chromatograph equipped with a flame ionization detector (FID) and capillary column VF-5 of 60 mm × 0.25 mm, 5% phenyl methyl siloxane. A temperature gradient of 70 to 240 °C was applied, with an increase of 3 °C per minute, maintaining the final temperature for 2 min. For the identification of the EO components, the relative retention times of standards and the corresponding Kovats indices were used. The quantification of the components was performed according to the areas of their chromatographic peaks.

These compounds were filtered (pore size 0.2 μm) (Fisherbrand, Shanghai, China) and stored at −20 °C in amber glass vials (Thermo Scientific, Madrid, Spain) until required.

5.2. Effectiveness of Plant Essential Oils on Fungal Growth and Aflatoxin Production

The effect of EOs at different concentrations on *A. flavus* S.44-1 growth and its ability to produce AFs were evaluated on CYA medium (45.5 g/L of modified Czapek–Dox agar (Pronadisa, Spain), 5 g/L of yeast extract (Pronadisa, Spain)). EOs were diluted in polyethylene glycol 400 (PEG (Acros, Geel, Belgium)) and added to the medium to obtain final concentrations of 10, 100, 500, and 1000 μg/mL. The same amount of PEG was included in the control plates instead of EO. CYA plates supplemented with EOs were inoculated with 1.5 μL (4 mm of diameter) of a 10^6 spores/mL suspension on the center of the plate, and incubated at 28 °C for five days. All the conditions were tested in triplicate.

Fungal colony diameters were measured daily in two directions at right angles to each other until the medium was fully colonized (five days). Growth parameters were calculated from a linear model obtained by plotting the diameter (mm) against time (day). The parameters determined were λ, representing the lag phase (days prior to mycelial growth), and $μ_{max}$, representing the maximum growth rate (mm/day), for control plates and each EO concentration tested.

AFs were extracted from the plates after six days of incubation, as described elsewhere [35]. Three agar plugs were removed from the centre, medium, and outer edge of the colony and toxins were extracted with 1 mL of methanol (Merck, Spain). Samples were stored at −20 °C until analysis. AFs were measured by high performance liquid chromatography (HPLC) using the protocol described below.

5.3. Effect of Satureja Montana and Origanum Virens Essential Oils Encapsulated in Niosomes on Fungal Growth and Aflatoxin Contamination

5.3.1. Procedure for Microencapsulation of Essential Oils

EOs extracted from *S. montana* and *O. virens* were encapsulated in non-ionic surfactant-based lipid vesicles (niosomes). These particles were prepared by Nanovex Biotechnologies S.L. (Oviedo, Spain) starting from 40 mL of each type of EO. Niosomes were obtained using the thin film hydration (TFH) method with homogenization and sonication to obtain niosomes with a good polydipersity index (PDI) and a particle size between 100 and 200 μm, with an EO concentration of 12 μL/mL.

The characterization of the niosomes was performed with a Zetasizer Nano ZS particle size analyzer (Malvern Panalytical Ltd., Malvern, UK), which uses dynamic light scattering (DLS) to determine particle size, and the M3-PALS technique to calculate the ζ-potential.

A nanoparticle tracking analysis (NTA) was performed using a nano sight particle tracking analyzer (Malvern Panalytical Ltd., Malvern, UK) to determine concentration and size distribution.

5.3.2. Effect of Niosome-Encapsulated Essential Oils on Fungal Growth and Aflatoxin Production on Maize Grains

Previously autoclaved maize grains were inoculated with *A. flavus*, strain A7. Then, 100 g of corn was immersed for 2 h in 100 mL of spore suspension 10^4 spores/mL to obtain a final concentration of 10^2 spores/g. Subsequently, the effect of niosome-encapsulated EOs was tested in a small-scale test in Petri dishes as well as in polypropylene woven bags simulating real storage conditions. At the beginning of the experiments, grain moisture was measured using a Hygropalm HP23A (Rotronic, Bassersdorf, Switzerland) and water activity was 0.95 in all cases.

Small-Scale Assays

Ninety millimeter Petri dishes were filled with crystalized potassium sulphate (Acros, Spain) to maintain a_w at 0.97 [36]. A 50 mm Petri dish containing 7 g of inoculated maize was placed inside the larger one. *Satureja montana* and *O. virens* EOs were applied directly to the grains or encapsulated in niosomes at 500 µg/g. Control assays, mock-inoculated with water, were also included. Incubation was performed at 28 °C and the effect of niosome-encapsulated EOs or those directly applied on maize grains was evaluated at 7 and 21 days.

After the incubation period, a sample of 3.5 g was taken from each treatment and a viable count was performed using serial decimal dilutions and inoculation on Rose Bengal with Chloramphenicol medium. Plates were incubated in darkness at 28 °C for two days. The fungal growth of maize grain was expressed as colony forming units per gram of maize (CFU/g).

Afterwards, another sample of 3.5 g was taken from each treatment, and shaken for 20 min with 35 mL of chloroform for AF extraction. AFB_1 was measured by thin layer chromatography (TLC), as described below.

Polypropylene Woven Bag Assays

One-hundred grams of inoculated maize was placed in small polypropylene woven bags. Subsequently, niosome-encapsulated EOs (*S. montana* and *O. virens*) were added at a dose of 500 µg/g and mixed. Bags were incubated at room temperature in darkness for 90 days in independent plastic boxes for each treatment ($40 \times 40 \times 30$ cm). Inoculated maize grains without EOs were used as control. Temperature and relative humidity were registered using a data logger El-USB-1 (Easylog; LASCAR electronic, Salisbury, UK) every 8 h until the end of the assay.

After the incubation period, the bags were cut open and the maize was diluted in 900 mL of sterile saline solution (9 g/L) containing 0.05% Tween 80. The mixes were incubated in an orbital shaker (140 rpm) at 4 °C for 60 min to release spores. Then, serial decimal dilutions and culture in Rose Bengal with Chloramphenicol were used to estimate fungal growth as CFU per gram of maize.

A sample of 14 g of corn was taken from each beaker, and shaken for 20 min with 35 mL of chloroform for AFB_1 extraction and subsequent evaluation by TLC, as described below.

5.4. Detection of Mycotoxins

5.4.1. Detection of Mycotoxins by High Performance Liquid Chromatography (HPLC)

After AFB_1 extraction with methanol, mycotoxin measurements were performed in the "Laboratorio Arbitral Agroalimentario" (Madrid, Spain) following its standardized protocols. AF was measured by HPLC on a reverse phase C_{18} column (Inertsil ODS3; 5 µm, 4.6 mm $\times$ 250 mm; GL Sciences, Tokio, Japan) at 40 °C in a Waters chromatograph 515 HPLC coupled with a fluorescence detector 474 (Waters, Milford, MA, USA) at excitation and emission wavelengths of 362 and 435 nm, respectively. The mobile phase contained water, methanol, and acetonitrile (60:20:20), and the flow rate was 1 mL/min. AF was eluted and quantified by comparison with a calibration curve generated from AF standards (OEKANAL®, Sigma–Aldrich, Steinheim, Germany). The detection limit of the technique was 2.5 ng/g.

5.4.2. Detection of Mycotoxins by Thin Layer Chromatography

After AFB_1 extraction with chloroform, samples were filtered using 0.45 µm syringe filters (Fisherbrand, Spain) and an aliquot of 1 mL was evaporated in a vacuum concentrator, Eppendorf™ Concentrator Plus with Pump and GB Plug (Fisher Scientific, Madrid, Spain).

Silica gel 60 chromatography plates (Merck, Germany) were used, and AFB_1 presence was determined according to the protocols described elsewhere [37,38].

Samples and AF standards were re-suspended with 500 µL toluene/acetonitrile (95:5) (Panreac, Spain). Then, 10 µL of each sample was spotted on the plate. Toluene/acetone/acetonitrile (1:1:1 (LabKem, Barcelona, Spain)) was used as a mobile phase. Toxins were visualized under ultraviolet light (Spectronics, Westbury, NY, USA).

5.5. Statistical Analysis

Statistical analysis was performed on the effect of EOs encapsulated in niosomes with StatsGraphics Centurion XVII V.17.2.04 program (Statpoint Technologies Inc., Warrenton, VA, USA). The Shapiro–Wilk and Levene tests were used to check normality and homoscedasticity, respectively. Data were analysed using analysis of variance (ANOVA).

When data did not meet normality and homoscedasticity criteria, a non-parametric Kruskall–Wallis test was performed. These analyses were performed using the software InfoStat/E 2011 (FCA, Córdoba, Argentina). This was necessary in the case of the fungal growth and aflatoxin production variables indicated in Section 5.2 of Material and Methods.

In all cases, the significance level was set at $p < 0.05$.

Author Contributions: All authors conceived the experimental design. M.G.-D., J.G.-S., and B.P. helped with laboratory analysis. M.G.-D. and J.G.-S. performed statistical analysis and wrote the original draft. B.P. and C.V. reviewed and edited the manuscript. All authors read and approved the final version of the document.

Funding: This research was supported by the Spanish Ministry of Science and Innovation, grant number AGL 2014-53928-C2-2-R, and Marta García-Díaz was funded through an FPI fellowship by the Spanish Ministry of Science and Innovation (BES-2015-074533).

Acknowledgments: The authors would like to thank the Agricultural Research Centre of Albaladejito for supplying the purified essential oils, as well as "Laboratorio Arbitral Agroalimentario" for the measurements of mycotoxins using HPLC. Nanovex Biotechnologies was our partner in the process of encapsulating essential oils.

Conflicts of Interest: The authors declare no conflict of interest.

References

1. Kensler, T.W.; Roebuck, B.D.; Wogan, G.N.; Groopma, J.D. Aflatoxin: A 50-year odyssey of mechanistic and translational toxicology. *Toxicol. Sci.* **2011**, *120*, S25–S48. [CrossRef] [PubMed]
2. Alshannaq, A.; Yu, J.H. Occurrence, toxicity, and analysis of major mycotoxins in food. *Int. J. Environ. Res. Public Health* **2017**, *14*, 632. [CrossRef] [PubMed]
3. Wu, F.; Stacy, S.L.; Kensler, T.W. Global risk assessment of aflatoxins in maize and peanuts: Are regulatory standards adequately protective? *Toxicol. Sci.* **2013**, *135*, 251–259. [CrossRef] [PubMed]
4. Gil-Serna, J.; Vázquez, C.; Patiño, B. Mycotoxins/Toxicology. In *Encyclopedia of Food Microbiology*; Academic Press: Cambridge, MA, USA, 2014.
5. Food and Agriculture Organization of the United Nations, Statistic Division. Available online: http://www.fao.org/faostat/en/#data/QC (accessed on 18 September 2019).
6. Battilani, P.; Toscano, P.; van der Fels-Klerx, H.J.; Jeggieri, M.C.; Brera, C.; Rortais, A.; Goumperis, T.; Robinson, T. Aflatoxin B$_1$ contamination in maize in Europe increases due to climate change. *Sci. Rep.* **2016**, *6*, 24328. [CrossRef] [PubMed]
7. European Commission. Regulation N° 165/2010 amending Regulation (EC) No 1881/2006 setting maximum levels for certain contaminants in foodstuffs as regards aflatoxins. *Off. J. Eur. Union* **2010**, *50*, 8–12.
8. Hussein, H.S.; Brasel, J.M. Toxicity, metabolism, and impact of mycotoxins on humans and animals. *Toxicology* **2001**, *167*, 101–134. [CrossRef]
9. Chulze, S. Strategies to reduce mycotoxin levels in maize during storage: A review. *Food Addit. Contam.* **2010**, *27*, 651–657. [CrossRef]
10. Lagogianni, C.S.; Tsisigiannis, D.I. Effective chemical management for prevention of aflatoxins in maize. *Phytopathol. Mediterr.* **2018**, *57*, 186–197.
11. Ji, C.; Fan, J.; Zhao, L. Review on biological degradation of mycotoxins. *Anim. Nutr.* **2016**, *2*, 127–133. [CrossRef]

12. Da Cruz Cabral, L.; Pinto, V.F.; Patriarca, A. Application of plant derived compounds to control fungal spoilage and mycotoxin production in foods. *Int. J. Food Microbiol.* **2013**, *166*, 1–14. [CrossRef]

13. Pandey, A.K.; Kumar, P.; Singh, P.; Tripathi, N.N.; Bajpai, V.K. Essential oils: Sources of antimicrobials and food preservatives. *Front. Microbiol.* **2017**, *7*, 2161. [CrossRef] [PubMed]

14. Kumar-Dwivedy, A.; Kumar, M.; Updhyay, N.; Prakash, B.; Kishore-Dubey, N. Plant essential oils against food borne fungi and mycotoxins. *Curr. Opin. Food Sci.* **2016**, *11*, 16–21. [CrossRef]

15. Da Silva, N.; Polis, L.; Faggion, J.; Yumie, C.; Galerani, S.A.; Grespan, R.; Botiao, S.; Augusto, C.; Abreu, B.A.; Machinski, M. Antifungal activity and inhibition of fumonisin production by *Rosmarinus officinalis* L. essential oil in *Fusarium verticillioides* (Sacc.) Nirenberg. *Food Chem.* **2015**, *166*, 330–336. [CrossRef] [PubMed]

16. Esper, R.H.; Gonçalez, E.; Marques, M.O.M.; Felicio, R.C.; Felicio, J.D. Potential of essential oils for protection of grains contaminated by aflatoxin produced by *Aspergillus flavus. Front. Microbiol.* **2014**, *5*, 269. [CrossRef] [PubMed]

17. Kedia, A.; Kumar-Dwivedy, A.; Kumar-Jha, D.; Dubey, N.K. Efficacy of *Mentha spicata* essential oil in suppression of *Aspergillus flavus* and aflatoxin contamination in chickpea with particular emphasis to mode of antifungal action. *Protoplasma* **2016**, *253*, 647–653. [CrossRef]

18. Ribeiro-Santos, R.; Andrade, M.; Sanches-Silva, A. Application of encapsulated essential oils as antimicrobial agents in food packaging. *Food Sci.* **2017**, *14*, 78–84. [CrossRef]

19. Mäes, C.; Bouquillon, S.; Fauconnier, M.L. Encapsulation of essential oils for the development of biosourced pesticides with controlled release: A review. *Molecules* **2019**, *24*, 2539. [CrossRef]

20. Donsí, F.; Annunziata, M.; Sessa, M.; Ferrari, G. Nanoencapsulation of essential oils to enhance their antimicrobial activity in foods. *LWT Food Sci. Technol.* **2011**, *44*, 1908–1914. [CrossRef]

21. Nesci, A.; Passone, M.A.; Barra, P.; Girardi, N.; García, D.; Etcheverry, M. Prevention of aflatoxin contamination in stored grains using chemical strategies. *Curr. Opin. Food Sci.* **2016**, *11*, 56–60. [CrossRef]

22. Da Silva, P.T.; Fries, L.L.M.; de Menezes, C.R.; Holken, A.T.; Schwan, C.L.; Wigmann, E.F.; Bastos, J.D.O.; da Silva, C.D.B. Microencapsulation: Concepts, mechanisms, methods and some applications in food technology. *Cienc. Rural* **2014**, *44*, 1304–1311. [CrossRef]

23. Prakash, B.; Kujur, A.; Yadav, A.; Kumar, A.; Singh, P.P.; Dubey, N.K. Nanoencapsulation: An efficient technology to boost the antimicrobial potential of plant essential oils in food system. *Food Control* **2018**, *89*, 1–11. [CrossRef]

24. Amoabediny, G.; Haghiralsadat, F.; Naderinezhad, S.; Helder, M.N.; Kharanaghi, E.A.; Arough, J.M.; Zandieh-Doulabi, B. Overview of preparation methods of polymeric and lipid-based (niosome, solid lipid, liposome) nanoparticles: A comprehensive review. *Int. J. Polym. Mater. Polym. Biomater.* **2018**, *67*, 383–400. [CrossRef]

25. Císarová, M.; Tancinová, D.; Medo, J.; Kacaniová, M. The in vitro effect of selected essential oils on the growth and mycotoxin production of *Aspergillus* species. *J. Environ. Sci. Health Part B* **2016**, *51*, 668–674. [CrossRef] [PubMed]

26. Perczak, A.; Gwiazdowska, D.; Marchwinska, D.; Jus, K.; Gwiazdowski, R.; Waskiewicz, A. Antifungal activity of selected essential oils against *Fusarium culmorum* and *F. graminearum* and their secondary metabolites in wheat seeds. *Arch. Microbiol.* **2019**, *201*, 1085–1097. [CrossRef]

27. Wang, L.; Jiang, N.; Wang, D.; Wang, M. Effects of essential oil citral on the growth, mycotoxin biosynthesis and transcriptomic profile of *Alternaria alternate. Toxins* **2019**, *11*, 553. [CrossRef]

28. Prakash, B.; Kedia, A.; Mishra, P.K.; Dubey, N.K. Plant essential oils as food preservatives to control moulds, mycotoxin contamination and oxidative deterioration of agri-food commodities—Potentials and challenges. *Food Control* **2015**, *47*, 381–391. [CrossRef]

29. Abbaszadeh, S.; Sharifzadeh, A.; Shokri, H.; Khosravi, A.R.; Abbaszadeh, A. Antifungal efficacy of thymol, carvacrol, eugenol and menthol as alternative agents to control the growth of food-relevant fungi. *J. Mycol. Med.* **2014**, *24*, 51–56. [CrossRef]

30. Gao, T.; Zhou, H.; Zhow, W.; Hu, L.; Chen, J.; Shi, Z. The fungicidal activity of thymol against *Fusarium graminearum* via inducing lipid peroxidation and disrupting ergosterol biosynthesis. *Molecules* **2016**, *21*, 770. [CrossRef]

31. Zeisig, R.; Cammerer, B. Liposomes in the food industry. In *Nano-and Microencapsulation for Foods*; Vilstrup, P., Ed.; Leatherhead: London, UK, 2001; pp. 101–109.

32. Maina, A.W.; Wagacha, J.M.; Mwaura, F.B.; Muthomi, J.W.; Woloshuk, C.P. Postharvest practices of maize farmers in Kaiti district, Kenya and the impact of hermetic storage on populations of *Aspergillus* spp. and aflatoxin contamination. *J. Food Res.* **2016**, *5*, 53–66. [CrossRef]

33. Perricone, M.; Arace, E.; Corbo, M.R.; Sinigaglia, M.; Bevilacqua, A. Bioactivity of essential oils: A review on their interaction with food components. *Front. Microbiol.* **2015**, *6*, 76. [CrossRef]

34. González-Salgado, N.; González-Jaén, M.T.; Vázquez, C.; Patiño, B. Highly sensitive PCR-based detection method specific for *Aspergillus flavus* in wheat flour. *Food Addit. Contam.* **2008**, *25*, 758–764. [CrossRef] [PubMed]

35. Bragulat, M.R.; Abarca, M.L.; Cabañes, F.J. An easy screening method for fungi producing ochratoxin A in pure culture. *Int. J. Food Microbiol.* **2001**, *71*, 139–144. [CrossRef]

36. Bernáldez-Rey, M.V. Desarrollo de Métodos de RT-PCR en Tiempo Real Para la Cuantificación de Mohos Toxigénicos Viables en Alimentos. Ph.D. Thesis, University of Extremadura, Badajoz, Spain, 2016.

37. Scott, P.M.; Lawrence, J.W.; van Walbeek, W. Detection of mycotoxins by thin-layer chromatography: Application to screening of fungal extracts. *Appl. Microbiol.* **1970**, *20*, 839–842. [PubMed]

38. Gimeno, A.; Martins, M.L. Rapid thin layer chromatography determination of patulin, citrinin, and aflatoxin in apples and pears, and their juices and jams. *J. Assoc. Off. Anal. Chem.* **1983**, *66*, 85–91.

© 2019 by the authors. Licensee MDPI, Basel, Switzerland. This article is an open access article distributed under the terms and conditions of the Creative Commons Attribution (CC BY) license (http://creativecommons.org/licenses/by/4.0/).

MDPI
St. Alban-Anlage 66
4052 Basel
Switzerland
Tel. +41 61 683 77 34
Fax +41 61 302 89 18
www.mdpi.com

Toxins Editorial Office
E-mail: toxins@mdpi.com
www.mdpi.com/journal/toxins

www.ingramcontent.com/pod-product-compliance
Lightning Source LLC
LaVergne TN
LVHW070106170726
843515LV00007B/1625